The Disorder of Mathematics Education

Hauke Straehler-Pohl • Nina Bohlmann
Alexandre Pais
Editors

The Disorder of Mathematics Education

Challenging the Sociopolitical Dimensions of Research

Springer

Editors
Hauke Straehler-Pohl
Freie Universität Berlin
Berlin, Germany

Nina Bohlmann
Freie Universität Berlin
Berlin, Germany

Alexandre Pais
Manchester Metropolitan University
Manchester, United Kingdom

ISBN 978-3-319-34005-0 ISBN 978-3-319-34006-7 (eBook)
DOI 10.1007/978-3-319-34006-7

Library of Congress Control Number: 2016942553

Printed on acid-free paper

This Springer imprint is published by Springer Nature
The registered company is Springer International Publishing AG Switzerland

Acknowledgments

This edited collection emerged from a small conference with the title *The Disorder of Mathematics Education. Working towards research of social and political relevance*. The conference was held at the Freie Universität Berlin from January 15th to January 17th, 2015 (www.ewi-psy.fu-berlin.de/dome). We would kindly thank the Department of Education and Psychology for financing the conference and all the participants' institutions for granting the subsidiary funds needed to attend the conference. Even more so, we particularly thank all those participants who invested own resources in order to be part of "DOME". A particular acknowledgement is reserved to Petra Gips, head of the Department of Education and Psychology, who trustfully allowed two young and half-baked scholars to organize everything in the way they desired. The success of the conference and the material existence of the subsequent book you hold in your hands were possible by the further help of Birgit Abel, Daria Fischer, Philipp Friedrichs and Birte Zoege.

All the chapters compiled in this book are not solely to be considered as works of the designated authors alone—even though they undoubtedly are to be made responsible for the originality, radicality and venturesomeness in the first place. The chapters are nevertheless inspired, enhanced and expedited by the collectivity that made up for the intense discussions during the conference. Besides the (present and absent) authors of this book, this collective includes: Gelsa Knijnik (Brazil), Maria Johansson (Sweden), Kenneth Mølbjerg Jørgensen (Denmark), Elizabeth de Freitas (USA) and Natividad Adamuz (Spain).

Finally, all papers have been reviewed at least twice. We thank all those who, by reviewing chapter drafts, have significantly contributed to this volume: Jehad Alshwaikh (Palestine/South Africa), Melissa Andrade-Molina (Denmark), Peter Appelbaum (USA), Christer Bergsten (Sweden), Lisa Björklund Boistrup (Sweden), Nina Bohlmann (Germany), Tony Brown (UK), Ditte Storck Christensen (Sweden), Elizabeth de Freitas (USA), Uwe Gellert (Germany), Eva Jablonka (UK), Kenneth Mølbjerg Jørgensen (Denmark), Gelsa Knijnik (Brazil), David Kollosche (Germany), Anna Llewellyn (UK), Sverker Lundin (Sweden), Mônica Mesquita

(Portugal), Alex Montecino (Denmark), Candia Morgan (UK), Alexandre Pais (UK), Aldo Parra-Sanchez (Denmark), Hauke Straehler-Pohl (Germany), David Swanson (UK), Paola Valero (Denmark/Sweden) and David Wagner (Canada).

The artwork on the book cover was designed by Daria Fischer.

Sylvia Violet and Jairo Dorado Cadilla were responsible for proofreading and editing the manuscript.

Contents

Contributors

Jehad Alshwaikh Faculty of Education, Birzeit University, Ramallah, Palestine

Faculty of Humanities, University of the Witwatersrand, Johannesburg, South Africa

Melissa Andrade-Molina Faculty of Engineering and Science, Aalborg University, Aalborg, Denmark

Peter Appelbaum Department of Curriculum, Cultures and Child/Youth Studies, Arcadia University, Philadelphia, PA, USA

Christer Bergsten Department of Mathematics, Linköping University, Linköping, Sweden

Lisa Björklund Boistrup Department of Mathematics and Science Education, University of Stockholm, Stockholm, Sweden

Nina Bohlmann Fachbereich Erziehungswissenschaft und Psychologie, Freie Universität Berlin, Berlin, Germany

Tony Brown Education and Social Research Institute, Manchester Metropolitan University, Manchester, UK

Ditte Storck Christensen Department of Education and Special Education, University of Gothenburg, Gothenburg, Sweden

Uwe Gellert Fachbereich Erziehungswissenschaft und Psychologie, Freie Universität Berlin, Berlin, Germany

Eva Jablonka Department of Education and Professional Studies, King's College London, London, UK

David Kollosche Humanwissenschaftliche Fakultät, Universität Potsdam, Potsdam, Germany

Anna Llewellyn School of Education, Durham University, Durham, UK

Sverker Lundin Department of Literature, History of Ideas and Religion, University of Gothenburg, Gothenburg, Sweden

Mônica Mesquita Policy and Governance, MARE - Marine and Environmental Sciences Centre, Caparica, Portugal

Alex Montecino Faculty of Engineering and Science, Aalborg University, Aalborg, Denmark

Candia Morgan Department of Curriculum Pedagogy and Assessment, University College London Institute of Education, London, UK

Institute of Education, University College London, London, UK

Alexandre Pais Education and Social Research Institute, Manchester Metropolitan University, Manchester, UK

Aldo Parra-Sanchez UNESCO Center of PBL in Engineering and Science, Department of Planning and Development, Aalborg University, Aalborg, Denmark

Hauke Straehler-Pohl Fachbereich Erziehungswissenschaft und Psychologie, Freie Universität Berlin, Berlin, Germany

David Swanson Manchester Institute of Education, The University of Manchester, Manchester, UK

Paola Valero Faculty of Engineering and Science, Aalborg University, Aalborg, Denmark

Faculty of Science, Stockholm University, Stockholm, Sweden

David Wagner Faculty of Education, University of New Brunswick, Fredericton, NB, Canada

Chapter 1
Welcome to the Jungle. An Orientation Guide to the Disorder of Mathematics Education

Hauke Straehler-Pohl, Alexandre Pais, and Nina Bohlmann

Abstract This introductory chapter builds on the assumption that the sociopolitical dimensions of mathematics education have been gradually recognised as an important part of mathematics education research. We problematise the process of institutionalising these dimensions as a firm strand of mathematics education research just as "philosophy of mathematics (education)", "history of mathematics (education)", "modelling and applications" or "geometry". This leads us to identify and conceptualise "disorder" as the foundation of the sociopolitical dimensions and accordingly to propose a shift from focussing diversity towards focussing disorder. Finally, we illustrate how the chapters of this book contribute to such an alternative self-conception of sociopolitical research in mathematics education.

The Promise of Sociopolitical Research in Mathematics Education

> Welcome to the jungle, we've got fun and games
> We got everything you want honey, we know the names
> We are the people that can find whatever you may need
> If you got the money honey we got your disease
> (Guns'n'Roses. "Welcome to the jungle")

Regardless of whether the attention to the sociopolitical dimensions of mathematics education is to be rated as a "shift of paradigm", a "turn" (e.g. Gutiérrez, 2013a, 2013b; Valero, 2004) or rather as the development of a new "branch" (Jablonka, Wagner, & Walshaw, 2013; Jablonka & Bergsten, 2010), such dimensions have been gradually recognised as an important part of mathematics education

---

H. Straehler-Pohl (✉) • N. Bohlmann
Fachbereich Erziehungswissenschaft und Psychologie, Freie Universität Berlin, Berlin, Germany
e-mail: h.straehler-pohl@fu-berlin.de

A. Pais
Education and Social Research Institute, Manchester Metropolitan University, Manchester, UK

H. Straehler-Pohl et al. (eds.), *The Disorder of Mathematics Education*,
DOI 10.1007/978-3-319-34006-7_1

research. It is about to become institutionalised as a firm strand of mathematics education research just as "philosophy of mathematics (education)", "history of mathematics (education)", "modelling and applications" or "geometry". Established conferences like ICME and CERME now incorporate in their programmes working groups exclusively dedicated to sociopolitical studies; the "Mathematics Education and Society" conference series has become an inherent part of the field (see for example Berger, Brodie, Frith, & le Roux, 2013; Gellert, Jablonka, & Morgan, 2010; Mukhopadhyay & Greer, 2015); on a regular basis, themes like "equity", "diversity", "social justice" and "critical education" are problematized in edited collections (see for example Alrø, Ravn, & Valero, 2010; Atweh, Graven, Secada, & Valero, 2011; Bishop, Tan, & Barkatsas, 2015; Black, Mendick, & Solomon, 2009; Clarkson & Presmeg, 2008; de Freitas & Nolan, 2008; Forgasz & Rivera, 2012; Herbel-Eisenmann, Choppin, Wagner, & Pimm, 2012; Skovsmose & Greer, 2012; Valero & Zevenbergen, 2004) and special issues of the most renowned journals (see for example Gutiérrez, 2013b; Meaney & Valero, 2014; Morgan & Kanes, 2014); and one of the four sections that compound the very recent Third International Handbook of Mathematics Education is dedicated to "Social, Political and Cultural Dimensions in Mathematics Education" (Clements, Bishop, Keitel-Kreidt, Kilpatrick, & Leung, 2013).

"The sociopolitical dimensions of mathematics education" has thus become a banner whereby researchers, who desire to contribute to the betterment of society, can situate their work within mathematics education. They do not longer need to do so at the margin of the field, as it was the case 30 years ago when the first elements of sociopolitical approaches started to appear as research areas such as ethnomathematics or critical mathematics education (see also Gellert in this volume). Those works are often critical of past and current approaches to mathematics education, and have the explicit aim to politicise the practices of teaching and learning mathematics as well as research itself.

Against this background, the term "sociopolitical" promises to bring new insights to the persisting question of why mathematics appears to be just for some, but not for all (see for example Gates & Vistro-Yu, 2003). We can conceive this promise as a finally sounding wake-up call; however, it can be conceived as well as the python Kaa from The Jungle Book (Walt Disney Productions, 1967) when singing:

> Trust in me, just in me
> Shut your eyes and trust in me
> You can sleep safe and sound
> Knowing I am around.

To succumb to the temptation and therefore embracing the sociopolitical label provides a sense of reassurance and clean conscience to the researcher. It supplies researchers with a coherent narrative wherein to situate their work amidst a field and a practice where failure (in school mathematics) is a generalised feature. Also, it enables researchers to see themselves as partisans against the negative effects of mathematics education, thus making it difficult to critically reappraise what might be their own role in these same effects. By not being at the margins of the field anymore—with implications for publishing, teaching, funding and travelling (to conferences, project meetings and the like)—the politically engaged researcher finds reassurance in the idea that, through their work, political concerns are being addressed in mathematics education. The fact that the apparent progress marked by research is hardly accompanied by an improvement of the teaching and learning of mathematics outside the realm of research (that is after the research caravan has moved on or where it does not dwell), often goes unremarked.

The politically engaged researcher can continue her or his research because the sociopolitical banner is there to offer her or him reassurance that he or she is on the right path. The aspiration of this volume is to show that the motive of "politicising" operates in the ambiguous field of tension between political activation and (unintended) political pacification.

Sociopolitical Research: A Diverse Forest?

There is nowadays in the field of mathematics education a considerable array of different approaches, theories and methodologies the politically engaged researcher can select from. One can opt for a postmodern approach, emphasising issues of power and identity (e.g. Gutiérrez, 2013a; Stinson & Bullock, 2012; Valero, 2015; Valero & Stentoft, 2010; Walshaw, 2004); one can also decide for a more traditional use of critical theory (e.g. Gutstein, 2006; Skovsmose, 1994), or for exploring the vast array of contemporary theories by bringing into the field the work of contemporary philosophers, linguists or sociologists and their cutting-edge research (Brown & Walshaw, 2012; Brown, Williams, & Solomon, 2016). However, the politically engaged researcher is also confronted with a quite narrow and pre-defined horizon to which he or she should align his or her movement. There appears to be an unquestionable assumption that "mathematics for all" is the only possible emancipatory prospect wherein to situate one's work, if the purpose is to be recognised for politically relevant research or to build up an identity as a political mathematics educator. In order to be of value or importance, ideas must contribute to the evolution of this sublime prospect.

It is our contention that such a narrowing of the political horizon to a regulative ideal—"mathematics for all"—curtails the ways in which researchers could conceive the "political" in mathematics education and disavows a more critical approach to the field's place in political economy. It narrows down the speculative *could* which is oriented to a yet to be thought future, to a normative *should* that is oriented to perpetuating the ideals of the present tense.

Instead of conceiving the systematic failure in providing mathematics for all as the result of particular obstacles that, once removed, would allow the fulfilment of the ideal, we challenge the reader to conceive these obstacles as being immanent to the field of mathematics education as such. That is, not only these obstacles cannot be removed but are there precisely to create the illusion that without them, mathematics for all will be possible. It is thus not the obstacles, but the illusion what maintains the status quo.

Fuelled by the definition of what should be, there is a clear danger for the socio-political dimensions of research to become both proceduralised and technicalised: fixed theories on how to conceive mathematics for social justice; fixed methodologies how to research it; and finally fixed pedagogies for how to apply it. As a result, while there is no doubt about the diversity of approaches to research the societal dimensions of mathematics education, this diversity, however, takes place within a relatively unified symbolic order; an order which is well aligned to the operating modes of global capitalism.

This volume rests on the possibility of finding the social and political relevance of mathematics education exactly where it appears to be contradictory, chaotic or even "messy". Instead of taking for granted the ideal of "mathematics for all", the contributions gathered in this volume seek to unsettle this ideal, by probing the way researchers use it as an empty signifier creating a sense of harmony between different research approaches. In this regard, the title of the volume can be seen as an indication that in order to revitalise our political imagination, we need to break with the alleged coherence or "order" of mathematics education.

From Diversity to Disorder

But how is it possible to break with this allegedly coherent order, particularly under the condition that this order is not monolithic, but already diversified? Moreover, how can this be possible, when the allegedly coherent order is not an authoritative mandate, but rather an "open" appeal that agents pursue in a self-determined, free and often enthusiastic way? Breaking with order *from within* appears as an indissoluble dilemma, as an impossibility. At the same time, any order of meanings that is not simply self-referential but relates to the empirical reality of lived experiences must include a constitutive moment of inner contradiction: a paradoxical moment whose recognition from within can be suspended temporarily for practical (or pragmatic) reasons, but that nevertheless remains finally unresolved and unresolvable.[1] Such paradoxical moments call into question the distinction between "inside" and "outside" of a symbolic order. However, this commonly assumed dichotomy has been effectively undermined by contemporary philosophers (see for example Deleuze's (1993) concept of the "fold", or Žižek's (2009) elaborations on the "parallax"). Following such attempts to undermine the only apparently unbridgeable division of inside and outside, the task would be, then, to systematically arrange an array of "inside" perspectives, so that they lay bare the intrinsic paradoxes in the *mode of collision*. Let's take Jorge Luis Borges' encyclopaedia of animals as an example, which has not least achieved fame through its discussion in the preface of Foucault's "The order of things" (2009, p. xvi. originally published in French in 1966):

[1] See the Gödel theorem (Gödel 1931) as an example that should be more or less familiar to most mathematics educators.

> This passage quotes a "certain Chinese encyclopaedia" in which it is written that "animals are divided into: (a) belonging to the Emperor, (b) embalmed, (c) tame, (d) sucking pigs, (e) sirens, (f) fabulous, (g) stray dogs, (h) included in the present classification, (i) frenzied, (j) innumerable, (k) drawn with a very fine camelhair brush, (l) et cetera, (m) having just broken the water pitcher, (n) that from a long way off look like flies."

According to Foucault, this encyclopaedia succeeds in breaking up all the ordered surfaces and all the planes with which we are accustomed to tame the wild profusion of existing things, and continuing long afterwards to disturb and threaten with collapse our age-old distinction between the Same and the Other. (ibid.)

It is *through* this apparently absurd and surreal order of categories that we are enabled to think meanings that were impossible to think beforehand. The fact that we can think these meanings which were formerly impossible is less the result of naming and fantasising a corresponding meaning (e.g. sirens) than the result of a contradictory order that distinguishes the categories by means of a classification:

> It is not the "fabulous" animals that are impossible, since they are designated as such, but the narrowness of the distance separating them from (and juxtaposing them to) the stray dogs, or the animals that from a long way off look like flies. What transgresses the boundaries of all imagination, of all possible thought, is simply that alphabetical series (a, b, c, d) which links each of those categories to all the others. (p.xvii)

Hence, it is through admittance to an order that we can reach a space beyond it—a disorder—without simply assuming a different, supposedly coherent "external" order. It is the specification of a (yet) strange and suspicious order that forces us to stop thinking things in the manner we are used to; it compels us to make familiar orders of meaning collide with unfamiliar ones. It is through this collision that orders reveal their contingency and lay bare their intrinsically "political foundation" (Žižek, 2000). Following this line of thought, we perceive "the disorder of mathematics education" *not as the absence of order*, nor as an allegedly original and natural state that precedes our current perception of the world. Rather, we perceive the disorder *as the intrinsic excess of order,* a not foreseeable surplus, an obscene downside that results from the process of ordering itself. As Pfaller (2011) demonstrates, any symbolic order necessarily includes the command for its very own violation.[2]

Thus, when we aim at breaking with the alleged coherence or "order" of mathematics education from within to lay bare the disorder beyond familiar perceptions of mathematics education, a relativist celebration of diversity appears as a dead end. Simply adding a new alternative perspective leaves the status quo perfectly intact. Only by facing the obscene downside of order can we dismantle the political contingency of the status quo, revealing its disordered foundation, injecting it back into the political arena.

[2] "We shall not forget that 'symbolische Ordnung' makes both in French and English semantic reference to the social system of rules [Regeln] and also to a command [Gebot]; hence, 'l' 'ordre symbolique' and 'the symbolic order' are not only denoting an order [Ordnung] but also a command, referring to the command [Anordnung] to transgress that order." (Pfaller 2011, p. 26, own translation)

Thus, we shall not conceive the "disorder of mathematics education" as neither a complement to the existing branch of "sociopolitical research" (e.g. "deconstructions of sociopolitical research in mathematics education"), nor as a sub-category of "sociopolitical research" (e.g. "poststructuralist approaches to the sociopolitical dimensions of mathematics education"), but integrate it into the (yet) absurd taxonomy of existent branches of mathematics education as a research field:

> (a) calculus, (b) algebra, (c) history of mathematics education, (d) mathematics education with zero gravity, (e) embodied mathematics, (f) gardening mathematics (g) workplace mathematics, (h) gala dinners at ME conferences, (i) sociopolitical dimensions, (j) et cetera, (k) dynamic geometrical software, (l) technologies, (m) disorder, (n) branches that from a long way off look like praying wheels, (o) mathematical modelling, (p) all kinds of animals.

For this reason the present volume gathers a selection of scholars (from PhD students to well-established professors), who seek on the one hand for ways of productively engaging themselves in the disordering of mathematics education as a research field, while on the other hand practising a strong commitment to theory (another name for the symbolic order). This strong devotion to theory, however, must be coupled with the recognition of the inbuilt fallibility of any theory. Instead of disavowing the messiness and the contradictions that emerge from researching mathematics education through a theoretical lens, the authors of this volume seek to make these contradictions visible. In this way, the present volume highlights precisely the problems, tensions and contradictions that are part of existing research.

How to Systematically Organise a Disorder (in Research Practice) …

If we want to perceive disorder not as the absence of order, but as the excess of acts of ordering, how can we practically create an academic space that maximises the likelihood of collisions that generate an unforeseen surplus? This volume has its origins in a conference that Hauke Straehler-Pohl and Nina Bohlmann hosted at the Freie Universität Berlin in January 2015. In the call for the conference, 27 international delegates were invited to relate their work in one way or the other to "the disorder of mathematics education". Instead of identifying a unified and unifying *problem*, the call just problematised the current state of sociopolitical research in mathematics education, thus leaving it open for delegates to interpret the "disorder" in their own fashion (see a slightly elaborated version of the initial call on http://www.ewi-psy.fu-berlin.de/dome). In this way, through relating a very own perspective on "disorder", each delegate contributed with their perception of what the problem of current sociopolitical research is. Some of the delegates have developed or are currently developing cutting-edge methodologies allowing mathematics education research to meet the claim of being sociopolitically relevant. Other contributors have produced a radical and re-politicised critique on the kind of research approaches that claim to be sociopolitically relevant. As a result, we celebrated a conference, where a range of different assessments of what the problem is, and how it could and should be approached, collided—in the spirit of sharing, discussing and contesting the developed approaches. In brief: to make the different assessments disputatious and hereby sound out the conditions under which an understanding of sociopolitically relevant mathematics education research can be developed beyond any abstract idealisation of the societal role of mathematics education. The delegates left the conference with the feeling that they had worked hard and constructively for 3 days in a common spirit; however, having produced a disorder that points towards a multiplicity of different and partly contradictory directions.

... and How to Retrospectively Order It

Against this background, if there is a common trace uniting the different contributions present in this volume, it is the assumption that any sense of unity always displays the structure of a defence against the contradictory nature of education and research. On the one hand, any political vision of mathematics education needs to be contextualised within a broader picture that transgresses the boundaries of educational institutions, like schools or universities. Albeit in different ways, contributors assume that mathematical knowledge, beliefs in and about mathematics, and a "mathematical mind-set" are not the sole result of an institutionalised education (schools and universities). A mathematical rationality is present in contemporary society and reproduces itself through technologies, social practices, media and other spheres of social life. The analysis of how current cultural, social and political practices enact mathematics in a panoply of different ways, allows contributors to criticise the apparent ideological coherence on what mathematics is and to criticise the common shared idea that more and better mathematics for all is an intrinsically benign goal. In order to make this compilation accessible for readers, we thus set out and classify the diverse papers in sections, aware that each classification highlights certain group-resemblances and neglects others, making visible certain contradictions by suppressing others. Thus, during the process of editing this volume, the titles and the compositions of the sections were in a state of permanent transformation—which might never stop transforming if we were to edit this volume for ever, trapping us in a vortex, like the eyes of the python Kaa. However, as a date of publication inevitably serves as a record of one single moment in time, we have come to the following classification of the contributions to this volume:

I. What bonds us to mathematics,
II. Disordering narratives of progress in mathematics education,
III. Disordering school mathematics,
IV. Disordering the role of the mathematics education researcher.

Part I: What Bonds Us to Mathematics

The first section of this volume challenges the assumption that the collective effort of making more and more students learn more and more mathematics is beneficial in itself. When focussing on different roles of actors in social practices (adults/parents, consumers, mathematics educators), instead of assuming a division between the proponents and opponents of mathematics, the chapters in this section analyse how opposition and endorsement are effective together. In this way, the chapters of this section analyse how the bond to mathematics is a result of this ambivalence.

Sverker Lundin and Ditte Storck Christensen ask why people who rarely use mathematics in their daily lives (adults), nevertheless, consent to the necessity of making (their) children learn it. The authors explore the role of compulsory schooling in the development of this ambivalent attachment to mathematics, where people learn to love and hate mathematics simultaneously. By sending children to school, where a "show" of mathematics-love is performed for the adults' gaze, adults delegate to children their disavowed love for mathematics, similar to for example the use of praying wheels.

Hauke Straehler-Pohl discusses available reflections regarding the flowering of mathematisation and demathematisation of social practices and relates them to current technological developments. He reveals mechanisms that allow people to enjoy this development despite (or rather because of) a sense of alienation that comes along with it, and draws conclusions concerning the further development of a mathematics education that critically reflects its role in society. He proposes to provide a legitimate space for students in the mathematics classroom to reject the demand to solve problems of social significance by means of mathematics.

Alexandre Pais asks for the reasons why so many mathematics educators prefer to create a reality so at odds with the one experienced by the vast majority of teachers and students worldwide. He shows how researchers, instead of recognising this mismatch as a symptom, utilise it to sublimate the reality of research at the cost of the reality of school mathematics. The author conceptualises the collective act of sublimation as a defence mechanism, leading mathematics education as a research field into a state of narcissism. As an alternative, Pais suggests a form of reality-therapy for mathematics education, one that invites researchers to seriously engage with its symptoms: students' systematic failure, absence of change, increasing of testing, etc.

If the chapters are right in their common conclusion that the desire for more mathematics education is constitutively built on the condition of ambivalence, it becomes clear that mathematics education requires a profound discussion about how it *does* perceive and how it *could* perceive progress.

Part II: Disordering Narratives of Progress in Mathematics Education

The second section therefor focuses on current narratives that shape the field of mathematics education (research) and give meaning to it. The chapters particularly centre upon those narratives that purport and address progress and development by means of mathematics. This section compiles chapters that deconstruct and/or reconstruct current narratives, revealing how they respond to historical and social developments, trends and requirements. In this way, the chapters lay bare the narratives' intrinsically political foundations and hence open them for political and scholarly debate.

Uwe Gellert opens this section with a chapter that comments on the argumentation by Pais in the preceding chapter. He argues for further reflections on the concrete demands of mathematical knowledge in contemporary society. The topic of universality of mathematical education is the pivot around which historical, functional, emancipatory and political considerations unfold. Gellert meticulously reconstructs how "mathematics for all" emerged as a response to tangible material and social imbalances and confronts this reconstruction with the critique developed by Pais (2012) on the topic of equity. In this way he lays bare the shortcomings both of the narrative of "mathematics for all" as well as of its critique. As a result, he argues for a search for mediating alternatives.

Aldo Parra-Sanchez critically analyses the assumption prevalent in ethnomathematics that its privileged focus of study should be the intersection of culture and mathematics. If we are to understand "emancipation" as a dominant narrative of progress in ethnomathematics, Parra-Sanchez reveals how the privileging of intersectional approaches rather tends to undermine than to fulfil this intention. As an alternative he suggests to shift attention away from the study of intersections towards the practice of "barters" that serve for researchers *and* researched to mutually inform and, in particular, to irritate each other.

Eva Jablonka and Christer Bergsten focus on the narrative of "good mathematics teaching" and its actualisation in teacher evaluation and curriculum design. The authors explore how "good mathematics teaching" easily slips into becoming an empty signifier. They reconstruct through three different examples the way in which the signifier "good mathematics teaching" feeds into a hegemonic narrative for persuading sponsors and policy makers to fund and promote research. In this process, meaning is created self-referentially, installing a self-perpetuating machinery for financing, defining, measuring and producing "progress".

Similarly, Candia Morgan addresses the junction between research, policy and practice. She focuses on the theoretical and ethical problems that arise from this encounter. However, instead of deconstructing one particular narrative of progress, she turns her attention to a general constitutive dilemma of research in mathematics education: "users" of research in politics and practice tend to recontextualise research to serve their own interests and to incorporate the results into alternative discourses that appropriate the users' pre-existing narratives. Not as a solution for the dilemma, but as a way to productively deal with it, Morgan suggests to step out of the role of researcher in order to engage in the social practices one seeks to affect.

Alex Montecino and Paola Valero produce a critical analysis of the way international entities such as OECD and UNESCO convey progressive ideologies about the importance of mathematics and the role of teachers as both products and agents of those ideologies. They systematically analyse the way teachers are portrayed in a multitude of different documents, and conclude that there is a strong tendency in current educational policies to transform teachers into agents of the market with the task of selling this precious piece of knowledge called mathematics.

Anna Llewellyn considers and critiques the role of technologies of power and surveillance, and governmentality in mathematics education research. The chapter deconstructs the fiction of the free, autonomous self, and discourses of progress as a key taken-for-granted truth of mathematics education research within the UK and other Western contexts. It is argued that the natural, developmental, free, child is (re)produced through both overt and covert surveillance and monitoring, from both schools and universities. Llewellyn thus calls into question the modern idea of progress, of which mathematics is one of the cornerstones.

Common to the chapters in this section is their concern with developing a practice of reflexivity (Bloor, 1976; Bourdieu, 2001) on mathematics education as a research field. That is to reflect the external circumstances that have shaped mathematics education as a field in its emergence, showing how it is socially and historically contingent. The next section will take advantage of this practice of reflexivity and move the focus back to the privileged subject of mathematics education: school mathematics.

Part III: Disordering School Mathematics

In the third section of this volume the authors centre their attention on the particular object of school mathematics. This happens in the spirit of avoiding the shortcomings of the taken-for-granted narratives of progress that have been deconstructed in the preceding section. All chapters address school mathematics as fundamentally situated in social, political and economic contexts, revealing its idiosyncrasies. Thus, this section puts the premises of the whole book into test: how can "disordering" practices of reflexivity help research to develop an alternative relation to its object?

David Kollosche casts a gaze on students' perceptions of mathematics from a socio-critical perspective. He does so by developing a framework that allows him to interpret students' perceptions as expressions of their developing subjectivities. He discusses how devotion to mathematics, suffering from mathematics, as well as seeing personal relevance and the opportunity to be challenged by mathematics, can be considered technologies of the self that students develop in response to mechanisms of subjection. These mechanisms, through privileging and sanctioning, make visible the dogmatic power-knowledge of school mathematics.

Jehad Alshwaikh and Hauke Straehler-Pohl focus the relationship between learning mathematics and the sociopolitical context where it "lives in" in Palestine. The authors work out how the construction of a passive mathematics learner and the simultaneous construction of mathematics as an abstract form of knowledge serve the status quo in Palestine. However, as the promotion of agency and relevance resemble all too much those narratives of progress deconstructed in the first and second section of this book, the authors decide to identify with this dilemma instead of avoiding it. They provide drafts for two hypothetical classroom activities that are designed as provocations for teacher education that simultaneously sensitise for the urgency of making mathematics relevant to Palestinian life and sensitise for the risks that come along with this.

While assessment is invariably conceived as a pedagogic strategy to enhance learning, Lisa Björklund-Boistrup in her chapter posits it instead as a governing apparatus. Reconstructing everyday assessment acts from Swedish mathematics classrooms, she construes four different discourses of assessment that position students differently in terms of power and in terms of opportunities to engage with mathematics. Relating these four discourses to each other, she construes an assessment dispositive that not only acts as a gatekeeper for some students, but also effects teachers' opportunities to transform their practice.

David Swanson refocuses debates on alienation and mathematics education around the unifying factor of the commodity form of production. Inspired by Walter Benjamin, he develops a methodology of montage that does not simply borrow from the arts, but serves as an experimental scholarly approach to truth. This methodology enables him to arrange excerpts from student interviews in a way that they reveal how the commodity form of production has translated down from a macro-structure to students' experiences. At the same time, the montage opens up an outlook for a different possible realisation of mathematics education.

Melissa Andrade Molina and Paola Valero focus—similarly to Kollosche—on how school mathematics instils technologies of the self in students, shaping what they call "the desired child". They illuminate how school geometry privileges certain conceptions of space while sanctioning others. The analysis of curricular material from Chile and OECD documents reveals how school geometry fabricates a desired form of subjectivity that requires students to detach from the "eyes of their bodies" in favour of a rationalised, a "sightless eye". The power-knowledge of school geometry thus subjugates one configuration of the body to another, instrumental one at the cost of alienating students.

Tony Brown's chapter suggests a similar effect of school mathematics when he analyses the relation of rationality and belief in learning mathematics. The chapter reveals how conceptualising beliefs as "irrational" distortions of "rational" mathematical thought is first of all a product of a tightened form of social management. The paper argues that rational mathematical thought necessarily rests on beliefs. Loosening the administrative grip so that the diversity of beliefs can play out in the mathematics classroom, Brown maintains, could release students' and teachers' own powers. In the process of creating this book, this chapter disappeared from the book's surface and reappeared in *Educational Studies in Mathematics* (doi: 0.1007/s10649-015-9670-7). All contributors to this book, including the author himself, consider it nonetheless a part of this book.

Common to all the chapters in this section is a suspension of the temptation to already base their analysis of school mathematics on normative pre-assumptions of what *ought-to-be* and focus on what *is* in the first place. Nevertheless, based on such analysis of the current state, the chapters offer speculative imaginations of what school mathematics *could be*.

Part IV: Disordering Role of the Mathematics Education Researcher

Developing research with the aim of transforming the status quo through speculative imaginations of *could-bes* implies walking on a thin line, always risking slipping into pacifying narratives of *ought-to-bes*. Furthermore, it often implies relating oneself to discourses within which one is positioned and with which one simultaneously seeks to break; it often implies taking advocacy for not only emancipating oneself but also one's environment, while being aware that emancipation always implies a subjective position. To put it shortly, it confronts the sociopolitically engaged researcher with challenges on the level of her or his subjectivity. The last section of the book is thus dedicated to exercises of researchers' self-reflexions about their role and place in research practice.

Peter Appelbaum's chapter addresses the dilemma of how mathematics educators who are positioned simultaneously as inside and outside of the field of mathematics education can productively deal with their desire to promote change. He builds on Deleuze's notion of the *fold* in order to destabilise the differentiation of insides and outsides, and suggests the development of nomadic topologies as seeds for change (of subjectivity, of the research field, and of broader sociopolitical contexts simultaneously). Reflecting on his own research biography, Appelbaum illustrates how developing such topologies can alter researcher's subjectivity by exploring five arbitrary phases towards enacting educational space as (artistic) studio.

David Wagner reflects on the way *he* positions *himself* when publishing research in mathematics education. He takes the recurrent question "Where is the maths?" as an example that is often strategically employed to artificially construct a division of mathematics education as research field, coercing researchers to either position themselves on the "side of politics/culture/etc." or on the "side of mathematics". Analysing his own research practice as an author in relation to a reader, he illuminates how differently developed storylines not only address different readers, but also bring different model readers into being. Wagner concludes with concrete suggestions for how researchers can reflect already in the process of writing on the positioning that can be associated with the texts they produce.

Mônica Mesquita closes this section and in doing so, she closes the book. Mesquita addresses the probably most fundamental question that the contributors to this volume grapple with: how is it possible to be a critical researcher while simultaneously struggling with surviving in a capitalist world-order? How can critical researchers "realise themselves"? Based on her own biography, she reflects on how this desire places the researcher in a "boundary space". However, Mesquita argues, critical researchers should not let themselves be paralysed by being "in" the boundary, as it is exactly in the boundary, where the production of yet to be thought knowledge has the potential to disturb the hegemony of the system. However, producing such form of new knowledge requires a certain posture and so Mesquita closes this book with an appeal of Etiènne Balibar:

> Let us be intolerant with ourselves and "pass on to another stage."

… If You Got the Money Honey, We Got Your Disease …

Finally, the title of the volume plays with a double meaning of the word "disorder". While it has become clear that we perceive mathematics as a chaotic realm of different meanings, whose (dis)order is contingent upon collective acts of ordering, the second meaning of the word humorously plays with the position of the contributing authors in the field of mathematics education. While it is certainly wrong to claim that the contributors suffer from exclusion or discrimination—some authors have reached quite powerful positions in the international field, continuously publishing cutting-edge articles in the most renowned journals—they share the feeling that they are enduringly regarded in a distanced, at times suspicious manner. Scholars with a "disorder" can thus be humorously understood as those who appear not to function in the way they are supposed to; a way that is not aligned with some of the most unquestioned assumptions in mathematics education, such as the idea that mathematics is important for the daily life of people or the enticing goal of "mathematics for all". Not seldom, researchers engaged in the "disorder" are insinuated because of their cruel "pessimism", as if their world-view was an infectious disease and not simply a political positioning. A political positioning that more often than not comes along with an optimism that a different world, actually *is* possible—an optimism that this world is possible in exactly the same reality we *live* in, and not in a *dream*-reality that we first need to construct by a more and more fine-tuned research machinery.

The aim of the present volume is not to achieve consensus but precisely to unravel the established consensus on the importance of mathematics and its role in education and the broader society. The chapters are thus to be understood as invitations to challenge fossilised beliefs, and it should not be too difficult to find opposing positions in the different contributions. This very same spirit animated the conference which preceded this volume. Our contention is that contrasting and disagreeing is a much more prolific method to address the current problems of mathematics education than constructing an apologetic narrative that, although narcissistically pacifying, leaves many of the contradictions of mathematics and its education untouched. Through this volume, we invite everyone in mathematics education to join the "folks with the disorder" and to dare to be dysfunctional at times.

References

Alrø, H., Ravn, O., & Valero, P. (Eds.). (2010). *Critical mathematics education: Past, present, and future. Festschrift for Ole Skovsmose*. Rotterdam: Sense.

Atweh, B., Graven, M., Secada, W., & Valero, P. (Eds.). (2011). *Mapping equity and quality in mathematics education*. Dodrecht: Springer.

Berger, M., Brodie, K., Frith, V., & le Roux, K. (Eds.). (2013). *Proceedings of the Seventh International Mathematics Education and Society Conference* (Vol. 1–2). Cape Town: Hansa.

Bishop, A., Tan, H., & Barkatsas, T. (Eds.). (2015). *Diversity in mathematics education. Towards inclusive practices*. Cham: Springer.

Black, L., Mendick, H., & Solomon, Y. (Eds.). (2009). *Mathematical relationships in education: Identities and participation*. New York: Routledge.

Bloor, D. (1976). *Knowledge and social imagery*. London: Routledge and Kegan Paul.

Bourdieu, P. (2001). *Science of science and reflexivity*. Chicago: University of Chicago Press.

T. Brown & M. Walshaw (Eds.). (2012). Special issue: Mathematics education and contemporary theory. *Educational Studies in Mathematics, 80*(1–2).

T. Brown, J. Williams & Y. Solomon (Eds.). (2016). Special issue: Mathematics education and contemporary theory. *Educational Studies in Mathematics*.

Clarkson, P., & Presmeg, N. (Eds.). (2008). *Critical issues in mathematics education*. New York: Springer.

Clements, M. A. K., Bishop, A., Keitel-Kreidt, C., Kilpatrick, J., & Leung, F. K.-S. (Eds.). (2013). *Third international handbook of mathematics education*. New York: Springer.

de Freitas, E., & Nolan, K. (Eds.). (2008). *Opening the research text: Critical insights and in(ter) ventions into mathematics education*. New York: Springer.

Deleuze, G. (1993). *The fold: Leibnitz and the Baroque*. Minneapolis, MN: University of Minnesota Press.

Forgasz, H., & Rivera, F. (Eds.). (2012). *Towards equity in mathematics education. Gender, culture, and diversity*. Heidelberg: Springer.

Foucault, M. (2009/1966). *The order of things: An archaeology of the human sciences*. London: Routledge.

Gates, P., & Vistro-Yu, C. (2003). Is mathematics for all? In A. Bishop, M. A. K. Clements, C. Keitel-Kreidt, J. Kilpatrick, & F. K.-S. Leung (Eds.), *Second international handbook of mathematics education* (pp. 31–74). Dordrecht: Kluwer.

Gellert, U., Jablonka, E., & Morgan, C. (Eds.). (2010). *Proceedings of the Sixth International Mathematics Education and Society Conference* (Vol. 1–2). Berlin: Freie Universität Berlin.

Gödel, K. (1931). Über formal unentscheidbare Sätze der Principia Mathematica und verwandter Systeme I. *Monatshefte für Mathematik und Physik, 38*(1), 173–198.

Guns'n'Roses. (1987). "Welcome to the jungle." Appetite for Destruction. Geffen. MCD.

Gutiérrez, R. (2013a). The sociopolitical turn in mathematics education. *Journal for Research in Mathematics Education, 44*(1), 37–68.

R. Gutiérrez (Ed.). (2013b). JRME equity special issue. *Journal for Research in Mathematics Education, 44*(1).

Gutstein, E. (2006). *Reading and writing the world with mathematics: Toward a pedagogy for social justice*. London: Taylor & Francis.

Herbel-Eisenmann, B., Choppin, J., Wagner, D., & Pimm, D. (Eds.). (2012). *Equity in discourse for mathematics education. Theories, practices, and policies*. Dodrecht: Springer.

Jablonka, E., & Bergsten, C. (2010). Theorising in mathematics education research: Differences in modes and quality. *Nordic Studies in Mathematics Education, 15*(1), 25–52.

Jablonka, E., Wagner, D., & Walshaw, M. (2013). Theories for studying social, political and cultural dimensions of mathematics education. In A. Bishop, M. A. K. Clements, C. Keitel-Kreidt, J. Kilpatrick, & F. K.-S. Leung (Eds.), *Third international handbook of mathematics education* (pp. 41–67). New York: Springer.

T. Meaney & P. Valero (Eds.). (2014). Socioeconomic influence on mathematical achievement: What is visible and what is neglected. *ZDM: The International Journal of Mathematics Education, 46*(7).

C. Morgan & C. Kanes (Eds.). (2014). Special issue: Social theory and research in mathematics education. *Educational Studies in Mathematics, 87*(2).

Mukhopadhyay, S., & Greer, B. (Eds.). (2015). *Proceedings of the Eighth International Mathematics Education and Society Conference* (Vol. 1–3). Portland: Ooligan.

Pais, A. (2012). A critical approach to equity in mathematics education. In O. Skovsmose & B. Greer (Eds.), *Opening the cage: Critique and politics of mathematics education*. Rotterdam: Sense Publishers.

Pfaller, R. (2011). *Wofür es sich zu leben lohnt: Elemente materialistischer Philosophie*. Frankfurt am Main: Fischer.

Skovsmose, O. (1994). *Towards a philosophy of critical mathematics education*. Boston: Kluwer Academic Publishers.

Skovsmose, O., & Greer, B. (Eds.). (2012). *Opening the cage: Critique and politics of mathematics education*. Rotterdam: Sense.

Stinson, D. W., & Bullock, E. C. (2012). Critical postmodern theory in mathematics education research: A praxis of uncertainty. *Educational Studies in Mathematics, 80*(1), 41–55.

Valero, P. (2004). Socio-political perspectives on mathematics education. In P. Valero & R. Zevenbergen (Eds.), *Researching the socio-political dimensions of mathematics education: Issues of power in theory and methodology* (pp. 5–24). Boston: Kluwer.

Valero, P. (2015). Re-interpreting students' interest in mathematics: Youth culture and subjectivity. In U. Gellert, J. Giménez-Rodriguz, C. Hahn, & S. Kafoussi (Eds.), *Educational paths to mathematics* (pp. 15–32). Heidelberg: Springer.

Valero, P., & Stentoft, D. (2010). The "post" move of critical mathematics education. In H. Alrø, O. Ravn, & P. Valero (Eds.), *Critical mathematics education: Past, present, and future. Festschrift for Ole Skovsmose* (pp. 183–196). Rotterdam: Sense Publishers.

Valero, P., & Zevenbergen, R. (Eds.). (2004). *Researching the socio-political dimensions of mathematics education: Issues of power in theory and methodology*. Boston: Kluwer.

Walshaw, M. (Ed.). (2004). *Mathematics education within the postmodern*. Greenwich, CT: Information Age.

Walt Disney Productions. (1967). *The jungle book*. Dir. Wolfgang Reitherman.

Žižek, S. (2000). *The Ticklish subject: The absent centre of political ontology*. London: Verso.

Žižek, S. (2009). *The parallax view*. Cambridge: MIT Press.

Part I
What Bonds Us to Mathematics

Chapter 2
Mathematics Education as Praying Wheel: How Adults Avoid Mathematics by Pushing It onto Children

Sverker Lundin and Ditte Storck Christensen

Abstract Why is mathematics education endorsed even by people who know little mathematics and almost never use it? We explore this question with reference to a theoretical framework inspired by anthropology and psychoanalysis. Our answer is that participation in mathematics education generates ambivalence towards mathematics, where it is at the same time loved and feared. The response to this ambivalence is to endorse mathematics per proxy, by sending children to school where they put up a "show" of mathematics-love. By means of a psychological mechanism involving what we will call *the naïve observer* this arrangement allows people to both *seem to love mathematics* and at the same time *keep it at arm's length.*

> It may indeed be the highest secret of monarchical government and utterly essential to it, to keep men deceived, and to disguise the fear that sways them with the specious name of religion, so that they will fight for their servitude as if they were fighting for their own deliverance, and will not think it humiliating but supremely glorious to spill their blood and sacrifice their lives for the glorification of a single man.
>
> Baruch Spinoza 1670, Theological-Political Treatise

The question posed by Spinoza, how it is possible that men fight for their own servitude, has been called "the fundamental problem of political philosophy" (Deleuze & Guattari, 2004, p. 29). It is this question that stands in the focus of the present article. As our theoretical point of departure, we will take how this theme is developed by the Austrian philosopher Robert Pfaller in his *The Pleasure Principle in Culture* (2014). He, in turn, is much inspired by the Lacanian variety of psychoanalysis, as expounded most popularly by the Slovenian philosopher Slavoj Žižek (e.g., 1999, 2008). While both Žižek and Pfaller have connected their interpreta-

S. Lundin (✉)
Department of Literature, History of Ideas and Religion, University of Gothenburg, Gothenburg, Sweden
e-mail: sverker.lundin@gu.se

D.S. Christensen
Department of Education and Special Education, University of Gothenburg, Gothenburg, Sweden
e-mail: ditte.storck.christensen@gu.se

H. Straehler-Pohl et al. (eds.), *The Disorder of Mathematics Education*,
DOI 10.1007/978-3-319-34006-7_2

tions to art, literature, and pop culture, often with a clear aim of understanding the ideological workings of modern capitalism, the empirical focus of this chapter is mathematics education.

The article is structured as an argument for the single thesis that mathematics education can be interpreted as an instance of desire for servitude. With Pfaller (2014), we will add an important twist, namely the presence of certain displacements within the group of oppressed: between the time and place of desire (for servitude) and time and place of actually serving; between that which is purportedly desired and the activity in which this desire finds expression. We will argue that mathematics education can be interpreted as an expression of a characteristically modern fight for mathematics. But while mathematics is associated with power and emancipation, these values are *inverted* when they find expression in educational activity.

We want to highlight this chapter's close connection to the article "Hating School, Loving Mathematics. On the ideological function of critique and reform in mathematics education" (Lundin, 2011). That article was focused on *engagement* with mathematics education from the inside perspective of pupils, teachers, researchers and reformers. This chapter instead explores the perspective of parents and politicians who see mathematics education from a calm distance but carrying the experience with them, described in "Hating School, Loving mathematics," of having once been immersed in it. Our question is: *What do they see in mathematics education that makes it so important to them?*

Our argument has three parts. First, we show that a typically modern sentiment towards mathematics is *ambivalence*; while most people agree that mathematics is "very important" and a necessary prerequisite for competent participation in professional and everyday life, they usually are not very good at mathematics themselves and not that keen on actually using it (see for example Kollosche in this volume). In a second step, we show that ambivalence is often connected to what we will call *norm-evasion*; a display of acknowledgment (in our case of the importance of mathematics) that serves to hide its opposite (in our case the sentiment that mathematics is not important at all) at the same time. We argue that mathematics education is such a superficial display of engagement, performed by children and youth, which releases adults from the pressure of having to engage with mathematics.

Our argument forms a circle, perhaps reminiscent of Pierre Bourdieu's (1990) theory of "reproduction". It is closed by an analysis of how participation in norm-evasion, because of its superficiality, leads to exactly the state of ambivalence to which norm-evasion functions as a well-adjusted response. Children, as they take part in mathematics education, learn that it is their obligation to know and to love mathematics, while at the same time they get installed a fearful respect in them that makes them want to keep mathematics at arm's length. The solution to this predicament is mathematics education. Paradoxically, the solution and the cause of the problem turn out to be the same (this type of circularity is also a recurrent theme in Žižek's work, e.g., 2008, p. 211).

In the last section of the article, we discuss what we call *the disorder* of mathematics education. We there introduce the possibility of dissatisfaction with mathematics education as a response to ambivalence to mathematics. For good or for evil, such dissatisfaction disrupts the harmonious circle of social reproduction and introduces the possibility of change. It introduces measures to make mathematics education correspond to the mathematical ideal of power and emancipation better, instead of just functioning as a means of superficial evasion of the pressure to conform to this ideal. While this ambition of making education "work" towards its declared ends is well-intended, it usually takes the form of increased regulation and control, possibly in combination with an expansion of the educational sphere. In its consequences, such faith in the possibility of emancipation by means of mathematics education may therefore actually increase the level of subordination demanded by children in the modern education system. We conclude by placing this situation, where increased efforts of emancipation lead to increased subordination, in the context of the "dialectic of enlightenment" (Adorno & Horkheimer, 1997).

Ambivalence Towards Mathematics

At the surface, modern society seems to acknowledge the importance of mathematics. It is claimed, not only by politicians and researchers, but also in public documents such as curricula, that mathematics is, to state it mildly, "important." This claim is backed up by the position of mathematics in the education system, where pupils spend a substantial amount of time studying it. Furthermore, (lack of) mathematical knowledge is a perennial topic of public debate. In these debates, the question is almost never *if* mathematics is important, but rather by what measures the state of mathematics in society can be most efficiently improved. This sentiment is aptly illustrated by a recent Swedish public report, with the telling title "High time for mathematics":

> Proponents of education, industry and society, powerfully and unequivocally express that knowledge of mathematics is important and that good, meaningful knowledge [of mathematics] is a prerequisite for self-confidence, democracy, [economic] growth and lifelong learning. Concerted efforts for a long-term, sustainable development of mathematics education is both demanded and welcomed by all social groups and on all levels of education. (Hög tid för matematik 2001, p. 11, our translation)

However, it is not difficult to find cracks in the surface appearance of universal support for mathematics. Why, for instance, does there seem to be such a pressing need for reform of mathematics education? It is readily acknowledged in the public debate that "many have blockings and anxiety before the subject" (Matematikdelegationen, 2004, p. 201, our translation) and as mathematics education researchers know, there is a whole research area concerned with "mathematics anxiety" (see Fries, 2013 for an overview). The fact that mathematics education functions as an instrument for the reproduction of society by "sorting" the students has been well established for decades (Dowling, 1998). Furthermore, many researchers claim that mathematical knowledge is of little relevance in everyday life (Dowling, 2010; Lave, 1988; Lundin, 2008; 2010; 2011; Pais, 2013; Walkerdine, 1988). Such claims also resonate in mainstream media (Hultén, 2010; Johnson, 2004).

As supported by this presence of dissenting views on the general value of mathematical knowledge, our contention is that the education system as such, with its centrally established curricula and mandatory participation, is founded upon the conviction that the free will of modern citizens cannot be trusted when it comes to mathematical knowledge: here the state needs to make the choices, not only for the children but also for parents who are distinctly not allowed to exclude mathematics from the upbringing of their children. Mathematics education is an attempt to *construct and sustain unity* in an area where such unity would otherwise not obtain.

The cultural arrangements of education can be compared with those pertaining to politics. In liberal democracies, the political sphere is supposed to function as a neutral arena where people with different opinions can meet and argue for their standpoint. In a totalitarian society, to the contrary, the political sphere is "filled," so to speak, with one particular doctrine that is considered to be "right" and which is therefore prescribed by the state. It is considered to be the only one *rational*, in the sense that the presence of dissenting views can be explained with reference to lack of knowledge and understanding.

In totalitarian societies, it may be possible to identify dissenting views in a similar way as we have above pointed to dissenting views regarding the value of mathematical knowledge in our liberal democratic society. But in both cases these alternative voices can never really find foothold in the public sphere. The reason why this is impossible is that such public recognition would necessitate a radical reconceptualization of the societies in question. In the case of liberal democracies, the view that mathematical knowledge is useless for everyday life would necessitate a radical reconsideration of the relationship between science, education, and society. In a totalitarian, say communist, society, the public recognition of the view that communism is just one particular political vision among many others, and perhaps even a deeply problematic one as well, would necessitate a radical reconsideration of the functioning of government and of the sphere of politics.

The point of this comparison between education and totalitarian politics is that in both cases dissenting views may be held by everybody, but as "dominated," so to speak, by an institutionalized "official story" that regulates what people do in public. In liberal democracies, school attendance testifies to a practical acceptance of the "official story" of the universal usefulness of mathematical knowledge. In totalitarian societies, people may need to *show* their support for the regime by attending for example elections, demonstrations, and celebrations, even when they keep dissenting views for themselves.

It is the dynamics of such situations, where two "truths" coexist on different levels that we are interested in. What happens when people are forced into acting *as if* they had thoughts, feelings, and desires that they personally "do not believe in"? With Pfaller (2014), we contend that it may result in the establishment, in the individual, of an unconscious conflict. It is such a conflict that amounts to what we also call *ambivalence*. When it has taken hold of someone, they start to see reality double. On the one hand, when they think with their "public" part of the brain, they consider the truth of the official story obvious. As mentioned, this is the understanding of the world usually guiding action. On the other hand, when people are asked about what they "actually think," it is as if they had another part of the brain that they could suddenly switch on, which concludes, with just as much conviction, that the claims of the official story are obviously false.

Moderns are ambivalent towards many things. The reason for this is that there is a host of norms for which it is deemed unreasonable to disagree, that is, issues on which there is an "official story" built into the workings of public life. Take the climate, for instance. For some years now, activities to decrease emissions of CO_2 stand high on the political agenda. Everybody seems to agree with the claim that these measures are "important." Nonetheless, in private, people seem very much to disagree, because the number of people who chose to go on charter trips by plane (that releases great amounts of CO_2) steadily increases. When left to themselves, in a similar way as how they neither like nor use mathematics, they do not seem to care very much for the climate. They may very well think that they care, and see themselves as caring. But just as obvious as it is to publicly endorse the fight for the climate, it is obvious that such considerations should not be allowed to interfere with the choice of holiday location.

Gensing and Reisin (2013) argue that there is a tendency today to bring public issues, for instance pertaining to gender, health, ethnicity, and the climate into the realm of an "official story" from which deviations are deemed irrational. Insofar as this happens, they are brought into a sphere of the modern world that has been the home of mathematics for a long time. A sign of this happening is that the proper standpoint becomes a topic of research and education. This does not mean that they have become issues of agreement. What happens is that they are transformed from being issues of disagreement to being issues of ambivalence. Insofar as it is possible to disagree in public, persons can have one view each that are more or less different from each other. When agreement is mandatory, each person tends to hold two views at the same time.

At the most general level, we think that ambivalence is the normal sentiment towards *any* norm that it is obligatory to comply with. Individuals are born into a collective where norms, as well as practices conforming to them, are already present. To be a subject in such situations means to have some distance towards these norms and those practices. A characteristic feature of all cultures—including modernity—is that people cannot always act according to their own personal ideas concerning what is right and true. People need to adjust and comply. The point, however, is that such compliance normally does not extend into the core of people's world-views and identities. Beside their compliance, people keep sentiments, thoughts, feelings, and desires that stand over and beyond the norms that they accept in practice.

The question that is not answered in this example is *how*, more specifically, people act as "good moderns" who love science and think mathematically. It turns out that, for mathematics as for issues such as gender and climate, there exists a set of commonly recognized acts of acceptance. Rather than just thinking or saying that we love mathematics, it is the performance of these acts that make it so clear to themselves who they are that they can be blind for how they also do, think, and desire otherwise. The strange logic of such practical production of freedom from norms is the topic of the next section.

Avoiding Mathematics Through Education

What happens when people find themselves in situations where they are expected to do things and have thoughts, feelings, and desires that they personally "do not believe in"? Most often, they act as if they agree; they keep up appearance. The French sociologist Luc Boltanski (2011) calls this a *pragmatic* way of relating to reality. Put shortly, reality is not perfect, and protesting all the time would lead nowhere but to unemployment, prison or perhaps an asylum. Compliance is usually automatic.

With Boltanski, we contend that this is how cultures usually work. Two things are important to note. First, that there is a gap between what we do and what we "want" or what we "think makes sense," and second: that this gap is not made into an object of reflection. Paraphrasing Kant (*The Critique of Pure Reason*, B131-32) one could say that all our actions are accompanied by an unconscious "I do not quite agree," making our "selves" so to speak *float above* what we actually do and what can be concluded about ourselves from this activity. The phenomenon we refer to is similar to, but probably not exactly the same as what Žižek (2008, p. 43) talks about in terms of how "ideological interpellation" is never complete. The point is that, when culture or ideology "addresses" a subject and says: this is how reality is; this is who you are; this is what it is rational for you to do, the subject never completely agrees, even as she acts as if she did. We take this *unconscious disagreement*, as we call it, to be constitutive of subjectivity; it is what makes us more than cogs in a cultural machinery.

Acting *as if* can be thought of as putting up a shield, to keep cultural norms at a distance, to establish a "personal space." By complying with what is expected from us automatically, not thinking about it, not being bothered, we make ourselves free to think "our own" thoughts and, as we will see, to some extent also do things that contradict the "official story" of the culture that we are part of. As the metaphor goes, we can hide our thoughts, desires, and even actions from the view under our shield of complicit action.

Consider the norms of research. They say that you should write and publish; they specify the form of your writing and often even what you should write about. Our shield, in this case, might consist of a steady stream of formally impeccable academic achievements, demonstrating that we are in fact, objectively speaking, doing proper research (cf. Žižek, 2008, p. 32). Under this formal surface, however, we may very well think more freely than the publications suggest, perhaps being quite critical of the "system" that we nonetheless continuously comply with in practice.

A completely different example of shielding an inner subjective core from demands of culture can be found in a simple handshake. A handshake is a display of respect (or friendship) that says very little of the actual feelings and thoughts of those involved. In the same way as we may be however free-thinking we want at the university as long as we "behave" according to academic standards, we may think and feel what we want about our colleagues, as long as we show respect, through the handshake and a polite "what's up?" (cf. Pfaller, 2014, p. 232). By conforming to a set of agreed-upon and simplified practical rules—for academic performance and politeness—we can sustain a personal space for thinking and feeling, beyond the reach of obligations connected to social life. We can disagree with norms as we please insofar as this does not disturb our "show" of adherence.

Acting as if, putting up a shield of superficial practical compliance, can be called norm-evading. It establishes a space beyond the reach of the norm. Crucial for our argument is that we also engage in these norm-evading activities when we are alone, as they protect us from a pressure to conform *coming from ourselves*. A host of useful examples of such activities have been presented by Žižek (2008) and Pfaller (2014): listening to canned laughter, the lightning of candles in church, the spinning of prayer wheels, video-recording, and many more. The point of such activities is that they make it seem, *to ourselves*, as if we had certain desires, thoughts, and feelings, and, paradoxically, even as if we did things that we do not in fact do.

It is this line of reasoning that we will now apply to the relationship between modern adults and mathematics. We can thus expect that there are certain actions that can be performed with relatively little effort that are universally *interpreted* as endorsement of mathematics, even though they do not involve any actual engagement with it. Our suggestion is that mathematics education provides adults with a range of possibilities for such actions. More specifically, modern adults make it seem, to themselves and to others, as if they cared about mathematics by attending to the performance of children participating in mathematics education.

In order to understand how this works, we must distinguish sharply between on the one hand the intensive, formalized, and time-consuming activity of children in school and on the other hand the rather less demanding activity of adults outside school needed to sustain this activity. Following Pfaller (2014, pp. 31 and 63) and Žižek (2008, p. 31), we suggest that this difference can be understood as homologous to the difference between Christians lighting candles in church and the ensuing burning of these candles, or similarly, Tibetan monks setting praying wheels in motion and the ensuing spinning of these wheels. In these examples, the candles and wheels should be understood as representing the participants, and as "acting-out," in their burning and spinning, the behavior that their respective culture demands from them. By *delegating* the demanded behavior to *things,* such rituals—that involve the operation of objects that "act"—generate relief for the participants. The same way a handshake demonstrates respect, participation in rituals of candle-lighting or wheel-spinning make it clear that the religious obligations have been properly met. Similar to the act of lighting a candle, we see the sending of children to school, and more generally, any engagement with mathematics education as a *replacement* for an engagement with mathematics. The children sit in school as candles, burning for mathematics *for* their caretakers. This explains, of course, the public gravity of this work of children. In school, the performance of children *represent* society as a whole; it is as if society were powerless in face of the *significance* of their performance, so that the only way to make us satisfied with ourselves regarding mathematics was through the manipulation of the performance of children.

This means that children are not primarily put to school for how it transforms *them* (what we understand in terms of *learning*). In our interpretation, mathematics education should be seen as a "rite of initiation." The British anthropologist Jean La Fontaine (1986, p. 104) explains that such rituals are "for those already initiated as much as for the novices." Using Pfaller's (2014, pp. 26–29) terminology, one can say that school attendance is a form of "civilized magic" by which moderns consecrate themselves as scientific. To repeat, it is thus not primarily the children who are "elevated" when they pass through mathematics education, as in the educational sociology of Pierre Bourdieu (1990). Our focus lies on how society—as a spectator of schooling—understands itself as becoming elevated by the performance of pupils. This is why adult society is so concerned with knowledge assessments: the knowledge found in the children represent, in a displaced form, the knowledge supposedly residing also in adults. It should be noted how this interpretation depends on the interpretation of knowledge assessments as establishing *once and for all* the capacities of individuals. Everybody *knows very well* that this is not the case—that children forget—but we interpret knowledge assessments *as if* they established, within each and every one, a "glassy essence" of knowledge and competence (cf. Rorty, 1979, pp. 15ff.).

A good question to ask at this point is why this works: How can Tibetan monks be released from the obligation to pray by the spinning artifacts that all sensible persons know very well cannot pray? How is it possible, as we claim, that moderns are released from their obligation to use mathematics by putting children to work at school? Pfaller (2014) introduces the idea of the *naïve observer* as a means of explanation. The point of this concept is that our patterns of reaction—of feelings of pleasure and pain, of triumph and shame—take place in culture *as if we were observed by a stupid version of ourselves* that somehow imputes her reactions and feelings onto our proper intelligent and reflective selves. Two things are characteristic of this naïve observer: Firstly, it has internalized the common sense view of things and reacts accordingly; Secondly, it accepts appearances at face value.

If something seems to be a certain way, the naïve observer takes it to actually be like that. This means that we, as noted above, have a certain split vision of reality, where we on the one hand can "see through" appearances and by means of reflection reach a more nuanced, comprehensive, and deep understanding of phenomena, but where we on the other hand carry a *capacity to ignore* with us, of how complex reality really is. It is this capacity that we use when we act pragmatically and it should be noted that it is instrumental for making culture possible (Boltanski, 2011; Winnicott, 2005 p. 4 and pp. 17–19). Reality is not how it should be; people are not what they pretend to be—in order to be able to interact, we must ignore this most of the time. People who insist on acting only on what they take to be the truth are utterly asocial and cannot participate in collective action.

It is also this capacity of sticking to surface appearances that makes it possible for us to derive pleasure from play. We can "bracket out" our power of reflection, so that we can act, think, and feel *as if* the world was in a way that we consciously know that it is not. For children, it comes naturally to pretend to be a hero or a princess, superman or a gnome. Pfaller (2011) shows that the same mechanism is in play for adults when they take on a grand pose when smoking, or when they dress up for a party: they pretend to be someone else for a while, whom they know very well that they are not. It is a show for the naïve observer, who pays back with acknowledgement and recognition. We contend that this is also the modus operandi of mathematics education—it is a manner of pretending to be grand, to be magnificently modern.

Mathematics education can be seen as a *compromise formation* that makes compliance with the norm of mathematics mandatory at certain times, in certain places, while it generates the possibility of evading this norm altogether at other times, in other places. Mathematics education is a part of modern society where compliance with the norm of mathematics is prescribed, enforced and meticulously monitored. It is, as researchers have noted (Palm, 2002), a simulation, but not of life as it is, but of modern life envisioned as founded upon mathematical knowledge. In the educational sphere, children function as candles to whom we delegate the hard work to of being properly modern; of celebrating and enjoying the growth of mathematics, of using it to get along in their daily life.

Mathematics education signifies, for us moderns, that we are properly modern (cf. Meyer, 1977). It is an engagement with mathematics per proxy, where children and youth perform the work of engagement. Adults act out their acceptance and adherence to the view that everybody should know mathematics through children in the displaced and miniaturized act of sending them to school. What children do is nothing less than producing the image of modern society as being properly modern, that is: a "knowledge society" or a "learning society." Mathematics education is thus not only for mathematics. It is for mathematics in the demarcated sense of signifying acceptance of the view that mathematics is important and should be known. But it is also a shield, for adults, against this view, a device that makes it possible for adults to not know mathematics, to not engage with it—to be, in a way, invisible in the eyes of mathematics; because its eye (we think of *The Lord of the Rings* here) is firmly directed at the children and their school performance. Thanks to mathematics education, we can consider ourselves to be modern, rational, scientific, and everything mathematics stands for—and at the same time be free to *not* think about it, except when we go through our mandatory ordeals as children and youth.

As a cultural entity, mathematical knowledge fits perfectly with Pfaller's characterization of a holy object (2011, p. 21). When an everyday life situation calls for the use of mathematics—for instance in the form of a simple calculation—this is usually not taken as a cause for celebration. Excluding mathematicians and teachers, it is more likely to resemble an encounter with a rare holy animal where you close your eyes and hope it will treat you kindly. As such a holy animal, mathematics is loved ritually but avoided and feared in everyday life.

In a way that is reminiscent of Donald Winnicott's analysis of *transitional objects* (2005), children taking part in mathematics education seems to make mathematics manageable in modern society. The translation of mathematics to an educational form can thus be seen as the making of a stuffed tiger that *represents* at the same time the presence of the real tiger that we identify with, and it being at a safe distance. In the form of mathematics education, we can "cuddle" with mathematics in a way that we would never dear to cuddle with mathematics itself (it would kill us instantly). In a way that is quite similar to how the blanket often standing in the center of Winnicott's analysis lends the child independence from the mother by representing her presence in a manageable form (p. 8), mathematics education lends society independence from mathematics while at the same time acknowledging its importance.

This function of the child sheds light, not only on educational theory—which through the concepts of learning and knowledge *makes sense* of the practice of replacement—but also on developmental psychology. It is well known that the notion of "the child," which rose to prominence as a topic of discussion in the early twentieth century (Berggren, 1995), bore many resemblances with the "primitives" that were at this same time explored and discussed in anthropology and philosophy. Just as allegedly primitive cultures functioned as points of reference for modernity, as past stages in its history of development, the child was *constructed*, one could say, as a stepping stone for a new kind of modern adulthood. Children were constructed as *in need* of the kind of systematic manipulation that was provided by the emerging systems of education. Piaget (e.g., 1957, p. 745 and p. 765) saw mathematics education as central to mental development. The construction of the child as an object of manipulation should, we contend, be understood as an indirect construction of the modern adult, who is the result—the end-point—of the individual development orchestrated in the education system. The educational theory of "ontogenetic recapitulation" (Hall, 1905) makes the connection between the historical development of modernity out of primitive culture and the individual development out of childhood explicit. The manipulation of the child within the education system *demonstrates* this development, in the form of a rite of initiation, for modern society. It makes this development manifestly present, in the activity of the children. It constitutes an imaginary "elevation" of modernity. In the educationally interpreted measurements of performance, this elevation is documented; with its effects in society, of sorting students into "stations" of different elevation, it is made into a functional part of cultural reality.

Becoming Ambivalent

How does it feel to function as an object of delegation? What does it feel like to spin and burn in the mathematics education classroom? It is a hell of an experience, many could tell you. Our contention is that it is this both intensive and extensive experience that normally generates exactly that complex relationship towards mathematics to which, later in life, delegation by means of mathematics education functions as a well-adjusted response.

Following the more detailed discussion in "Hating School, Loving Mathematics" (Lundin, 2011), mathematics education can be described as a game. It has rules that you are obliged to follow. Unlike some other games like football and poker, mathematics education also attributes meaning to its rules. There is a narrative or "imaginary world" in mathematics education, in which the game supposedly takes place and in which it make sense (cf. Dowling, 1998; Lave, 1992). In his *Ritual and Religion in the Making of Humanity* (1999), Roy Rappaport presents a definition of the "ritual form" (pp. 23ff.). According to this definition, ritual practice is largely determined by others than the performers themselves; it often takes place in spaces separate from other cultural activities, following its own rhythm and schedule; it is usually formalized, punctilious, carefully supervised and controlled. Clearly, mathematics education can thus also, from a slightly different perspective, be described as a *ritual*. Just like a game, ritual activity takes place *as if* the world was a particular way. Roy Rappaport calls this *enactment of meaning* (1999, pp. 107ff.): The meaning of the ritual is not "told," it is acted out; the meaning becomes apparent, to the performers themselves as well as potential spectators, in the structure of the activity itself. Rappaport calls this meaning the *message* transmitted by ritual performance.

What is the message about mathematics transmitted by the structure of mathematics education? On the most fundamental level, mathematics education has the structure of a series of questions, answers and evaluations: Pupils get questions, respond to them, and get informed about the quality of their response. In the narrative lending meaning to this practice, mathematics appears in at least three guises:

- *Firstly*, mathematics appears as the cause of successful performance. This meaning of mathematics positions it as an object of desire: Wanting to be successful, wanting to have power, corresponds, in the game, to wanting to have mathematical knowledge.
- *Secondly*, mathematics appears as a judge. According to the imaginary of mathematics education, it is not up to the teachers to decide what is right and wrong. All decisions and effects are derived from a supposedly universal logic inherent in mathematics itself. Wanting to be successful thus also corresponds to *conforming* to the "rule" of mathematics.
- *Thirdly*, mathematics appears as the cause of mathematics education itself. It is because mathematics supposedly reflects the fundamental structure of both the natural and social world and is practically useful in professional and everyday life that mathematical knowledge is given such importance.

This threefold message about mathematics is transmitted through enactment of mathematics education; it is implicit in its structure. This means that, when children participate in mathematics education, they learn that mathematics is: a powerful resource, an immutable judge and fantastically important.

Crucial to add, however, is that children also—even while they are learning this— are fully aware of the limited applicability of the imaginary of schooling to everyday life. Children often do not understand why they are obliged to learn mathematics. From an anthropological perspective, this is not surprising. La Fontaine (1986) has pointed out that it is quite normal that participants in maturity rites are not aware of the meaning of their own performance. Children participating in such activities understand that what they do somehow plays an important function for their transition into adulthood, but the cultural significance of their activity largely remains a mystery to them. The information received by children participating in rites of initiation are often minimal and reduced to instructions on how to behave (p. 103), making the experience of participation little more than an "experience of the ritual" (ibid.). This account fits well with Laves (1992) observation that children normally know the difference between the world of schooling and life outside of school very well. They know what they are doing when they become skilled in solving the problems encountered in the school setting very well, but they do not understand why they are doing it.

When people are immersed in play, they become possessed by what Johan Huizinga (1998) calls "holy seriousness" (cf. Lundin, 2011). They thus start to think, feel, and desire according to the logic of the game, even though they are well aware that it is "just a game." Pfaller (2014) has explained not only how the very knowledge of the difference between the game and reality is constitutive of the power of the game, but also why it does not really matter if you want to play the game or not. Thus, even if you are *forced* to play, you cannot resist having the imaginary of the game installed in yourself (pp. 242–244). This way of understanding the effects of schooling nicely complements the ritual theory of Rappaport (1999 p. 119) who talks about the inevitability of receiving the message transmitted by one's own performance.

From the perspective of the naive observer that we have described in this chapter, we continue to play the game of mathematics education even as we leave school, only at a much lower intensity, by sending children to school, well aware that what they learn there is generally quite useless. We know this as adults in a similar way as we already knew it as children. Just as we had no choice but to comply in school, we have no choice but to comply as adults: the difference lies only in what it is that we are obliged to do.

Just as we learn that mathematics education is just a game, we also learn that it is *not* just a game. This piece of knowledge is connected to the third guise of mathematics described above, as the ultimate cause of mathematics education itself. As such, mathematics appears as the queen of science, fundamentally opposed to the superficiality of schooling. Put shortly, we learn that there is something in mathematics that lies beyond our own powers of comprehension that justifies the idea of mathematics education, independently of how this idea happens to be implemented as actual reality (cf. Lundin, 2010, 2011). Even though we ourselves do not exactly "know" this justification, we are convinced of its existence. The presence of such *faith* in the idea of mathematics education complicates the picture we have painted so far.

The Disorder of Mathematics Education

In our analysis so far, mathematics education appears to be a rather well-functioning institution. As announced in the introduction, we have described a circle reminiscent of Pierre Bourdieu's (1990) theory of social reproduction. In the center of this description stands the *naive observer*, which both holds the institution together and keeps it in motion. Belief in mathematics education has the form of an "(unconscious) fantasy structuring our social reality" (Žižek, 2008, p. 30), without anyone actually *having* this belief in mathematics education themselves. While everybody sees through the claim that mathematical knowledge is useful in everyday life, social reality is kept together by everybody acting *as if* they believed. Quite obviously however, this is not the whole story. Even when reflecting, most people have a certain *faith*—if not in the usefulness of what is actually learned in mathematics education—more likely in the *possibility* of learning something useful, and in the *necessity* of such learning taking place.

As mentioned in the previous section, we take such faith to be part of the message transmitted by the enactment of mathematics education. We cannot but assume there to be some kind of power residing in mathematics that explains the massive presence of mathematics education in modern society. One could perhaps assume that the fact that we do not understand this power-meaning residing in mathematics would diminish its trust-worthiness, but as Žižek (2008, p. 35) notes, it is the other way around: "this traumatic, non-integrated character of [the assumed foundation of social reality] is a positive condition of its [authority]" over us. We do not endorse mathematics education because it works—we know very well that it does not. We endorse it because we (think we) know that it *must* work; the assumption that it can work is necessary to make for modern reality sense. Insofar as everybody started to "see through" mathematics too, in the same way as they routinely see through the actual workings of mathematics education, "the very texture of the social field" (p. 34) would disintegrate. We are thus doubly attached to mathematics education: on the one hand, we are compelled to endorse it in practice, because it releases us from an unconscious pressure to engage with mathematics personally, on the other hand we are compelled to have faith in its "idea" because we need this idea for our world to make sense. The question we want to pose at the end of this chapter is how this attachment to mathematics education should be normatively evaluated. We will by no means provide a conclusive answer. What we hope to do is to contribute to a clarification of what is at stake.

Let us start by bringing out some central characteristics of the circular motion propelled by the naive observer. It is quite clear that acting on the supposed belief of others for good or for bad amounts to a general acceptance of the cultural order as it is. It amounts to a taken for granted resignation in face of the power of reality. With Sloterdijk (1987), it can be called *cynical*; Boltanski (2011) calls it *pragmatic*; according to Pfaller (2014) and Rappaport (1999), it is simply normal.

A central point of this chapter however, is that this attitude does not amount to a *wholesale* acceptance of reality as it is. To the contrary, it includes a moment of play that allows for a partial escape from reality "as it is." Reality, accepted in this way, is not quite what it seems to be. Modern society seems to care very much for science and mathematics, but in fact adult society is virtually free of both—thanks to the ingenuous mechanism of delegation. The logic of the naive observer thus establishes, through miniaturization and delegation, within the symbolic order, spaces of freedom. It can thus be said to amount to a compromise between acceptance and rejection.

We find it important that this logic of "the other supposed to believe"—to introduce another way of talking about the naive observer—also entails a particular kind of transcendence, present in the midst of cultural reality. The acceptance of culture as it is makes it very clear that reflective reason reaches beyond it. The situation established by this logic is close to what Kant recommended in *What is Enlightenment?* (1784): that culture is accepted in practice (Kant calls this the "private" use of reason), so that freedom can reign in the realm of (public) theoretical reflection. An important point here is that the logic of the naive observer does not in itself entail an articulation of what, more exactly, it is that everybody must comply with. The norm is instead primarily present implicitly, in the structure of practice. This *lack of image*, as one could call it, leaves the field open for different interpretations of why we do what we do and what we could do otherwise.

Tentatively, we suggest that the logic of the naive observer entails a certain moderation in the policing of cultural obligations. This moderation can be illustrated by the story in the Bible about a child falling in a well on the Sabbath when it is not allowed to work (Luke 14:5; Matthew 12:11). Jesus uses this as an example of an exception where we should temporarily and locally put the cultural obligations aside to solve an immediate problem. When we do this, we use our full capacity of reflective reason and act in a way that is clearly different from how we behave normally. Jesus tells us that we should be aware of our capacity to transcend culture and use that capacity when needed. The point is that, while the logic of the other supposed to believe entails acceptance in general, it does not entail acceptance always and everywhere and if new obligations are invented, people have a "ground" so to speak, outside culture that can be taken as a point of departure for protests.

When people act on the supposed existence of a justification of mathematics education, they act differently. A first thing to note about this logic is that it opens up for change. It aims to bring reality in accordance with its own idea; it aims to make mathematics education work. This is of course the good thing with this attitude towards reality, which is missing in the reproductive logic of the naive observer. Boltanski (2011) discusses the dynamics of such improvement in terms of *moments of reflection* that break the habitual pragmatism of social life. Arguably, they play a crucial and benign function in modern social life, interrupting processes of corruption of institutions, such as democracy and human rights, trying to bring them "back on track."

In the case of mathematics education however, the situation seems to be slightly different. The first problem is that the very idea of a well-functioning mathematics education is—if our analysis is correct—a result of the enactment of mathematics education itself, that is, the result of the performance of an act filling the function of mitigating the impact of mathematics on social life. From this perspective, the ambition to "improve" mathematics education seems misguided. Not only do people not actually want the actually useful mathematical knowledge that such ambitions of improvement aim for; such knowledge is probably not even possible.

There are also other problems with faith-driven attempts at improvement. It is not okay, according to the logic of faith in mathematics, to just go on with mathematics education as it is. Using the terminology of Johan Huizinga (1998), the person with faith is a *Spielverderber,* who prevents others from deriving pleasure from just "playing along" with culture. This prohibition is connected to the fact that the ideal, around which enthusiastic but dead serious reformers gather, cannot be seen through; it is very difficult to have distance towards it since it appears as a manifestation of reason. The logic of faith is thus not only an opening up of the horizon, but also a limitation of it. The presence of publicly acknowledged truth in culture makes it difficult to think otherwise.

While it certainly seems like a bad idea to us to discredit all attempts at improvement driven by faith in ideals (as e.g., Hayek, 2007; Popper, 2013), we think it is important to recognize its dangers. The desire for subordination noted by Spinoza comes in different varieties. While powerful visions of a better world may fill a positive function of creating distance from reality as it is, they can also result in obligatory obsession that adds insult to injury (see the discussion in Butler, Laclau, & Žižek, 2000).

Tentatively, we will suggest that a *healthy culture* is a culture where refusal to participate in rituals is recognized as a sort of normal exception. Rituals are thus restricted, by means of reasonable protests, to particular times and places, where they can fulfill their peculiar function of keeping norms in place, while at the same time restricting their power. In such cultures, ritual obligations are respected but from a perspective that transcends them.

In our culture, unfortunately, reason is increasingly subordinated to the logic of certain rituals, not least that of learning. We seem to have little control of their construction. It is very difficult for anyone to say no to them, even in the face of idiocy and suffering (Graeber, 2015; Paulsen, 2014). To paraphrase our epigraph, modern people fight for more knowledge as if they were thus fighting for their own deliverance. As indicated by our analysis, this seems to be a problem inherent in a modernity priding itself of reason (cf. Adorno & Horkheimer, 1997; Illich & Cayley, 2005; Peukert, 1989; Weber, 1992). What one should do today is to insist on the necessity of distance towards the modern rituals, the possibility of refusing to participate and of the rights of an even more powerful reason transcending them to make them into objects of reflection. We should recognize that the rituals are there for our sake, to keep the norms that we cherish at a proper distance—not for the purpose of their universal realization.

References

Adorno, T. W., & Horkheimer, M. (1997). *Dialectic of enlightenment*. London: Verso.

Matematikdelegationen. (2004). *Att lyfta matematiken: Intresse, lärande, kompetens: Betänkande*. Statens offentliga utredningar 2004:97. Stockholm: Fritzes offentliga publikationer.

Berggren, H. (1995). *Seklets ungdom: Retorik, politik och modernitet 1900-1939*. Stockholm: Tiden.

Boltanski, L. (2011). *On critique: A sociology of emancipation*. Cambridge: Polity.

Bourdieu, P., & Passeron, J.-C. (1990). *Reproduction in education, society and culture*. London: Sage.

Butler, J., Laclau, E., & Žižek, S. (2000). *Contingency, hegemony, universality: Contemporary dialogues on the left*. London: Verso.

Deleuze, G., & Guattari, F. (2004). *Anti-oedipus: Capitalism and schizophrenia*. London: Continuum.

Dowling, P. (1998). *The sociology of mathematics education: Mathematical myths/pedagogic texts*. London: Falmer.

Dowling, P. (2010). Abandoning mathematics and hard labour in schools. A new sociology of knowledge and curriculum reform. In C. Bergsten, E. Jablonka, & T. Wedege (Eds.), *Mathematics and mathematics education: Cultural and social dimensions: Proceedings of MADIF 7* (pp. 1–30). Linköping: Svensk förening för matematikdidaktisk forskning (SMDF).

Fontaine, J. S. L. (1986). *Initiation*. Manchester: University Press.

Fries, M. (2013). *Mathematical anxiety. A literature review regarding research 2008-2013*. Malmö högskola, Lärande och samhälle.

Gensing, P., & Reisin, A. (2013). *Der Präventivstaat: Warum Gesundheits-, Kontroll- und Verbotswahn Freiheit und Demokratie gefährden*. Köln, Germany: Lingen Verlag.

Graeber, D. (2015). *The utopia of rules: On technology, stupidity, and the secret joys of bureaucracy*. Brooklyn: Melville House.

Hall, S. G. (1905). *Adolscence, its psychology and its relations to psychology, anthropology, sociology, sex, crime, religion and education*. London: D. Appleton and Co.
von Hayek, F. A. (2007). *The road to serfdom: Text and documents*. Chicago: University of Chicago Press.
Huizinga, J. (1998). *Homo ludens: A study of play-element in culture*. London: Routledge.
Hultén, E. (2010). Mattebluffen. *Aftonbladet*. Retrieved January 31, 2010.
Illich, I., & Cayley, D. (2005). *The rivers north of the future: The testament of Ivan Illich*. Toronto, ON; New York: House of Anansi Press; Distributed in the U.S. by Publishers Group West.
Hög tid för matematik. (2001). Göteborg: Nationellt centrum för matematikutbildning, Göteborgs universitet.
Johnson, C. (2004). Inte så himla viktigt att kunna matematik. *DN debatt*. Retrieved October 31, 2004.
Lave, J. (1988). *Cognition in practice: Mind, mathematics and culture in everyday life*. Cambridge: Cambridge Univ. Press.
Lave, J. (1992). Word problems: A microcosm of theories of learning. In P. Light & G. Butterworth (Eds.), *Context and cognition: Ways of learning and knowing* (pp. 74–92). Hillsdale, NJ: L. Erlbaum Associates.
Lundin, S. (2008). *Skolans matematik: En kritisk analys av den svenska skolmatematikens förhistoria, uppkomst och utveckling*. Uppsala: Acta Universitatis Upsaliensis.
Lundin, S. (2010). The missing piece. An interpretation of mathematics education using some ideas from Zizek. In C. Bergsten, E. Jablonka, & T. Wedege (Eds.), *Mathematics and mathematics education: Cultural and social dimensions: Proceedings of MADIF 7* (pp. 168–178). Linköping: Svensk förening för matematikdidaktisk forskning (SMDF).
Lundin, S. (2011). Hating school, loving mathematics. The ideological function of critique and reform in mathematics education. *Educational Studies in Mathematics, 80*(1), 73–85.
Meyer, J. (1977). Formal structure as myth and ceremony. *American Journal of Sociology, 83*(2), 340–363.
Pais, A. (2013). An ideology critique of the use-value of mathematics. *Educational Studies in Mathematics, 84*(1), 15–34.
Palm, T. (2002). *The realism of mathematical school tasks: Features and consequences*. Umeå: Umeå University.
Paulsen, R. (2014). *Empty labor: Idleness and workplace resistance*. Cambridge: Cambridge University Press.
Peukert, D. J. K. (1989). *Max Webers Diagnose der Moderne*. Göttingen: Vandenhoeck and Ruprecht.
Pfaller, R. (2014). *On the pleasure principle in culture: Illusions without owners*. London: Verso.
Pfaller, R. (2011). *Wofür es sich zu leben lohnt. Elemente materialistischer Philosophie*. Frankfurt am Main: S. Fischer Verlag.
Piaget, J. (1957). The myth of the sensory origin of scientific knowledge. *Psychology and Epistemology: Towards a Theory of Knowledge*, 45–62.
Popper, K. (2013). *The open society and its enemies*. Princeton: Princeton University Press.
Rappaport, R. A. (1999). *Ritual and religion in the making of humanity*. Cambridge: Cambridge University Press.
Rorty, R. (1979). *Philosophy and the mirror of nature*. Princeton, NJ: Princeton University Press.
Sloterdijk, P. (1987). *Critique of cynical reason*. Minneapolis: University of Minnesota Press.
Walkerdine, V. (1988). *The mastery of reason: Cognitive development and the production of rationality*. London: Routledge.
Weber, M. (1992). *The protestant ethic and the spirit of capitalism*. London: Routledge.
Winnicott, D. W. (2005). *Playing and reality*. London: Routledge.
Žižek, S. (1999). *The ticklish subject: The absent centre of political ontology*. New York: Verso.
Žižek, S. (2008). *The sublime object of ideology*. London: Verso.

Chapter 3
De|mathematisation and Ideology in Times of Capitalism: Recovering Critical Distance

Hauke Straehler-Pohl

Abstract This essay takes an ethical dilemma posed by current technological developments as a point of departure for discussing the dialectic of mathematisation and simultaneous dematematisation as a social phenomenon and a challenge for society. This dialectic of *de|mathematisation* is discussed in relation to a conception of ideology that has been developed within the field of mathematics education as well as in relation to an alternative conception recently imported to mathematics education from the field of political psychoanalysis. By analysing media reports, advertisement campaigns and a sociological study on dating portals, the analysis reveals shifts that have recently occurred in de|mathematisation and its ideological framing. This embedding in popular culture allows for a better understanding of the interplay between de|mathematisation and the sphere of political economy within late capitalism. Finally, the author provides an outlook on how the conclusions drawn could contribute to a further development of a mathematics education that critically reflects its role in society.

Introduction

> "Let's assume you are sitting in a car controlled by a computer. One of those cars Google is currently developing. Let's also assume you are driving this car on a two-lane road in a big city. There is a cyclist to your right and a motorcyclist to your left. Let's assume now that a group of children has suddenly entered the road. The measuring-device of the computer comes to a diagnosis: it is too late for breaking. What should the computer do? Run over the children? Turn to the left and ram against the motorcyclist? Or turn to the right, where the cyclist is pedaling?" (Die ZEIT 33/2014, own translation)

What would you decide? There are three options, each of which will definitely do harm to human beings. Surely, you would decide not to run over the children. But would you actively decide to endanger the life of the motorcyclist instead? If we had the choice, we would surely prefer not having to take that decision. This short passage

H. Straehler-Pohl (✉)
Fachbereich Erziehungswissenschaft und Psychologie, Freie Universität Berlin, Berlin, Germany
e-mail: h.straehler-pohl@fu-berlin.de

H. Straehler-Pohl et al. (eds.), *The Disorder of Mathematics Education*,
DOI 10.1007/978-3-319-34006-7_3

from a German news magazine poses an ethical dilemma: A situation in which a decision must be taken and only negative options are available. We may find comfort in the fact that this is just a hypothetical situation—"let's assume"—and furthermore, it is science fiction—"Google is currently developing". Even more comforting appears to be the fact that, even if this hypothetical scenario should manifest itself in future reality, we will not be the ones having to decide what to do. The question is "what should the computer do?" Ironically, agency appears to be with the computer. The computer will have to decide whom to do harm to. But as we know very well, computers do not act because they have an own mind, but because they have been programmed in a certain way. Hence, humans are not fully released from the ethical dilemma and the respective guilt that would result from harming an innocent human being. Humans could decide what the computer should do in such a situation. Once the "Google car" would be about to enter the market, a group of humans certainly *must* develop an algorithm that will provide the data required to take the decision between running over the children, or ramming against the motorcyclist, or knocking down the cyclist, even if this particular or a comparable situation were not actually "real".

In this chapter I will take this ethical dilemma as a point of departure for revisiting the analyses that have been developed in mathematics education regarding the phenomena of mathematisation and demathematisation until 2007. I will relate these findings to examples from media reports, advertisement campaigns and dating portals in order to illustrate shifts that have occurred since. Relating these phenomena to a critique of ideology will allow me to draw conclusions concerning further developments of a mathematics education that critically reflects its role in society. For this purpose, the concepts of "ideology of certainty" (Borba & Skovsmose, 1997) and "evolutionism" (e.g. Pais, Fernandes, Matos & Alves, 2012; Straehler-Pohl & Pais, 2014) will be discussed and delineated in their relation to capitalism. This essay, thus, builds on the flourishing introduction of political psychoanalytic theory into the field of mathematics education (e.g. Brown, 2008, in this volume; Davis, 2005; Lundin, 2012; Lundin and Christensen in this volume; Gerofsky, 2010; Mesquita in this volume; Mesquita, Restivo & D'Ambrosio, 2011; Pais, 2012, 2015, in this volume; Pais & Valero, 2012, 2014; Straehler-Pohl, Gellert & Bohlmann, 2014; Straehler-Pohl & Gellert, 2015; Straehler-Pohl & Pais, 2014). The essay reconstructs the role of critical distance towards mathematisation and concludes suggesting how it could unfold more of its desired emancipatory effects. A stronger and more explicit anchoring within a critique of capitalism and the promotion of self-confidence in rejecting mathematics-based argumentation within the mathematics classrooms are proposed as necessary ingredients of a truly critical mathematical literacy.

The Formatting Power of Mathematisation

The introductory example is just one of many possible examples of how mathematics is increasingly penetrating all different aspects of our contemporary world. Mathematics appears as the ultimate meta-language that stakeholders use and rely on when describing the world, trying to predict its development and arguing for the validity of an argument. Already in 1986, Davis and Hersh (1986, p. 17) drew our attention to this development: "The social and material worlds become more and more mathematised". As the appealing title of their book suggests, "Descartes' dream" is about to materialise in the run of capturing each single aspect of our lives by mathematisation. They suggest that "we should observe these developments critically, as they could do damage to all of us" (ibid). This critical attention is due to the observation that mathematisation is not just an innocent tool, which can be used to describe the world and, based on this description, provide valuable predictions. It is simultaneously altering our lives. This is evident, for example, in a definition of mathematisation proposed by Skovsmose (2014, p. 442):

> Mathematization refers to the formatting of production, decision-making, economic management, means of communication, schemes for surveilling and control, war power, medical techniques, etc. by means of mathematical insight and techniques.

The usage of the word "formatting" is indicative of the profound effects that Skovsmose attributes to mathematisation. When a practice is mathematised, this does not simply affect a change in our behaviour, as we have an additional source of information that we can rely on when making decisions. Instead, mathematisation intervenes more deeply in the very structure of practice, re-organising the basic conditions our actions refer to.[1]

This critical drawing of attention to the formatting power of mathematisation is clearly distinct from a view on "mathematisation as a didactic principle" (Jablonka & Gellert, 2007, p. 2). Such a view is prevalent in the variety of approaches on "mathematical modelling" in schools (e.g. de Lange, Keitel, Hutley & Niss, 1993; Houston, Blum, Huntley & Neill, 1997), where mathematics is proposed as a means for structuring a problem of reality. In this perspective, mathematics is sharply distinct from reality itself. Generated mathematical solutions can then be transferred back into reality and evaluated by reference to reality. Circular approximation brings the description closer to what is described. Such a view portrays mathematics as a means for the description and prediction of reality. The focus on formatting undermines this artificial dualism: through the availability of the mathematical model, the original problem of reality changes itself. Description and the described approximate each other, the described object itself also transforming towards the description. As we have seen in the discussion of the introductory example of the "Google car", the reality of driving a car *after* the market introduction of autonomous cars would actually be a different reality of driving than it was (or is) before. Car drivers will be concerned with different actions, routines and thoughts and confronted by different questions arising from their practice. Furthermore, the responsibility for certain problems of car driving will be delegated to actors who solve these problems in the absence of their immediate experience: not as car drivers, but as computer scientists.

[1] See Skovsmose (1994) or Keitel, Kotzmann and Skovsmose (1993) for a more detailed account of the "formatting power" of mathematics and mathematisation.

In his definition, Skovsmose selectively steers our attention to social practices that could commonly be considered intrinsically political and hence ideologically suspicious on first glance. It is easy to imagine stakeholders having an interest in intentionally influencing the algorithms at work in decision-making, economic management and the like for their own purposes. Davis and Hersh (1986, p. 125ff.), however, make us aware that the formatting power of mathematisation can be understood much broader, including its application on practices that are often perceived as non-political and non-ideological on first glance. As such, they analyse the "computerisation of love". Davis and Hersh (1986, p. 128) observe how dating brokers use prescribed categories to produce a "very rough mathematical model of the candidate with all her/his needs and wishes". These models are further used for a (yet manual) preselection of potential candidates. While we can understand such preselection as a form of prescription, it is still far from the deep meaning of the structural intervention in the practice of love, that the term "formatting" suggests. The mathematically operated prescription may limit and refine the infinite number of potential candidates in the same way as our job, hobbies or the choice of a favourite pub do in so-called Western societies. It may have a (rather arbitrary) influence on *who* people love, but less on *how* people love. Indeed, Davis and Hersh (1986, p. 130) themselves constrain that "from [hitherto] available statistics, we can conclude that the success of the computer is significantly lower than that of the traditional [broker] competitor: Despite that the traditional broker is on the pullback, while the computer is [...] on the advance". Accordingly, in the 1980s, the stakeholders' interest in operating the "computerisation of love" was rather an indication of putting a shiny gloss of the technology that made dating agencies fashionable instead of actually unfolding a material impact on the practice of love. The "computerisation of love", as observed by Davis and Hersh, is illustrative of an illusive power that people attributed—and, that is a question of this essay, maybe still do—to mathematics, but that mathematics actually did not hold up to. The "computerisation" of love is thus illustrative of an ideology that inhibits the recognition of the "real" limited inherent power of mathematics, an irrational hope that can be instrumentalised against people's rational concerns.

The Ideology of Certainty

Mathematics is commonly attributed a reputation of being a neutral and objective advisor, blind to any kind of ideological bias, similar to the personification of justice. It is said that mathematics do not have any interest of its own and it is hence apolitical and non-ideological. Supposedly, mathematics is the neutral ground governed by raw numerical facts, where arguments can be judged against objective measures. Borba and Skovsmose (1997, p. 17) reverse this assumption and attest that the power of mathematics itself is founded on an "ideology of certainty" that is characterised by a "view of mathematics as an 'above-all' referee, as a 'judge', one that is above humans, as a non-human device that can control human imperfection".

A crucial manifestation of that ideology concerning mathematisation is what Jablonka and Gellert (2007, p. 8) call the "myth of the infallibility of technology. When airplanes crash or nuclear power stations run into problems, it is often attributed to human error". This can be illustrated by the case of the "Ariane 5 flight 501" (Le Lann, 1997), a space rocket which crashed in 1996 on its maiden voyage. The crash was caused by a computational rounding error that in turn caused an unstoppable avalanche of errors in the control system of the rocket. At the core of this problem lays the fact that, due to the mundane need to store a finite binary representation of a number, computers cannot process any real number, but solely an approximation whose accuracy is determined by the number of digits. Any finitely fixed number can just approximately match the material entity that it represents. The fact *that* computational errors occur is thus intrinsic to the computational system. The possibility of the crash of the space rocket is an integral problem of the technology, which can only be controlled and minimised to a certain degree—never complete—by humans. The public report of the inquiry board nevertheless concluded that "poor S/W [software] Engineering practice is the culprit" (Le Lann, 1997, p. 339); thus the crash was officially attributed to human error. However, Le Lann's critical re-analysis reveals that the mere fact that errors must occur is indissoluble. As mathematics is a human construction, it cannot transgress human imperfection, it cannot escape history.

The certainty and infallibility that humans expect from mathematising social practice is thus an illusion that is held in suspension through neglecting the irreducible difference between human practice and its mathematical description. It is a fantasy. Through Žižek (1992), we can understand occasions like the crash of the Ariane 5 flight 501 as a *symptom* of the ideology of certainty, a "real" kernel that resists full symbolisation, an inextinguishable contradiction. Why is it that the appearance of such symptoms does not effectively alter our relation to the ideology of certainty? According to Žižek, ideology itself provides the fantastic material that allows to repress confrontation with the symptom and erase it from our direct perception of reality. We therefore only experience it in a distorted form, enacting what Žižek (2008, p. 12) calls "fetishistic disavowal": "We know very well that we will never solve all problems by means of mathematics, but still … (we apply mathematics to all problems)".

In this way, fetishistic disavowal allows for a retrospective misrecognition of the symptom, which nevertheless persists "off the radar" and continuously enters our perception in a distorted form.

The Dialectics of De|mathematisation

Conceptualising the certainty and infallibility of mathematisation as an ideology and its failed materialisations as symptoms immediately evokes the question about the mechanisms that facilitate their repression.

One of such mechanisms appears to be inherent to a process that Jablonka and Gellert (2007) call "the dialectic of mathematisation and demathematisation".

Even though technologies are introducing mathematisations into more spheres of life, the use of mathematics is simultaneously becoming less visible. It simply became easier to overlook the presence of mathematics, and hence to repress the symptoms. Demathematisation was recognised as a social phenomenon that occurred parallel to the phenomenon of mathematisation in the late 1980s by scholars like Keitel (1989) and Chevallard (1989, reprinted in 2007), who drew the attention to what they called "implicit mathematics":

> Implicit mathematics are formerly explicit mathematics that have become "embodied", "crystallized" or "frozen" in its objects of all kinds—mathematical and non-mathematical, material and non-material –, for the production of which they have been used and "consumed". (Chevallard, 2007, p. 58)
>
> Implicit mathematics makes mathematics disappear from ordinary social practice. (Keitel, 1989, p. 10)

According to Chevallard (2007, p. 58), the "social 'implicitation' of mathematics into objects" is taking place through the usage of mathematical knowledge in the production of objects. While it may take a high amount of mathematical knowledge to originally produce an object, this object can thereafter be produced with a dramatically lower mathematical effort, as mathematical rules of procedure that formerly had to be constructed can now be followed without having to repeat the initial mathematical "work". From now on, mathematics can be "frozen" or "crystallised" in this object. Within the object, the mathematical work can occur automated. Furthermore, once this object has been produced, it can be used as a tool in the process of designing—with the help of mathematics—new objects. As a brief example, digital technology for measuring time requires the development of an according algorithm and so does technology for measuring distance. Digital technology for measuring speed can make use of the former two algorithms as if they were objects instead of developing a complete algorithm from scratch. This invokes "an infinite regress" (ibid.). It implies that the consumption of mathematics becomes easier. In its implicit form, formerly complex and exclusive mathematical knowledge becomes accessible and applicable for consumers without any mathematical expertise of their own. Accordingly, Chevallard (2007, p. 57, emphasis in original) argues that "in contradistinction to societies as organised bodies, all but a few of their members can and do live a gentle and contended live *without any mathematics whatsoever*". Mathematisation itself brings the comfort of liberating the individual from the necessity of knowing mathematics.

While the analysis of Chevallard reveals that the process of mathematisation reinforces demathematisation, Keitel's (1989, p. 9) analysis of the historical emergence of the mechanical clock reveals how the process of demathematisation in turn reinforces the process of mathematisation:

> The mechanical clock extends the domain of quantification and measurability. Applying measure and number to time means measuring and quantifying all other areas, in particular those where time and space relate to one another. The measurability of time pushes forward the development of the natural sciences as (empirical) sciences of measurement (and hence objective sciences) and mathematics as the theory of measurement.

After the mechanical clock, any attempt seriously referring to an alternative, subjective conceptualisation of time appears at least odd, if not absurd—regardless it was a practical necessity hitherto. Be it on the level of sciences, public institutions or individuals living their lives, time as a measurable and measured entity has become the natural, the "real" perception of time.

> Thus, the mechanical clock changed the relation between mankind and reality far beyond its original domain of application. It initiated the creation of a second nature totally reconstructing the first, exclusively admitting objective, mathematical laws, devaluing the authority of individual (subjective) experience or insight. (ibid.)

Although we might be able to rationally deconstruct our perception of time, being the result of a historically contingent misjudgement—a collective illusion—this does not liberate us from the compulsion to align to that illusion in our actions. "The illusion is not on the side of knowledge, it is already on the side of reality itself, of what the people are doing" (Žižek, 2008, p. 29f.). We cannot simply deconstruct the "second nature" by peeling away all the illusive layers and get to our supposedly original and innocent "first nature". Mathematically spoken, "first" rather stands for the moment x and "second" for the moment $x+1$. The way we perceive "xth nature" as "first nature" is the necessary retrospective effect of constituting subjectivity within ideology (Žižek, 2006).

As we can see, mathematisation does not only reinforce demathematisation as a consequence of implication, but simultaneously demathematisation reinforces mathematisation as a consequence of naturalisation.

> We have a circularity: The more mathematics is used to construct a new reality, the better it can be applied to describe and handle exactly this reality. (Fischer, 1993, p. 118)

Thus, even though mathematisation and demathematisation are apparently antagonistic phenomena, they stand in a dialectical relation and thus can be seen as two sides of one and the same coin. Hence, the notation *de|mathematisation* signifies both mathematisation and its apparent antagonist within one dialectic relation. The circularity that Fischer attests to de|mathematisation describes the relation of de|mathematisation to its symptom quite well. Expanding mathematisation postpones the confrontation with the symptom, promising some kind of unstoppable progress towards a perfect condition. With Lacan (2008), such faith can be described as "evolutionism": the belief in a supreme good resulting from expanding mathematisation in a final goal of progress that guides its course from the very beginning. If we just keep following the tracks we are on, we will enter a benign state in the end.

Enjoying the Immersion in the Ideology of Evolutionism

According to Žižek (2008), the effectiveness of ideology should not be understood on the level of consciousness but should instead be understood on the level of our actions. An ideology does not owe its prevalence to the fact that people believe in it, but rather because they act *as if* the ideological narratives were true, regardless whether they believe in them or not. What makes people perform the ideology of evolutionism? Žižek's canonical answer is that they develop (often ambiguous) ways to enjoy their immersion in ideology. What enjoyment then do people gain from adhering to evolutionism despite encountering symptoms like the crash of "Ariane 5 flight 501"?

As long as machines "know" the necessary mathematics for the individual (that is: in her stead), the individual is relieved of the burden of knowing mathematics[2]: "technology facilitates the use of mathematics in social or technical situations precisely by liberating the user from the details of the mathematics involved" (Jablonka & Gellert, 2007, p. 11). The narrative of a comfortable future allows for effectively avoiding the confrontation with the symptoms of the ideology of certainty. As mathematisations "become materialised, they become part of our reality and most of the time we do not ask where they come from or what they are—there is no necessity for doing so" (Jablonka & Gellert, 2007, p. 7). We find relief in delegating the hard mathematical "work" and willingly addict ourselves to a fantasy of comfort. We can record *comfort* as one of the affects that contributes to the enjoyment of evolutionism. A recent commercial of one of the world-leading car producers illustrates the effectiveness of the fantasy of comfort on the level of the individual (see the story line in Table 3.1).[3]

The actual piece of technology, the collision prevention assist (CPA), has been made possible by a tremendous amount of formerly explicit mathematical work in which applied mathematicians designed algorithms that process all the numerical data that detectors capture in real time. These data, we may suspect, must be related in a three-dimensional vector space, modelling all nearby objects and their movements. In order to predict potential collisions and display warnings accordingly, these movements must be modelled in potential trajectories. The quality of the CPA, we may suspect, depends to a significant degree on the quality of the compliant algorithms. However, even though the advertised innovation is the CPA, the spot does not at all thematise the mathematical qualities that facilitate and make up for this innovative piece of technology. The CPA is not at all presented as a device that helps an already careful driver to drive even safer, but rather to allow him to completely refrain from paying attention. He can drive blindfolded through the street, as the analogy drawn to the boy suggests, he could even enjoy an ice cream. However, contrary to the blindfolded boy, he has a technological guardian that ensures him not to hurt himself. He can "first of all, simply trust the black box" (Jablonka & Gellert, 2007, p. 8), lounge back and enjoy. While the blindfolded boy hits the pole, the driver escapes this fate.

[2] See Lundin and Christensen in this volume for another mechanism that modern culture has developed in order to delegate the direct encounter with mathematics to someone else: school mathematics.

[3] See the video on http://www.youtube.com/watch?v=KqPGlr52xDw.

Table 3.1 Storyline of the commercial

	A young, blindfolded boy with an ice cream in his hand walks on the pavement. When he heads for a pole, makes all of a sudden a 90-degree turn, shortly before he hits the pole. He continues to walk, already approaching the next pole. In the meantime, an older boy on a chopper bicycle enters the road on a crosswalk and a car turns around the corner.
	While his co-driver is looking out the window in dreams, the driver avertedly stares at the young boy, not noticing that the older boy had stopped—without any visible cause—his chopper bike in the middle of the road.
	Right in time the advertised technology gives a sound signal and displays a red triangle in the console of the car. The driver stops the car with a full brake and the car stops right in front of the older boy with the bike. The younger, blindfolded boy with the ice cream hits the pole with his head.
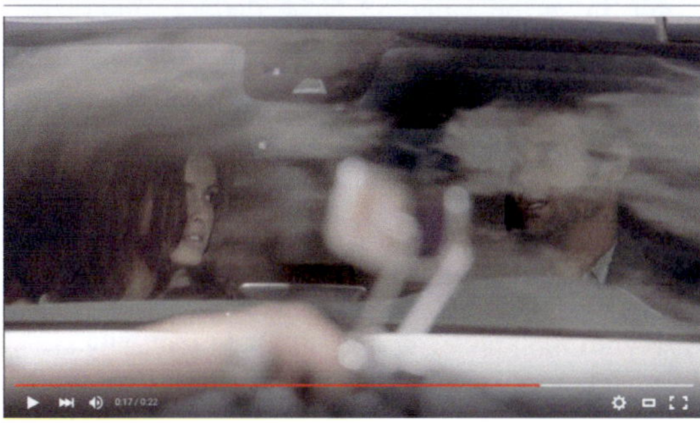	The driver and his co-driver breathe deeply and look at each other in relief, the older boy leaves the crosswalk. The last shot shows only the car on the street and the lettering "Issues a warning before collisions. Supports braking. The new B-class, serially with COLLISION PREVENTION ASSIST".

At the same time, comfort is a suspiciously egoistic motive, particularly when it is connected to the delegation of responsibility. It is an affront for the Kantian enlightened and self-determined subject. Delegating responsibility for the sake of comfort is likely to induce a sense of guilt. While *guilt* is a second affect involved in the immersion in evolutionism, it appears not yet clear how guilt contributes to enjoyment. However, as Žižek (1992) maintains, guilt induces the desire to extinguish this guilt. Probably for this reason, the commercial offers a second narrative, one of collective security: the CPA saves the life of an innocent child (the older boy on the bike). It allows to unconditionally enjoy an object (driving a nice car) and at the same time do something for the common good (contributing to safe streets). Renouncing the comfort of delegating responsibility to the CPA in this way itself appears as an irresponsible and egoistic act of defending the illusion of one's own self-determination.[4] The individual can delegate her very private responsibility for the supposed sake of collective responsibility. In this way, mathematisation does promise not only individual *relief*, but also a collective "relief from strain" (Fischer, 2007, p. 68) that allows to collectively adopt "certain mechanisms without a common idea about its whole, and, as a consequence, without any responsibility for it" (p. 69). The supposed sake of collective responsibility is of course an illusion. In this way, the individual's act of suppression of the symptom can be linked to a collective act of suppression of the symptom to what Fischer calls the "unconsciousness of society" (ibid.). Evolutionism is then the perfect conjunction of egotism and altruism. It simultaneously induces the apparently contradictory affects of comfort, guilt and relief and hereby creates a *perpetuum mobile* of enjoyment.

We see that the problem of people's ideological bond to de|mathematisation is not one of information, of a false consciousness of society about the real effects of de|mathematisation, as Borba and Skovsmose (1997) suggested with their conception of the ideology of certainty. At the dawn of the century, this appeared as a straightforward assumption, as at that time even well-recognised media very seldom "educate[d] us to the fact that mathematics is formatting a good portion of today's life and to point out where this is occurring" (Davis, 2007, p. 195). However, this has changed dramatically in the last years. The *algorithmisation* of social life has become a frequently occurring topic in serious media.[5] Nevertheless, enterprises whose productivity originates in the exploitation of the dialectic of de|mathematisation, like Google and Facebook, are as powerful today, both economically and politically, as they have never been before. Often we can observe a striking discrepancy between peoples' critical awareness on the level of consciousness and their actions. Why does critical distance not effectively translate into critical action?

---

[4] See http://www.youtube.com/watch?v=ytVdBLMmRno for a parody that most brilliantly mocks society's willingness to addict itself to a faith in evolutionism that posits mathematisation as a sacred force beyond history.

[5] Take the article, "How algorithms rule the world", which appeared in the guardian and which asks, "The NSA revelations highlight the role sophisticated algorithms play in sifting through masses of data. But more surprising is their widespread use in our everyday lives. So should we be more wary of their power?" (Hickmann 2013), as a more or less random English-speaking example.

Critical Distance as Part of the Interpellation of Evolutionism

Analysing another example from an advertising campaign for cars may take us one step further in better understanding this gap. While borrowing on the magical aura of mathematics in the 1980s seemed to be beneficial (see the Davis and Hersh's observation on the computerisation of love quoted in the beginning of this essay), mathematisation does not appear to be a selling point for advertising anymore. Quite the contrary: the campaign "escape the map"[6] by the same car producer that advertised the CPA even addresses directly to all those who feel alienated by proliferating de|mathematisation. It is an interactive campaign in which consumers can participate via social media and smartphones in a narrative aiming to liberate the beautiful Marie who is trapped in Google Street View: "Destined to a life of solitude within the confines of Streetview [sic!], Marie needs your help to escape. It's a strange world inside Streetview [sic!], and its effects on the human mind can be surprising; a place where life is anything but ordinary—a world full of glitches and digital vortexes" (Mercedes, 2011), the campaign's website explains. In the beginning of the costly produced, 4-min-long film around which the campaign is built, Marie asks rhetorically: "You don't want to end up like that?" The vehicle, that helps Marie to successfully escape out of the de|mathematised world, is—of course—the advertised car. Instead of advertising the rich technological equipment that the car has nevertheless, the narrative posits it as a weapon in the struggle against alienation in a more and more estranged, de|mathematised world. It appears as if the ideological interpellation with the one hand effectively enacts the ideology of evolutionism by selling mathematically sophisticated technologies and with the other hand critiques the very same ideology. Following Žižek's account of ideology this is no coincidence and it is naive to believe that the one hand does not know what the other hand is doing. According to Žižek (2008, p. 137), a certain degree of disbelief, or critical distance, is necessarily inbuilt in any effective ideological interpellation:

> So it is precisely this lack in the Other which enables the subject to achieve a kind of 'dealienation' caned by Lacan separation: not in the sense that the subject experiences that now he is separated for ever from the object by the barrier of language, but that the object is separated from the Other itself, that the Other itself 'hasn't got it', hasn't got the final answer—that is to say, is in itself blocked, desiring; that there is also a desire of the Other. This lack in the Other gives the subject—so to speak—a breathing space, it enables him to avoid the total alienation in the signifier not by filling out his lack but by allowing him to identify himself, his own lack, with the lack in the Other.

A complete, total identification with the ideology of certainty is an insult of the autonomous, self-aware and rational subject. It would imply that mathematics *actually* could solve all the dilemmas that the human itself is doomed to suffer. It would reverse the "optimal" relation between humankind and technology: humanity's welfare would be better entrusted to the hands of technology than to the hands of

[6] See https://www.youtube.com/watch?v=Z8rLyYYzEOo for a video around which the campaign unfolds.

humans. Further, this would in its last consequence lead to a rigid determinism and the end of the uniqueness of the human individual.

As Žižek maintains, in order to get rid of such uncanny effect of alienation, in order to "de-alienate", the subject does not need to take a full reversal and fully renounce the ideology of evolutionism. Instead, it can be assumed that mathematics itself "hasn't got the final answer"—that certainty itself is bothered with doubt. Allowing this breathing space may dismantle the ideology of certainty, but it leaves evolutionism intact. In this way, including critical distance towards the ideology of certainty as an immanent feature of the ideology of evolutionism can be understood as a mechanism of defence. It is critical distance that allows us to enact the fetishistic disavowal: "I know very well that I feel alienated by this mathematised world, but still … (I consume all the mathematised commodities)". By disavowing the discrepancy between consciousness and action, critical distance allows for even more enjoying the *perpetuum mobile* of comfort, guilt and relief.

In order to work effectively, ideology even requires people to not fully believe in the social fantasies they are performing. "The notion of social fantasy is therefore a necessary counterpart to the concept of antagonism: fantasy is precisely the way the antagonistic fissure is masked. Fantasy is a means for an ideology to take its own failure into account in advance" (Žižek, 2008, p. 142). Understanding critical distance as an inherent part of the ideology of evolutionism helps to comprehend why critical distance does not necessarily contribute to interrupting the process of de|mathematisation effectively: it takes its very own failure into account in advance. In this way, critical distance can even serve to fuel the enjoyment of evolutionism.

What Is Wrong with Critical Distance?

The observations above suggest that, within the contemporary media, it makes more sense to refer to an ideology of evolutionism than to an ideology of certainty. The ideology of evolutionism even instrumentalises critical distance—uncertainty—as part of its fantasy in order to avoid a total alienation that would finally lead to a disintegration of ideology. In this way, the ideology of certainty is able to immunise itself against its own failure, its symptoms. Evolutionism urges the subject to conceive failures as empirical obstacles, as marginal or even technical problems of a nevertheless good system (cf. Pais et al., 2012, p. 29).

At this point a critical reader may wonder about what could be wrong with critical distance. Is the author of this essay not himself supposing the existence of a "bad" system, intentionally (mis)interpreting all the instances of emergence of critical distance as arbitrary "failures" within such bad system? It might be argued that when people willingly enforce de|mathematisation by their actions despite a consciousness about all the negative effects that occur with it, there must be enough positive effects that people anticipate to overweigh their critical distance. At stake here is how we set the ruler of critique. When should we be more than just suspicious and reflect about countermeasures, when should we celebrate uncertainty as openness for a future that is coming towards us?

In her exhaustive research on dating portals, the American sociologist Eva Illouz (2007, p. 142, own translation) suggests a model of critique that tries to avoid the trap of both positive and negative evolutionism.

> I suggest to call this approach to social practice an 'impure' critique, a type of critique that seeks to stroll on a fine line between those practices that serve the subjects' wishes and needs—no matter how crude they may appear to us –, and those practices that clearly prevent subjects from reaching their aims.

This implies that a critique of the practice of a fetishistic disavowal in the form of "I know very well that I feel alienated by this mathematised world, but still … (I consume all the mathematised commodities)" has to seriously take into account the practitioner's perspective. As Pfaller (2002, p. 165, own translation) puts it: "The key is the posture the practitioners themselves adopt to follow their passions".

In her book "Gefühle in Zeiten des Kapitalismus" [*Emotions in times of capitalism*], Illouz provides an analysis of online dating portals that shows how significant the "computerisation" of love has changed since the observations of Davis and Hersh (1986). In the times of Davis and Hersh (1986, p. 126), the occurring mathematisation followed a quite static rubric of easily codifiable information, "income, height, smoker/non-smoker, education, preferred hobbies, favourite music … It [the computer] does not ask whether you are patient or not, considerate or brutal, tolerant or narrow-minded, impulsive or cautious". It did not allow for what Illouz (2007, p. 122, own translation) calls an "ontological presentation of the self" that was prevalent at the time of her research, and which "presupposes a movement that points towards a [supposed] solid inner core (who am I and what do I want?)". The Internet formalises the search for a partner in analogy to an economic transaction, "it turns the self into a packed product that competes with others on an open market, which is only regulated by the law of supply and demand" (p. 132, own translation). Through the technological progress of the Internet, the problem of searching for a partner becomes immediately interwoven with the problem of effectivity (ibid.), encouraging the participants to develop technologies of the self in order to increase their exchange value. This, Illouz (2007, p. 134 f., own translation) suspects, implies a radical transformation in the concept of love, which—in the so-called Western world—has hitherto been characterised by being (a) an unexpected, irrational and inexplicable event; (b) a unique physical experience that shakes our bodily certainties; (c) a non-instrumental value and (d) an expression of the uniqueness of the object of love. Thus, we can conclude that, in contrast to 1986, mathematisation has actually started to format the social practice of love. "The internet appears to have lifted the process of the rationalization of emotions and of love on a level that even Critical Theorist could not dream of" (p. 136, own translation). Illouz does not claim that there was an age before the Internet in which the (transcendental) romantic ideal of love really corresponded to the actual social practice of love. The actual social practice of love has always been interfered with economic considerations. One might argue that this was even more the case before the Cultural Revolution of the late 1960s. The social practice of love thus always required some form of fetishistic disavowal in the form of "I know very well that I chose my partner because of the social and economic security she offers, but still

… (I believe that I actually love her romantically)". For the participants in Illouz' study, however, such disavowal appears to serve the subjects' wishes and needs much less than it served practitioners' needs in non-mathematised off-line practices of love. Illouz (2007, p. 168, own translation) assumes that the reason for this change is the shift of ideology that occurred in late capitalism:

> If ideology is that, which allows us to live an internally contended life of contradictions, then I am not sure, whether the ideology of capitalism can still provide this. Possibly, the capitalist culture has reached a new stage: while the industrial and even the advanced capitalism simultaneously facilitated and required a split self, that could smoothly float between the domain of strategic to the domain of emotional interactions, from economic to emotional, from egoistic to cooperative, the internal logic of contemporary capitalism acts differently. Not only is the cost-utility analysis in the meantime applied to almost all private and domestic interactions, it appears to have also become more difficult to change from one register of practice (e.g. the romantic) to another (e.g. the economic).

While late capitalism is—of course—not able to liberate us from the contradictions that are constitutive of subjectivity, namely "a split self", it has discontinued the supply of strategies that facilitate a contended life within this split. This ideological constellation transforms critical distance from an emancipatory capacity into an invasive deadlock that finally rather sustains the dominant ideology than undermines it by fuelling the *perpetuum mobile* of enjoyment.

Recovering Critical Distance Towards De|mathematisation

The excursus to the domain of the mathematisation of love has helped us to better understand under what conditions critical distance towards mathematisation can unfold either an emancipatory or a preservative function. Late capitalism, with its unifying logic of evolutionism, undermines the emancipatory potential of critical distance. Without facilitating the space for a split self, the demand for critical distance becomes an invasive requirement, demanding to interpret any friction between competing principles as empirical obstacles. This accounts *a fortiori* in times of the so-called Internet of everything, where any social practice—including participants' behaviours, their wishes and desires, and the tools they employ—are mathematically captured and transferred into a unified network where everything can be commodified and disclosed with an exchange value.

If Illouz is right in her suspicion that a transition within the ideology of capitalism minimises the space for a split self, and if Žižek (2004, p. 3) is right in his assumption "that while it [late capitalism] remains a particular formation, it overdetermines all alternative formations, as well as all noneconomic strata of social life", is there any possibility that mathematics education can provide such a space, where critical distance opens options instead of foreclosing them? Is it that, within capitalism, a critical school mathematics education is doomed to "perform the role of what Freire (1998, p. 508) calls 'superficial transformations', designed precisely to prevent any real change in the core features of schooling" (Pais et al., 2012, p. 32)? Is it that we are "obliged to have faith in the necessity of mathematics education and in the importance of that which we do not understand" as Lundin (2012, p. 82) suggests in his analysis of the ideological function of critique and reform in mathematics education?

I fully agree with Pais and Valero (2012, p. 20) that "if the purpose is the high ideals of democracy, social justice and equality", we should not expect the solution from neither mathematics education research nor school mathematics. However, we should also not neglect the possibility that schools are a cultural and political achievement that provides at least a minimum of "relative autonomy" (Apple, 2002; Bernstein, 1990). While such a concept may be strange to a Žižekian theory of ideology, it appears not at odds with it. As I have described in detail elsewhere (Straehler-Pohl & Gellert, 2015), the "relative autonomy" of pedagogic devices, such as schools, is closely related to the idea of an intrinsic and indissoluble impossibility of the device to fully reproduce itself, similar to the logic of Žižek's symptom. In the same way Lacan (2007) maintains for language, any ideological interpellation cannot be anything other than a demand that necessarily fails to take full hold on the subject. I therefore claim that schools—because and not despite of their artificial character, where reality is rather emulated than real (Lundin, 2012)—have at least a minimal potential to be a catalyst for the failure of the capitalist interpellation. Relative autonomy surely implies that schools themselves cannot provide the solution to the dilemmas of capitalism. However, due to their nature as a simulated "game", they can provide the space for a split self within a mathematics classroom. This would at least sustain the potential of seeing capitalism as one ideology among others instead of mistaking it for reality. In the contemporary ideological climate, this appears as already an ambitious and important endeavour (Fisher, 2009).

However, this would require an even more radical break with the ideology of evolutionistic de|mathematisation than the one already proposed by critical mathematics educators. A first, and surely necessary step, is the dialogic development of what Keitel, Kotzmann and Skovsmose (1993, p. 272) have called "reflective knowing" and all the levels that they have proposed for its development. "Reflective knowing" in mathematics occurs through "addressing its own status" as knowledge. In this way, mathematical knowledge can be experienced as a contingent human construction among others, privileging certain values while neglecting others. It is reflective knowledge that includes a critical distance towards its own status. However, as this essay suggests, in order to emancipate, rather than entrap, students, school mathematics must provide a space where mathematical knowledge is not simply hypothetically experienced as one alternative among others, which are forgotten once they have been critically discussed. This would rather make "the ethical system of mathematics education seem *necessary*, regardless of how misdirected and destructive this system happens to be at the moment" (Lundin, 2012, p. 82, emphasis in original) than making it contingent. Therefore, I claim that in a necessary second step, school mathematics itself should provide a legitimate space for students to *completely reject the demand to solve problems of social significance by means of mathematics*. This would imply a space for the so-called split self that allows for a "dealienation" outside the ideology of evolutionism (instead of an alleged dealienation within, see above). One may object that a mathematics classroom—where problems are not solved by means of mathematics, but instead mathematics being rejected—would no longer be a mathematics classroom (but political education or an ethics classroom for example). However, if we follow an anthropological understanding of school mathematics, where school mathematics is self-referentially defined and continuously reproduced by its practice and not by a supposed substance of (school) mathematics (cf. Brown, 2010; Lundin, 2012), anything that addresses the status of mathematical knowledge should be legitimately considered as school mathematics. Finally, I claim that not only the intrinsic features of mathematisation, such as "ethical filtration" (Skovsmose, 2008; de Freitas, 2008), should be addressed in school mathematics, but also that, furthermore, the *contemporary entanglement of de|mathematisation with capitalism should be explicitly brought on the agenda of mathematics classrooms*.

Acknowledgments I would like to thank Candia Morgan, Nina Bohlman and Uwe Gellert for their helpful comments for developing this essay.

References

Apple, M. W. (2002). Does education have independent power? Bernstein and the question of relative autonomy. *British Journal of Sociology of Education, 23*(4), 607–616.

Bernstein, B. (1990). *The structuring of pedagogic discourse: Class, codes and control* (Vol. IV). London and New York: Routledge.

Borba, M. C., & Skovsmose, O. (1997). The ideology of certainty in mathematics education. *For the Learning of Mathematics, 17*(3), 17–23.

Brown, T. (Ed.). (2008). *The psychology of mathematics education: A psychoanalytic displacement*. Dordrecht: Sense.

Brown, T. (2010). Truth and the renewal of knowledge: The case of mathematics education. *Educational Studies in Mathematics, 75*(3), 329–343.

Chevallard, Y. (2007). Implicit mathematics. In U. Gellert & E. Jablonka (Eds.), *Mathematisation and demathematisation. Social, philosophical and educational ramifications* (pp. 57–66). Rotterdam: Sense (Orig. pub. 1989).

Davis, P. J. (2007). The media and mathematics look at each other. In U. Gellert & E. Jablonka (Eds.), *Mathematisation and demathematisation. Social, philosophical and educational ramifications* (pp. 187–196). Rotterdam: Sense.

Davis, P. J., & Hersh, R. (1986). *Descartes' dream: The world according to mathematics*. San Diego: Harcourt Brace Jovanovich.

Davis, Z. (2005). *Pleasure and pedagogic discourse in school mathematics: A case study of a problem-centred pedagogic modality*. Unpublished dissertation thesis. Cape Town: University of Cape Town.

Fischer, R. (1993). Mathematics as a means and as a system. In S. Restivo, J. P. van Bendegem, & R. Fischer (Eds.), *Math worlds: Philosophical and social studies of mathematics and mathematics education* (pp. 113–133). New York: SUNY Press.

Fischer, R. (2007). Technology, mathematics and consciousness of society. In U. Gellert & E. Jablonka (Eds.), *Mathematisation and demathematisation. Social, philosophical and educational ramifications* (pp. 67–80). Rotterdam: Sense.

Fisher, M. (2009). *Capitalist realism. Is there no alternative?* Hants: John Hunt Oublishing.

Freire, P. (1998). Cultural action for freedom. *Harvard Educational Review, 68*(4), 471–521.

de Freitas, L. (2008). Critical mathematics education: Recognizing the ethical dimension of problem solving. *International Electronic Journal of Mathematics Education, 3*(2), 79–95.
Gerofsky, S. (2010). The impossibility of 'real-life' word problems (according to Bakhtin, Lacan, Žižek and Baudrillard). *Discourse: Studies in the Cultural Politics of Education, 31*(1), 61–73.
Hickmann, L. (2013 July 1). How algorithms rule the world. Guardian. Retreived from: http://www.theguardian.com/science/2013/jul/01/howalgorithms- rule-world-nsa on September, 8th 2015.
Houston, S. K., Blum, W., Huntley, I., & Neill, N. (Eds.). (1997). *Teaching and learning mathematical modelling: Innovation, investigation and applications*. Chichester: Albion.
Illouz, E. (2007). *Gefühle in Zeiten des Kapitalismus*. Frankfurt: Suhrkamp.
Jablonka, E., & Gellert, U. (2007). Mathematisation—Demathematisation. In U. Gellert & E. Jablonka (Eds.), *Mathematisation and demathematisation. Social, philosophical and educational ramifications* (pp. 1–18). Rotterdam: Sense.
Keitel, C. (1989). Mathematics education and technology. *For the Learning of Mathematics, 9*(1), 103–120.
Keitel, C., Kotzmann, E., & Skovsmose, O. (1993). Beyond the tunnel vision: Analysing the relationship between mathematics, society and technology. In C. Keitel & K. Ruthven (Eds.), *Learning from computers: Mathematics education and technology* (pp. 243–279). Berlin: Springer.
Lacan, J. (2007). *The other side of psychoanalysis: The seminar of Jacques Lacan: Book XVII*. London: Norton.
Lacan, J. (2008). *The ethics of psychoanalysis: The seminar of Jacques Lacan: Book VII*. New York: Taylor and Francis.
de Lange, J., Keitel, C., Hutley, I., & Niss, M. (Eds.). (1993). *Innovation in maths education by modelling and applications*. Chichester: Ellis Horwood.
Le Lann, G. (1997). An analysis of the Ariane 5 flight 501 failure—A system engineering perspective. In *Proceedings of the 1997 Workshop on Engineering of Computer-Based Systems* (pp. 339–346). IEEE.
Lundin, S. (2012). Hating school, loving mathematics: On the ideological function of critique and reform in mathematics education. *Educational Studies in Mathematics, 80*(2), 73–85.
Mercedes Benz (2011). Escape the map with Marie and Mercedes Benz. Retrieved March 03, 2015 from http://news.mercedes-benz.co.uk/innovations/escape-the-map-with-marie-and-mercedes-benz.html
Mesquita, M., Restivo, S., & D'Ambrosio, U. (2011). *Asphalt children and city streets. A life, a city, and a case study of history, culture, and ethnomathematics in São Paulo*. Rotterdam: Sense.
Pais, A. (2012). A critical approach to equity in mathematics education. In B. Greer & O. Skovsmose (Eds.), *Opening the cage: Critique and politics of mathematics education* (pp. 49–93). Rotterdam: Sense.
Pais, A. (2015). Symbolising the real of mathematics education. *Educational Studies in Mathematics, 89*(3), 375–391.
Pais, A., Fernandes, E., Matos, J. F., & Alves, A. S. (2012). Recovering the meaning of "critique" in critical mathematics education. *For the Learning of Mathematics, 32*(1), 28–33.
Pais, A., & Valero, P. (2012). Researching research: Mathematics education in the political. *Educational Studies in Mathematics, 80*(1–2), 9–24.
Pais, A., & Valero, P. (2014). Whither social theory? *Educational Studies in Mathematics, 87*(2), 241–248.
Pfaller, R. (2002). *Die Illusionen der anderen. Über das Lustprinzip in der Kultur*. Frankfurt a.M: Suhrkamp.
Skovsmose, O. (1994). *Towards a philosophy of critical mathematics education*. Dordrecht: Kluwer.
Skovsmose, O. (2008). Mathematics education in a knowledge market: Developing functional and critical competencies. In E. de Freitas & K. Nolan (Eds.), *Opening the research text: Insights and in(ter)ventions into mathematics education* (pp. 159–188). New York: Springer.
Skovsmose, O. (2014). Mathematization as social process. In S. Lerman (Ed.), *Encyclopedia of mathematics education* (pp. 441–445). Heidelberg: Springer.

Straehler-Pohl, H., & Gellert, U. (2015). *Pathologie oder Struktur? Selektive Einsichten zur Theorie und Empirie des Mathematikunterrichts*. Heidelberg: Springer.

Straehler-Pohl, H., & Pais, A. (2014). Learning to fail and learning from failure: Ideology at work in a mathematics classroom. *Pedagogy, Culture and Society, 22*(1), 79–96.

Straehler-Pohl, H., Gellert, U., & Bohlmann, N. (2014). Reasons to believe: Mathematics and the reality of consumerism. In G. Aldon, B. Di Paola & C. Fazio (Eds.), *Mathematics and reality. Proceedings of CIEAEM 66, Quaderni di Ricerca in didattica, 24*(1), 186–189.

ZEIT (2014, August 23). Zu kurz gesprungen. Was tun selbst fahrende Autos, wenn sie beim Ausweichen wählen müssen: Hinein ins Motorrad oder in die Fußgängergruppe? (by Adrian Lobe). Retrieved November 20, 2014 from http://www.zeit.de/2014/33/autonomes-fahren-auto-strassenverkehr

Žižek, S. (1992). *Enjoy your symptom. Jacques Lacan in Hollywood and out*. New York: Routledge.

Žižek, S. (2004). *Organs without bodies: Deleuze and consequences*. London: Routledge.

Žižek, S. (2006). *The Parallax View*. Cambridge: MIT Press.

Žižek, S. (2008). *The sublime object of ideology*. London: Verso (Orig. pub. 1989).

Chapter 4
The Narcissism of Mathematics Education

Alexandre Pais

Abstract Why does mathematics education research create a reality so at odds with the one experienced by the vast majority of teachers and students worldwide? This chapter is part of an ongoing venture that seeks to analyse the ideological belongings of contemporary educational research, by focusing in the particular case of mathematics education. Here, the author displays some elements of Pfaller's materialist approach to philosophy and Žižek's ideology critique to analyse common shared assumptions of researchers when conceiving the influence of their work in practice. It is argued that mathematics education research needs to shift its perspective and recognise in its symptoms—students' systematic failure, absence of change, increasing of testing, pernicious political and economic influences, etc.—the violent expression of the disavowed part of itself.

Introduction

Mathematics education research makes sense in itself. Researchers have at their disposal a panoply of well-grounded theorisations and extensively tried methodologies, used in their scientific endeavours to understand and improve mathematics education. From the "inside", mathematics education research appears as a prolific, growing and reasonably fundable area of scientific enquiry. It also enjoys an aura of importance derived from the place mathematics occupies in the scientific, technological and economic development. Moreover, when animated by social, cultural and political concerns, mathematics education research is perceived as a crucial element against racism, poverty, lack of democracy, reproduction of class inequalities, and other social harms. Within the realm of its own syntax, mathematics education research makes sense and deserves to grow.

This semblance of significance is however challenged by the concrete reality of schooling. We know all too well how mathematics is often considered a meaningless, useless and boring school subject; cause of anxiety for many students as well

A. Pais (✉)
Education and Social Research Institute, Manchester Metropolitan University, Manchester, UK
e-mail: a.pais@mmu.ac.uk

H. Straehler-Pohl et al. (eds.), *The Disorder of Mathematics Education*,
DOI 10.1007/978-3-319-34006-7_4

as teachers. Within the popular imaginary, no other school subject raises so opposite feelings of love and hate. Such discrepancy is itself symptomatic of the perilous place that mathematics occupies in today's society. School mathematics has been also an important social tool to systematically measuring (and positioning) individuals, schools and countries (Keitel, 2013). It is used as an instrument of governmentality and bio-politics (Popkewitz, 2004; Pais & Valero, 2012), and it shows a strong tendency to reproduce class, race and gender inequalities (Clements, Keitel, Bishop, Kilpatrick, & Leung, 2013). It is a privileged mean to seize people into the production of surplus value, thus an introduction into capitalist economics (Baldino & Cabral, 2013). All this is well known. Nevertheless the spirit of research is one where mathematics is worthy of love (Boaler, 2010), an essential tool for technology, society, and a means towards critical citizenship (Skovsmose & Valero, 2008).

Why does research create a reality so at odds with the one experienced by many students and teachers worldwide? This question is at the centre of the infamous gap between research and practice (Sriraman & English, 2010). In the introductory chapter of the very recent *Third International Handbook of Mathematics Education*, Clements (2013, p. x, xi) poses a crucial question for all of those involved in mathematics education research: "Why has there not been a marked improvement, given the large amount of mathematics education research conducted around the world, and over a very long period of time, with respect to such fundamentally important curriculum matters?". This situation is problematic since mathematics education as a field of research is not only oriented to describing and analysing practice, but (and perhaps more importantly) to prescribe or at least identify good practice (Jablonka, Wagner, & Walshaw, 2013, p. 47). As I explore elsewhere (Pais & Valero, 2012), the discrepancy between the sophistication of research and the lack of change in school mathematics is often displaced from research and posited on the way governments, schools and teachers fail to "acquire" and implement the knowledge originating from academia. In research, everything appears to run smoothly; we know the best methods, theories and strategies. The problems of implementation rest in the school settings. Lundin (2012) has recently discussed the fallacy of this line of argumentation. What he calls the *standard critique* of mathematics education consists of describing the current state of affairs of school mathematics as suffering from a variety of malfunctions, and the role of mathematics education research to fix them, by providing direct recipes for the practitioners' work (e.g. Sriraman & English, 2010, p. 27). The problem with this argumentation is that it eschews research from a critical analysis of its own role in the creation of the very same gap that it so eagerly strives to close. As argued by Klette (2004, p. 3), the problem of change in mathematics education reform is not just a problem of "application" but may well be an embedded part of research itself. She argues that the "denial of change" is being constructed from the beginning, in the theoretical, methodological and conceptual ways in which research is done.

The purpose of this chapter is to probe the way in which the community of mathematics education researchers perceives itself in relation to its object—the teaching and learning of mathematics.[1] When a field begins to pose questions not only about its

[1] Some would argue that the object of mathematics education research cannot be reduced to the teaching and learning of mathematics—notwithstanding this being constantly stated in some of the

internal achievements, but also about the external circumstances, which render possible its existence as a field, this is usually called reflexivity (Bloor, 1976; Bourdieu, 2001). Such has been the case for many social sciences that at a certain point of their development, turn back upon themselves to investigate their own ways of working (e.g. Clifford, 1988), and the impact they have not in itself (that is, in the theoretical and methodological evolution of the field), but for its object of study (the teaching and learning of mathematics). The focus of this chapter is thus mathematics education research, and the examples analysed come from research rather than from concrete episodes of teaching and learning mathematics. It is my contention that such an approach, although not directly aimed at providing some kind of insight for action, can help redefine the coordinates we use to make sense of the problems of the field.

Ideal, Object and Example

Today, perhaps as never before, a banner catches together the community of mathematics education researchers. Because of the recent criticism made on the ideological mechanisms at work when researching equity (Pais, 2012; Pais & Valero, 2012), we might expect researchers to be more cautious when assuming the slogan of "mathematics for all". This ideal is constantly foregrounded as the ultimate horizon guiding our engagement in the field (Clements et al., 2013). A slogan such as "mathematics for all", functions as a *master-signifier* (Žižek, 2012, following Lacan, 2007), a banner upon which we all agree, uniting the field, thus offering a space, whereby different perspectives, theories and methodologies, can "work together". Notwithstanding all the evidence that mathematics is not for all, this ideal is posited as an achievable goal, and emphasis is given to the exploration of successful experiments, where students seem to learn meaningful mathematics for their lives. To develop and broadcast successful experiences seems to be the aim of research (Gutiérrez, 2010; Presmeg & Radford, 2008; Sriraman & English, 2010).

What should be rendered problematic here is the inability of research to break with sources like common sense. Let us take as an example the research on the use-value of mathematics, the critique recently made by Lundin (2012) and myself (Pais, 2013) to the ideology sustaining this research. Common sense says that

most important publications of the field (e.g. Clements, 2013; Presmeg, 2013). Their argument rests on a set of research work that has been occurring in mathematics education that is instead focused on developing an analysis of the cultural, social and political landscapes that animates mathematics education. However, the fact remains that most of these studies do so with the ultimate goal of improving the teaching and learning of mathematics. As stated by Christine Keitel (2013, p. 1), in the introductory chapter of the section of the recent Third International Handbook of Mathematics Education (Clements et al., 2013) dedicated to the Social, Political and Cultural Dimensions in Mathematics Education, this research—on the social, political and cultural dimensions—has the goal of "informing mathematics education researchers as they strive to achieve more equitable and effective environments in which the teaching and learning of mathematics occurs."

people do not use (school) mathematics in their daily lives.[2] Research often confirms this unimportance of mathematics for mundane activities (e.g. Brenner, 1998; Jurdak, 2006; Williams & Wake, 2007). However, instead of questioning the presupposition that people need mathematics for their mundane or professional activities, research takes to itself the task of improving the utility of mathematics. This is done by means of developing deeper analysis and positive experiences, whereby students actually transfer mathematics from and into school: people do not use mathematics, but (because they should) we need to continue developing efforts to change this situation. A paradigmatic case is provided by Jurdak's research (2006). After concluding that "the activity of situated problem solving in the school context seems to be fundamentally different from decision-making in the real world because of the difference of the activity systems that govern them" (p. 296), and that students "define their own problems, operate under different constraints, and mathematics, if used at all, plays a minor role in their decision making" (ibid.), Jurdak still insists on the importance of confronting students with real-life situations: "simulations of such authentic real life situations as embedded in situated problem solving may provide a plausible option to develop appreciation of the role, power, and limitations of mathematics in real world decision-making" (ibid.). He adds, "though quite different in real life from that in school, the process of mathematization *is essentially the same* and having experience in it in a school context may impact on mathematization in real life" (p. 297, my emphasis). However, it is impossible to find support in the research reported in Jurdak's text for such statements. The belief that the exploration of real-life situations in school will impact on the way in which people use mathematics in real life is based on a "leap of faith" (Lundin, 2012, p. 8).

When confronted with the difficulties in transfer, Jurdak proceeds by eliminating the obstacles, so that the higher goal of making mathematics useful for people's lives can be kept. Instead of assuming the impossibility of transfer (Evans, 1999; Gerofsky, 2010), the researcher ends up creating an ideology whose purpose is precisely to disavow such impossibility. This is an example of how the theory speaks louder than its object. When confronted with a situation that contradicts the theory, instead of questioning the theory, the researcher ends up keeping the theory and disavowing the difficulties posed by apparent educational, political or economic contradictions, which make the real of schools.

When a science lacks an object that can question theory from within, this is a sign of what Pfaller (2007, p. 40), following Bachelard, calls the *initial narcissism of theory*. Mathematics education research is narcissistic because, lacking a concrete object, sees nothing but itself—its own expectations, presuppositions and prejudices. A narcissist approach disavows those matters that do not fit into its own image. When confronted with obstacles to the teaching and learning of mathematics that cannot be controlled by research—poverty, inequality, economic constraints, and governmental

---

[2] Elsewhere (Pais, 2011, 2013) I discuss in depth the fallacy supporting the idea that people use mathematics in their daily lives, as well as the dialectic at play when confronting the mathematics people learn in schools with the mathematics people use in their daily activities (as professionals, consumers, lovers, etc.).

decisions, but also students' refusal to assume the symbolic mandate conferred upon them—researchers tend to forsake them for the sake of research. Instead of conceiving these "external" circumstances as the very arena in which the true nature of research's inner potentials is to be "tested", researchers conceive them as empirical impediments, thus keeping the presuppositions of research intact (Pais, 2012).

Providing science with an object, and breaking with its initial narcissism, is a thoroughly materialist task (Pfaller, 2007, p. 41). A materialistic approach to theory means to engage seriously with those matters that precisely do not fit into a given theoretical explanation. To paraphrase Pfaller (2007, p. 42), materialist researchers do not hesitate to play the role of the black sheep that speaks out the dirty truth nobody wants to acknowledge.

Within the materialist tradition,[3] the *example* plays a specific role. In most of mathematics education research, examples, or more generally, data from the classroom, are used to illustrate, support, or show what could be done in order to achieve a meaningful mathematics instruction (Pais, 2016). However, could an example function not as an "exemplification" or "illustration", but more precisely, as a *symptom*, as a counter-example to the entire theorisation? From a materialistic point of view, an example serves to undermine a given universality. According to Pfaller (2007, p. 38), the role of an example is not to illustrate or exemplify a general idea, but "to displace it; drag it away from its initial position, to "estrange" it". By caricaturizing another example, something appears, which was foreign to the idea that this initial example exemplifies: "it makes visible a theoretical structure in the original idea which, before, was not easy to discern or which was even hidden by another structure that appeared evident". As I argue elsewhere (Pais, 2014), it is only the exploration of disruptive examples of failure that can prevent theory from painting reality pink and becoming an idealist, "apologetic" narrative. The study of practice, more than corroborating a theory, should serve to question the theory from within. This is a way to break with the initial narcissism of mathematics education. The exploration of "dirty" examples allows us to not only question the idea that "mathematics for all" is achievable, but also the entire theoretical frame that sustains this ideal.[4]

The Primacy of Practice

What can classroom examples say to us about the ideal of "mathematics for all?" As any teacher knows, in a class of thirty students there will always be some (often many) who fail. The crude reality tells us that the ideal is at least an illusion (when not straightforward bait). In order to enable success, however, researchers set and organise classroom data in

[3] Here the reference is the work of people such as Marx, Freud, Wittgenstein, Althusser, Lacan and Žižek.

[4] In previous research (e.g. Pais, Fernandes, Matos, & Alves, 2012; Straehler-Pohl & Pais, 2014) I explore a set of "dirty" examples to criticise the entire discourse on the beatitude of mathematics education.

a way that can corroborate a priori assumptions. As I analyse elsewhere (Pais, 2016; Pais & Valero, 2012), in Luis Radford's (2006, 2008) theory of cultural objectification, for instance, the examples used to support the theory are all reports of successful experiences, whereby pupils always acquire (objectify) the mathematical content set by the teacher. The research environment is arranged in a way as to avoid friction and allow a meaningful mathematics learning to occur; and the classroom examples are chosen to fit the theory. Radford's theory of objectification is however "estranged" at the very moment we try to imagine it implemented in low-streaming schools in Germany (Straehler-Pohl & Pais, 2014), schools in post-apartheid South Africa (Skovsmose & Valero, 2008), ghetto schools in the USA (Gutstein, 2003) or even a public European school struggling with imposed forms of mathematics that do not match the safeness and aseptic schooling characteristic of Radford's research settings (Brown, 2011). In these, seldom do students "unite" (Radford, 2006, p. 54) with the culture of mathematics in the way envisaged by Radford's theory. Contrary, what often occurs is precisely a refusal to identify with the mathematical successful learner envisaged by the curriculum (Pais, 2016).

"Mathematics for all" displays the structure of a *fantasy* (Žižek, 2012). Its purpose is not to make sense of the world in a perfect way, but rather to conceal the impossibility of making sense of it by establishing an analogy between two incompatible polar terms: "The function of the whole structure is to conceal the original imbalance" (Maningler, 2012, p. 44). Although we know that mathematics is not for all, that it serves other purposes than the ones related with knowledge and competences, that many students find it meaningless or even traumatic, we have to rely on the illusion that mathematics can indeed be for all, that it can be an adventure into knowledge, a pleasurable and useful subject for students. The shocking evidence that mathematics is nothing of this does not inhibit from partaking in the illusion that it can indeed be so. As a result, instead of asking why it is not, we keep researching how it can be.

In a recent conversation with a colleague around these issues, he claimed that although we know very well that mathematics is not for all, we should refrain ourselves from saying it out aloud. Admitting that mathematics is not for all will potentially diminish its importance in schooling, (who says "chemistry for all?") with direct consequences for our work as researchers. It is because mathematics plays such a relevant role in society and schooling that we, as a research community, enjoy privileged funding and working opportunities. What this discourse renders evident, however, is how research is about nothing but itself. It seems that research is not about improving school mathematics, but about using the miserable state of school mathematics to give researchers conditions to develop their work.

For research to break with this "epistemological obstacle" (Bachelard, 2002) it needs to seriously take its object of study—the teaching and learning of mathematics—as "it is" instead of how it "ought to be" (Pais & Valero, 2014). This implies moving from questioning "what can a school do if it wants to engage all of its students actively and productively in relevant mathematics learning?" (Clements, 2013, p. ix), to questioning why schools cannot systematically engage all of its students actively and productively in relevant mathematics learning, notwithstanding the declared will of all involved? In other words, instead of seeing research as a mean to change practice, perhaps researchers should take practice itself—as it happens in most schools, outside the fixed environments designed by researchers—as a mean to change research theories, methodologies and approaches.

Research's Beautiful Souls

Although a fantasy, with few resemblances with the concrete circumstances of schooling, the reality of research has real effects (Lundin, 2012; Pais, 2013). It creates an entire research industry, outlines school curricula, prescribes classroom work, and is the main informant for the constitution of international assessment instruments like PISA and TIMSS. Researchers often see these instruments as corrupting positive developments originated from research: "[t]here is a concern that TIMSS, PISA, and other international testing programs will have a standardizing effect on school mathematics that will cramp promising developments arising from the "social turn" in research" (Clements, 2013, p. ix). We have thus two opposite positions. On the one side, we have governments and international agencies privileging economic interests and suspicious political agendas in education; and on the other side, we have researchers who are perceived as struggling against this educational reductionism. However, could it be that these two positions are not opposite but part of the same whole, each one performing a complementary role? Would it be possible to develop an instrument like PISA without all the knowledge produced by mathematics education research in the last three decades around the importance of mathematics for professional and mundane activities? PISA, the ultimate examination designed to evaluate students' use of mathematics in everyday activities, partakes and takes advantage of the research ideology that asserts the use-value of mathematics (Pais, 2013).[5] Moreover, PISA is embedded in the same discourse used by researchers to justify the importance of mathematics education for scientific, technological, social and economic development. Seen from this perspective, instruments such as PISA are not corrupting research but *actualising* core research suppositions.

The usual reproach to this argumentation consists in saying that, although research and policy may partake in the same discourse concerning the importance of mathematics for today's society, political and economic instances manipulate this discourse to achieve other purposes than the ones envisage by researchers. But can we exempt ourselves from the political and economic world order in which we live? What is school mathematics outside the practice of schooling? A researcher's dream of how things should be, if only … This posture resembles that of Hegel's "beautiful soul", a figure that withdraws from any committed action in order to retain the purity of intention and not to get dirty hands: "when the subject assumes

[5] Such a "commoditization" has been happening with ideas coming from ethnomathematics (Pais, 2011) and critical mathematics education (Pais et al., 2012) two fields that are highly critical towards existing school mathematics.

the position of a judge exempt from what he is passing a judgment on" (Žižek, 2000, p. 228). The contradiction is that in the attempt to remain pure, the beautiful soul is nevertheless involved in the system that it pretends to reject.

In Mathematics More Than Itself

We know too well how international comparative instruments like PISA and TIMSS put pressure on teachers and schools to reduce the educational process to a promotional process (Biesta, 2009). Teachers tend to tailor their instructional practices to the format of the test out of concern that if they design their teaching differently, their students will fail. Although they might know all the didactical novelties and methods to promote learning in a way meaningful to the students, if what counts is to pass the test, that is how they will "educate" their students (Lerman, 1998). It is as if something get "stuck" to mathematics when it goes into schools that "mortifies" what could be a meaningful and pleasurable experience.

What is this "something" that becomes attached with mathematics thus colouring its presence in school? As I explore elsewhere (Pais, 2014), it is the credit that mathematics embodies when in school that colours its functioning. The school's credit system (Vinner, 1997) functions as Lacanian's *object a*, that element whose exclusion (from research) constitutes and sustains research. Object a is both subtracted from the reality of research (thus lacking, namely, lacking an economic and ideological reading of mathematics education) and its excess—it cannot be processed, it is disavowed, recognised as important but beyond the scope of research.[6]

It might be difficult for a researcher to acknowledge that school mathematics is more about credit than knowledge or competences. However, for most students, what makes mathematics desirable is not mathematics itself, but precisely what in mathematics is more than itself: the object cause of desire, the school credit attached to this school subject.[7] The same happens to many teachers: they want to teach mathematics, but they want even more their students to pass, so they destroy mathematics—doing routine exercise, meaningless "real" problems, etc.—for the sake

[6] For example, see Abreu, Bishop, and Presmeg (2002 p. 4).

[7] As I explore elsewhere (Pais, 2016), the fascination towards the importance of mathematics results from something that gets attached to school mathematics, which then starts to colour its entire dynamic. Remember when a teacher proudly brings into the classroom a particular application of mathematics, a bit of history or some other curiosity, and students immediately ask: "will this appear in the exam teacher?" Teachers are compelled to say yes, if keeping students' interest in the agenda. Or imagine the feeling of betrayal that a teacher feels when a student openly admits that he or she does not want to "like" mathematics, but only to pass the exam. A student that says to the teacher: "train me the best you can, so that I can do the exam, and never again go through mathematics!" Something is coupled with mathematics (the object a, the credit system) that stands for its functioning. This something that structures students' desire to learn mathematics is the credit associated with this school subject. It is the object cause of desire (Lacan, 2007), which makes both teachers and students "enjoy" this school subject.

of the credit given by the exam. As Brown (2011) shows in his research with teachers, this happens even if the teacher is fully aware that promoting a "teaching to pass" is a deficient way of learning mathematics.

Research becomes possible from the moment it subtracts from reality the economic role of school mathematics, thus creating an imaginary world where mathematics can be an adventure into knowledge, the ultimate problem solving technology or the most crucial component of critical citizenship. Economy is subtracted from research (Pais, 2014), and this absence simultaneously allows research to flourish and thwarts any endeavours to actually change practice. However, if the object of mathematics education is the teaching and learning of mathematics as it happens in a panoply of different contexts (but mostly through schooling), research cannot disavow what are the concrete conditions of this teaching and learning. By disavowing these conditions in favour of a prototypical reality (Skovsmose, 2011), mathematics education research ends up speaking about nothing but itself.

If mathematics education research assumes the importance of changing the teaching and learning of mathematics, and wishes to break with the internal narcissism of research, it cannot afford disavowing what are the concrete circumstances of today's schooling. The disavowing of the economy of schooling by research (Pais, 2014) comes back through the back door in the form of instruments like PISA and TIMSS, or in the crude reality of all of those who year after year fail to succeed in this school subject. Research fails to analyse its own role in the same reality it laments. And instruments like PISA make explicit something that is already present in research but in a disavowed form. To get way from this deadlock, mathematics education research needs to shift its perspective and recognise in its symptoms—students' systematic failure, absence of change, increasing of testing, pernicious political and economic influences, etc.—the violent expression of the disavowed part of itself.

Acknowledgments I am grateful to Ditte and Uwe for their tenacious criticism.

References

Abreu, G., Bishop, A., & Presmeg, N. (2002). *Transitions between contexts of mathematical practices*. The Netherlands: Kluwer Academic Publishers.

Bachelard, G. (2002). *The formation of the scientific mind*. Manchester: Clinamen Press.

Baldino, R., & Cabral, T. (2013). The productivity of students' schoolwork: An exercise on Marxist rigour. *Journal for Critical Education Policy Studies, 11*(4), 1–15.

Biesta, G. (2009). Good education in an age of measurement: On the need to reconnect with the question of purpose in education. *Educational Assessment, Evaluation and Accountability, 21*(1), 33–46.

Bloor, D. (1976). *Knowledge and social imagery*. London: Routledge and Kegan Paul.

Boaler, J. (2010). *The elephant in the classroom: Helping children learn and love maths*. London: Souvenir Press.

Bourdieu, P. (2001). *Science of science and reflexivity*. Chicago: University of Chicago Press.

Brenner, M. (1998). Meaning and money. *Educational Studies in Mathematics, 36*(2), 123–155.

Brown, T. (2011). *Mathematics education and subjectivity: Cultures and cultural renewal.* Dordrecht: Springer.
Clements, M. A. (2013). Past, present and future dimensions of mathematics education: Introduction to the third international handbook of mathematics education. In M. A. Clements, A. Bishop, C. Keitel, J. Kilpatrick, & F. Leung (Eds.), *Third international handbook of mathematics education*. New York: Springer.
Clements, M. A., Keitel, C., Bishop, A., Kilpatrick, J., & Leung, F. (2013). From the few to the many: Historical perspectives on who should learn mathematics. In M. A. Clements, A. Bishop, C. Keitel, J. Kilpatrick, & F. Leung (Eds.), *Third international handbook of mathematics education*. New York: Springer.
Clifford, J. (1988). *The predicament of culture: Twentieth century ethnography, literature and art.* Cambridge: Harvard University Press.
Evans, J. (1999). Building bridges: Reflections on the problem of transfer of learning in mathematics. *Educational Studies in Mathematics, 39*(1), 23–44.
Gerofsky, S. (2010). The impossibility of "real-life" word problems (according to Bakhtin, Lacan, Žižek and Baudrillard). *Discourse: Studies in the Cultural Politics of Education, 31*(1), 61–73.
Gutiérrez, R. (2010). The sociopolitical turn in mathematics education. *Journal for Research in Mathematics Education, 41*, 1–32.
Gutstein, E. (2003). Teaching and learning mathematics for social justice in an urban, Latino school. *Journal for Research in Mathematics Education, 23*(1), 37–73.
Jablonka, E., Wagner, D., & Walshaw, M. (2013). Theories for studying social, political and cultural dimensions of mathematics education. In M. A. Clements, A. Bishop, C. Keitel, J. Kilpatrick, & F. Leung (Eds.), *Third international handbook of mathematics education*. New York: Springer.
Jurdak, M. (2006). Contrasting perspectives and performance of high school students on problem solving in real world situated and school contexts. *Educational Studies in Mathematics, 63*(3), 283–301.
Keitel, C. (2013). Introduction to section A: Social, political and cultural dimensions in mathematics education. In M. A. Clements, A. Bishop, C. Keitel, J. Kilpatrick, & F. Leung (Eds.), *Third international handbook of mathematics education*. New York: Springer.
Klette, K. (2004). Classroom business as usual? (What) do policymakers and researchers learn from classroom research? In M. Høine & A. Fuglestad (Eds.), *Proceedings of the 28th Conference of the International Group for the Psychology of Mathematics Education* (Vol. 1, pp. 3–16). Bergen, Norway: University College.
Lacan, J. (2007). *The other side of psychoanalysis: The seminar of Jacques Lacan book XVII* (1st ed.). New York: Norton & Company. 1991.
Lerman, S. (1998). The intension/intention of teaching mathematics. In C. Kanes (Ed.), *Proceedings of Mathematics Education Research Group of Australasia* (Vol. 1, pp. 29–44). Gold Coast: Griffith.
Lundin, S. (2012). Hating school, loving mathematics: On the ideological function of critique and reform in mathematics education. *Educational Studies in Mathematics, 80*(1), 73–85.
Maningler, P. (2012). Acting out the structure. In P. Hallward & K. Peden (Eds.), *Concept and form (volume 2): Interviews and essays on the Cahiers pour l'Analyse*. London: Verso.
Pais, A. (2011). Criticisms and contradictions of ethnomathematics. *Educational Studies in Mathematics, 76*(2), 209–230.
Pais, A. (2012). A critical approach to equity in mathematics education. In O. Skovsmose & B. Greer (Eds.), *Opening the cage: Critique and politics of mathematics education* (pp. 49–91). Rotterdam: Sense Publishers.
Pais, A. (2013). An ideology critique of the use-value of mathematics. *Educational Studies in Mathematics, 84*(1), 15–34.
Pais, A. (2014). Economy: The absent centre of mathematics education. *ZDM: The International Journal on Mathematics Education*. doi:10.1007/s11858-014-0625-8.

Pais, A. (2016). At the intersection between the subject and the political: a contribution to an ongoing discussion. *Educational Studies in Mathematics*, *92*(3), 347–359.
Pais, A., Fernandes, E., Matos, J., & Alves, A. (2012). Recovering the meaning of "critique" in critical mathematics education. *For the Learning of Mathematics, 32*(1), 29–34.
Pais, A., & Valero, P. (2012). Researching research: Mathematics education in the political. *Educational Studies in Mathematics, 80*(1–2), 9–24.
Pais, A., & Valero, P. (2014). Whither social theory? *Educational Studies in Mathematics, 87*(2), 241–248.
Pfaller, R. (2007). Interpassivity and misdemeanours: The analysis of ideology and the Žižekian toolbox. *International Journal of Žižek Studies, 1*(1), 33–50.
Popkewitz, T. S. (2004). The alchemy of the mathematics curriculum: Inscriptions and the fabrication of the child. *American Educational Research Journal, 41*(1), 3–34.
Presmeg, N. (2013). Editorial. *Educational Studies in Mathematics*, *84*(3), 279–280.
Presmeg, N., & Radford, L. (2008). On semiotics and subjectivity: A response to Tony Brown's "signifying 'students', 'teachers,' and 'mathematics': A reading of a special issue". *Educational Studies in Mathematics, 69*(3), 265–276.
Radford, L. (2006). The anthropology of meaning. *Educational Studies in Mathematics, 61*(1–2), 39–65.
Radford, L. (2008). Culture and cognition: Towards an anthropology of mathematical thinking. In L. English (Ed.), *Handbook of international research in mathematics education* (2nd ed., pp. 439–464). New York: Routledge, Taylor and Francis.
Skovsmose, O. (2011). *An invitation to critical mathematics education*. Heidelberg: Springer.
Skovsmose, O., & Valero, P. (2008). Democratic access to powerful mathematical ideas. In L. D. English (Ed.), *Handbook of international research in mathematics education* (2nd ed., pp. 415–438). New York: Routledge.
Sriraman, B., & English, L. (2010). Surveying theories and philosophies of mathematics education. In B. Sriraman & L. English (Eds.), *Theories of mathematics education: Seeking new frontiers*. Heidelberg: Springer.
Straehler-Pohl, H., & Pais, A. (2014). Learning to fail and learning from failure: Ideology at work in a mathematics classroom. *Pedagogy, Culture and Society.*. doi:10.1080/14681366.2013.877207.
Vinner, S. (1997). From intuition to inhibition—mathematics education and other endangered species. In E. Pehkonen (Ed.), *Proceedings of the 21th Conference of the International Group for Psychology of Mathematics Education* (Vol. 1, pp. 63–78). Helsinki: Lahti Research and Training Centre, University of Helsinki.
Williams, J., & Wake, G. (2007). Black boxes in workplace mathematics. *Educational Studies in Mathematics, 64*(3), 317–343.
Žižek, S. (2000). Da capo senza fine. In J. Butler, E. Laclau, & S. Žižek (Eds.), *Contingency, hegemony, universality: Contemporary dialogues on the left* (pp. 213–262). London: Verso.
Žižek, S. (2012). *Less than nothing*. London: Verso.

Part II
Disordering Narratives of Progress in Mathematics Education

Chapter 5
Revisiting Mathematics for All: A Commentary to Pais's Critique

Uwe Gellert

Abstract Doubts have been expressed whether research and development in mathematics education really support improvement of the processes of teaching and learning mathematics at school. The critique says that programmatic endeavours, such as "mathematics for all", tend to end up in rhetorical claims that conceal the structural conditions of inequity of institutionalised instruction. In this chapter, which is inspired by several publications of Alexandre Pais, I argue for further reflections on the demands of mathematical knowledge in contemporary society. The topic of universality of mathematical education is the pivot around which historical, functional, emancipatory and political issues unfold.

Introduction

This chapter is a reaction and a comment on a recent critique of research on equity that, although present in other researchers' work too, has been elaborated in most detail by Alexandre Pais. In these introductory remarks, I roughly summarise Pais's argument before sketching the reason for my engagement. Although not a requisite, the reader might benefit from reading Pais's chapter in this volume as it develops the critique further.

In Pais's critique, classroom studies and developmental work, which intends to improve mathematics classroom practices and teacher education activities, appear as presumably well-meaning but misguided attempts to solve a structural problem: not everyone is becoming a high achiever in mathematics. Despite decades of research on the mathematics curriculum, mathematics teacher education strategies, learning theories, teaching aids, and on mathematical activities in the classroom, no significant transformation of mathematics education practices in school seems to be made—at least in terms of equity. Pais's critique describes mathematics education research itself as an obstacle to equity and social justice. Because research in mathematics education ignores, or even disregards, the structural reasons for systematic failure in mathematics,

---

U. Gellert (✉)
Fachbereich Erziehungswissenschaft und Psychologie, Freie Universität Berlin, Berlin, Germany
e-mail: uwe.gellert@fu-berlin.de

H. Straehler-Pohl et al. (eds.), *The Disorder of Mathematics Education*,
DOI 10.1007/978-3-319-34006-7_5

it contributes to the maintenance of structural inequalities and social injustice. From this position, research in mathematics education does not contribute to clarify, let alone to solve, the problem. It is rather part of the problem itself as long as it does not recognise its perpetuating effects on the very practice, which it tries to ameliorate. Finally, the question emerges if the community of researchers on mathematics education really desires to solve the problem of inequity and injustice, as this community has made itself comfortable with; and being occupied with repetitions and replications. We confirm the "complexity" of the issue and call for more research. Consequently, the critique invites the reader to step back from the routines of doing research on equity in mathematics education, to pause and to reflect.

In this chapter, I accept this invitation and engage with the tenets and the arguments of the critique. By seizing the suggestion to scrutinise some basic assumptions of research in mathematics education, I would like to delve into a (dialectical) discussion about possible aims, goals and desires that are the drivers of research activities. It is my impression that part of the critique is justified, necessary and constructive. However, a tendency to oversteer seems to be a general pattern of human counteracting, thus the critique might have overshot the mark. My reaction tries to engage with the arguments—and will certainly not escape this general pattern, either.

Procedure

The critique of research in mathematics education on equity has been discussed intensively at the conference "Disorder of Mathematics Education", which provided the face-to-face basis to this volume. As these conference conversations are hardly accessible to the reader, I mainly rely on Alexandre Pais's (2012) chapter "A Critical Approach to Equity", in which the critique is exposed in detail, without prejudice to the fact that others advance a similar critique, although based differently (e.g. Lundin, 2012; Straehler-Pohl & Pais, 2014). In Pais (2012, p. 58), the "slogan" (p. 57) or "motto" (p. 58) "mathematics for all" is used as an exemplary case for the social fantasy that is "concealing the crude reality that, as any mathematics teacher knows, mathematics is not for all."

To begin with, and in order to be able to discuss this exemplary case, I reconstruct in the third paragraph "mathematics for all" as a research topic and as an influential programme to investigate and counteract structures of social injustice in developing and (post-) industrial countries. UNESCO's publications of selected papers from the International Congresses on Mathematical Education 5 and 6 (Damerow, Dunkley, Nebres, & Werry, 1984a; Keitel, Damerow, Bishop, & Gerdes, 1989) serve as the main resources for this reconstruction.

Although "mathematics for all" has been and still is a very powerful slogan, discussions about what mathematics ought to be taught to which students, in order to provide mathematical education for all, occur much earlier in the history of mathematics education. They are not an invention of the decade of the 1980s. There are traces of discussions, even disputes, about the appropriate mathematical instruction throughout all the history of institutionalised compulsory mathematical education. Of course, the intensity of the discourse varied depending on the time and the place. Since mathematics education became a globalised endeavour during the twentieth century, the discourse became increasingly global. In the fourth paragraph, I mention some of these disputes over the appropriate mathematical knowledge to be transmitted to different groups of students.

In the fifth paragraph, I draw extensively on Pais's (2012) critique of "mathematics for all" before closing with some general comments.

"Mathematics for All"

"Mathematics for all" has reached universal awareness among researchers in mathematics education during the 1980s, although attempts to provide mathematical education to "all and everyone" had already been targets for decades before. Indeed, at the time the influential Commission Internationale pour l'Étude et l'Amélioration de l'Enseignement des Mathématique (CIEAEM) held its 31st meeting in the year 1979 under the title "mathematics for all and everyone." It might be noted, however, that Howson, Keitel, and Kilpatrick (1981) in their thorough review of curriculum innovation during the 1960s and 1970s, do not mention "mathematics for all" as a key notion of mathematics curriculum reform. The worldwide focus on "mathematics for all" was mainly triggered by a UNESCO publication of the work of theme group 1 titled "mathematics for all" at the 5th International Congress on Mathematical Education (ICME) (Damerow et al., 1984a). The introductory chapter of that publication situates the programme "mathematics for all" within a global context. This global context was marked by two developments not long since: the establishment of universal elementary education and the establishment of massive secondary education in industrialised countries. In addition, curriculum development and implementation under the umbrella of the "New Maths" movement had been an unsatisfactory experience in most places. The introductory chapter poses four questions (Damerow et al., 1984b, p. 3):

– What kind of mathematics curriculum is adequate to the needs of the majority?
– What modifications to the curriculum or alternative curricula are needed for special groups of learners?
– How should these curricula be structured?
– How could they be implemented?

"Mathematics for all" is thus posited as a study of the curriculum for school mathematics that takes the most important global developments in mathematics education at the time into account. It is a research programme with an explicit political basis. Damerow and Westbury (1984, p. 23) polarise two political perspectives for mathematics within the canon of general education:

> Do we keep, for example, the highly selective frameworks and methods of traditional mathematics education but give up the privileged position of the subject as part of the core of general education? Or do we seek to keep mathematics at the core of the curriculum but find a way of teaching the subject to all students?

The "selective frameworks and methods of traditional mathematics" refer to selectivity on three levels: selectivity of the school system, selectivity in classroom interaction, and cultural selectivity in the case of the transfer of elite European mathematics curricula to developing countries (Damerow, Dunkley, Nebres, & Werry, 1984b). The dichotomy set by Damerow and Westbury might be considered an issue that could possibly be overcome (CIEAEM, 2000, p. 5). However, it emphasises the political nature of the issue as it construes the distribution of knowledge at its core. Some papers in the UNESCO publication depict possible (political) ways in which the mathematics curriculum might evolve towards "mathematics for all": De Lange (1984) reports on the developmental work within the Institute for the Development of Mathematics Education (IOWO) at Utrecht, which can be seen as an example of didactic engineering; Jensen (1984) shows how teachers' instructional practices in different forms of schooling in Denmark adjusted to the expansion of higher education before their modified curricular practice finally became official regulation.

At ICME-6, the discussion about "mathematics for all" continued in form of a Fifth Day Special Programme on "Mathematics, Education, and Society", whose outcome again was published by UNESCO (Keitel et al., 1989). Among the many papers collected in that volume, several ones report on the particularities of compulsory mathematics education in developing countries. These papers open up the scope to the discussion as they point to social situations characterised by the specific tensions of political post-colonialism and educational imperialism/liberation. Broomes (1989, p. 21) calls for attention to the special mathematical demands of rural economies. He emphasises the need for "educational regeneration" in the sense of a "community involvement strategy" of bringing educational institutions and curriculum closer to community life in order for the outcomes of schooling to be directly significant to the life and work in these communities. This is a call for a socio-communitarian productive education. Kanté (1989, p. 78), reporting the case of the Republic of Ivory Coast, reveals the constraints that developing countries face when trying to implement up-to-date mathematics curricula:

> Yet, for some 25 years now, the average yearly number of mathematics graduates of the university able to teach the subject in all grades of the secondary school has not surpassed three, while the deficit of secondary mathematics teachers currently is about 900.

Naumann (1989, p. 80) shows figures of student enrolment in Senegal. According to these figures, the call for "mathematics for all" can be regarded as aiming at higher access rates to mathematics education on a physical level: "For the country as a whole, the 1985 gross enrollment ratio [percentages of the official school-age population at each education level] was 55 % for the first, 13 % for the second, and 3 % for the third level", with classes tending to be very large (up to 69 pupils). Physical access to education has also been an issue during the 1980s in Papua New Guinea (Souviney, 1989). Government plans intended to raise the number of primary school students from 280,000 in 1979 to 450,000 in 1992, requiring an augmentation of mathematically educated primary school teachers from 9000 in 1979 to 14,000 in 1992. To sum up, in many developing countries "mathematics for all" was first of all directed towards comprehensive enrolment in primary, and more seriously, in secondary mathematics classrooms. Gates and Vistro-Yu (2003) show how "mathematics for all" was firmly embedded in UNESCO and UNICEF initiatives that culminated in a "World Declaration on Education for All: Meeting Basic Learning Needs" (UNESCO, 1990). Parallel to the issue of physical access to schooling, which largely is a politico-economic challenge, the emphasis on "*mathematics* for all" focused the debate on the appropriate mathematics curricula that reconcile the diverse post-colonial political realities with at that time still materially existing colonial traditions of institutionalised schooling.

What Curriculum for Whom?

Because compulsory (mathematics) education is an invention of the nineteenth century, it makes sense to start the historical retrospection of education and "mathematics for all" at this point. From the early nineteenth century on (the time varied among the different countries) schooling became institutionalised and teacher education and curricular decisions were gradually controlled by the state. Parallel to these important developments, the mathematical education of the offspring of the upper social strata, who always received some kind of mathematical education, was a matter of debate. In the eighteenth century and before, mathematical education (wherever and whomever it existed for) formed part as an integrated component of academic education, relating to both science and philosophy. In the aftermath of massive military conflicts in central Europe at the beginning of the nineteenth century, the government and administration of the defeated Prussia constructed a *system* of public education. On the one hand, this system was based on the dominant educational theory of the time, Neohumanism, in which the Greek ideal of mathematics "as an instrument of mental demarcation against utility and application in labour and trade" (Damerow & Westbury, 1985, p. 183) was a significant principle. On the other hand, the mathematical and scientific formation of the officer corps was recognised as a condition for the effectiveness of the French revolutionary troops. The foundation of the École Polytechnique in Paris in 1794, with Gaspard Monge among the founding fathers and as the school's professor of mathematics, profoundly influenced the debates on the kind of mathematics and science education for the new system of public education. As Jahnke (1986, p. 87) concludes, "it is probably due to the French influence that mathematics enjoyed an extraordinarily high esteem within the educational theory of Neohumanism" in Prussia. In this situation, the debate over the mathematics curriculum was first of all a debate on mathematics for (higher) secondary education. As a result of the controversial discussion, school mathematics was not simply defined as academic mathematics on a lower level in the sense of a preparation to the study of pure mathematics. Many mathematicians argued for a mathematical education that should reflect the recent achievements of mathematical science, but their position was challenged by parents from the influential industrial class ("Wirtschaftsbürgertum") who wished their children to be introduced to a more practical mathematics. Headmasters of prestigious secondary schools (Gymnasium) in Berlin as well as some school inspectors argued in favour of more "elementary courses aligned to practical requirements to parallel the scientific course" (Jahnke, 1986, p. 87). In 1829, for instance, the Prussian Ministry of Education assigned the task of constructing a mathematics curriculum for the Prussian Gymnasium to the influential mathematician August Crelle. When Crelle argued for school mathematics to be fully determined in methods and content by scientific mathematics and for excluding "common arithmetic lessons" altogether from the school curriculum, he received fierce opposition from the Ministry's committee. For the majority of the committee,

> The independence of a school subject from the related science and its characteristic features, is due to the fact that in the school subject everyday knowledge and scientific knowledge are mediated. Hence it is not appropriate to exclude "common arithmetic" from the curriculum as an unscientific subject, but it is necessary to find ways and means of linking "common arithmetic" and higher mathematics, everyday knowledge and scientific knowledge. (Jahnke, 1986, pp. 90f.)

The Gymnasium clearly was a school for the Prussian elite. However, even for that social minority school, mathematics was not configured as a specialisation in, and preparation for, pure mathematics. Despite the fact that the Greek ideal of a mathematics free of immediate use would have been fitting well to a school canon (see Kollosche, 2014 for a different perspective), in which the classic languages Latin and Greek once dominated, a more mundane orientation to mathematics was finally installed. The institutional mechanism for dealing with these oppositions, which are essentially conflicts between the "old elite" ("Bildungsbürgertum") and the Wirtschaftsbürgertum, was streaming: the splitting of secondary education in three school forms. Whereas in the traditional Gymnasium mathematics instruction still focused on Euclidean geometry and logical thinking, the new forms of the Realgymnasium and the Oberrealschule took on a more practical approach. With the intention to mediate contradictory orientations, the structure of the school mathematics curriculum was centred on the concept of the "mathematical operation" and the "principle of permanence", that is, the extension of the number concept. Infinitesimal calculus was constructed as formal school algebraic theory, and school algebra was constructed as the study of the formal properties of the arithmetical operations. This macro-structure of the school mathematics curriculum—higher school mathematics founding on elementary arithmetic—outlasted nearly two centuries since.

The curricular question during the nineteenth century and at the beginning of the twentieth century was not focused on education for all children. It was rather the question of what kind of mathematical training was to be applied to which elite in a modern country: "mathematics for all" elites. The reforms at the beginning of the twentieth century in France (Émile Borel) and Germany (Felix Klein), in which the emphasis on the concepts of function, functional reasoning, continuity and derivatives as well as links to physics and to application were predominant, focused on the education of the future students of the universities and other higher education centres (Gispert & Schubring, 2011).

While the debate about the appropriate mathematical education for the old and the new elites was lively, the mathematical education for the majority of the children at elementary schools was concurrently restricted to training in computational skills and directly linked to their fields of direct application. Beginning from the twentieth century, traditional elementary mathematics education was challenged by principles of progressive pedagogy (New Education Fellowship, Reformpädagogik). For the case of "general education" in the USA, Keitel (1987, p. 398) reconstructs how the resulting mathematical programme was an amalgam of "a trivial though dogmatic social-needs orientation on the one hand, and the child-orientedness of the Progressive Movement on the other". She displays the table of contents of a textbook of the year 1937 for grade 8 (taken from Jones, 1970, p. 220):

1. Mathematics and the Community.
2. The Merchant and the Community.
3. The Bank and the Community.
4. Taxes and other Community Funds.
5. Community Planning.
6. The Community and its Neighbors.
7. An Inventory of the Year's Work.

At the same time in Germany, a similarly trivial (mathematically) though differently dogmatic (ideologically) social-needs orientation is reflected in the mathematics education programme of National Socialism, as can be illustrated by the table of contents of an arithmetic textbook of the year 1940 for grade 8 (cf. Ullmann, 2008, p. 281):

1. Adolf Hitler takes over a desolate heritage.
2. Adolf Hitler—the savior.
3. What has been achieved by the first 4-year plan.
4. Germany must live even if we pass by.
5. Get yourself healthy for your people.
6. Computation in insurance.
7. Monetary transaction.
8. The German Reich Post.
9. The German Reichsbahn in the sign of reconstruction.
10. Geometry of space.

Both tables of content suggest that the direct subordination of mathematics education for the majority to the ideologically determined "needs" of the community, or of the nation, resembles to what de Lange (1984) has called "mathematics for all is no mathematics at all". The determination and justification of the mathematics curriculum exclusively from the field of its social applications replaces any mathematical structures by the fugacious and ideological contingencies of the political will of the moment.

The debate on the appropriate mathematics curriculum received a substantially new stimulus after 1950. The idea that mathematics—in any form whatsoever: pure and strictly oriented at academic mathematics or more applied and practical—is a knowledge not only to be transmitted to the elite stratum but to all children from all social classes, produced a radical shift and extension of the debate. It was no longer considered sufficient that the majority of children received training in elementary arithmetic computational skills. The idea became powerful because of the support from many influential though ideologically different sides: from mathematicians inspired by the French Bourbakistes trying to advance a mathematics curriculum oriented to mathematical structures (partly organised in the CIEAEM), from psychologist inspired by the notion of "mental structures", and from agencies for economic development (OEEC, later renamed as OECD) promoting the massification of secondary, and later, higher education, particularly science education (Gispert & Schubring, 2011). As a consequence of the considerable differences of perspective, the concrete reform attempts took on various forms too. Keitel (1987) contrasts the US National Science Foundation-financed School Mathematics Study Group (SMSG), who developed and promoted a sophisticated science-oriented programme for all students, with the approach of Unified Science and Mathematics in the Elementary School (USMES), which focused on mathematical applications in "real life". Keitel (1987, p. 398) scathingly summarises the experience with these and similar projects: "We must acknowledge that all attempts to make a science-oriented mathematics instruction a 'mathematics for all' have either failed or met almost unsurmountable difficulties." She perceives the fundamental and starting questions of these attempts as the reason for failure. Instead of asking: *What is important from a mathematical perspective?* Or: *What is needed for social and economical development?* It would make more sense to begin with reflections about what children can do and what teachers can teach. In Keitel's view, taking these two last questions as the starting point is one of the reasons why Freudenthal's work under the motto "mathematics for all and everyone" received international recognition and made the IOWO/OW&OC/Freudenthal Institute in Utrecht a landmark on the mathematics education world map. Quoting, again, Keitel (1987, p. 395):

> What is elementary in cognitive development and/or in the discipline of mathematics? What is socially useful or necessary and how can it be identified? Questions like these are complex and cannot be answered absolutely. To put them at the beginning of curriculum development often impedes, restricts, and even obstructs a comprehensive view of the curricular task. Starting as Freudenthal does opens up a vast area for undetermined observation, experimentation, tentative development of materials, trial, revision, and evaluation.

Consequently, for Freudenthal (1991, p. 179), "it [the contents of "mathematics for all"] will not be the same thing for each particular learner; there is a great deal of diversity not only for contents, but also or even more so, of breadth and depth of understanding." It might appear as an irony, despite this call for the individual being the key to "mathematics for all", the work of Freudenthal was, and still is, used by the OECD's Programme for International Student Assessment (PISA) as a key reference, although incorrect and inappropriate (Gellert, 2006), for promoting "Mathematical Literacy". The political is lurking where the director of the Freudenthal Institute and the chair of the PISA's Mathematical Functional Expert Group are the same person. Despite of variations across the world (Stacey et al., 2015), based on their political power, OECD is influential in implementing the mathematics of PISA as a forced and de facto new "mathematics for all".

A Personal Response to Alexandre Pais's Critique on "Mathematics for All" and Equity

Having prepared my argument in the two preceding sections, I can now turn to my purpose of responding to Pais's critique on "mathematics for all". In what follows, my personal response is structured by three headings, which refer to what I perceive as the main components of Pais's critique. Under the first heading, I react on a discussion of mathematics education as a selection device. The second heading "mathematics and society" discusses the arguments on the mathematics used in jobs and professions and on the role mathematics plays for access to socially powerful positions. In the last part, I comment on Pais's conceptualisation of the nucleus, the periphery and the boundaries of mathematics education research.

Mathematics Education as a Selection Device

The following passage from Pais (2012, p. 51) unfolds its premise and the gist of its argument:

> My premise is that exclusion and inequity within mathematics education, and education in general, are integrative parts of schooling and cannot be conceptualised without understanding the relation between scholarised education and capitalism as the dominant mode of social formation. As mentioned, it is common to find research that presupposes the idea that the problem of inequity transcends school and mathematics education. However, little research has been done that explicitly tries to understand exclusion as an integral part of schooling; that is to say, as something consubstantial to schooling itself.

Schooling not only being about the transmission of knowledge is a well-known fact within functional theories of education and schooling. These theories start with the premise that school, as an institution, operates to facilitate the stability of society. In early theories, e.g. in the context of capitalist America, achievement is the main mechanism for maintaining social structures. These theories conceive achievement in schools as based on merit. The function of schooling is to underpin capitalism through the distribution of success, allowing those with the highest merit to fill high-level positions in the capitalist society (Davis & Moore, 1945). Parsons (1959, p. 298) classical account of a structural functional perspective to education integrates the topic of social structures with the formation of personality. The paper conceives this integration as a problem of socialisation and selectivity:

> Our main interest, then, is in a dual problem: first of how the school class functions to internalize in its pupils both the commitments and capacities for successful performance of their future adult roles, and second of how it functions to allocate these human resources within the role-structure of the adult society.

As a matter of fact, Parsons's theory is uncritically accepting the existing inequalities of the structures of society. Equality does not yet figure as a kind of ideological supplement to the educational programme:

> The problem of inequity appears not as a contingency of a good system, but as a necessity of the same system that posits equity as a goal to strive for. Inequality is a necessity of capitalist economies, while equality functions as the necessary ideological supplement concealing the obscenity of what is going on. (Pais, 2012, p. 55)

More recent functional theories of education are explicitly influenced by Parsons's ideas. These theories are certainly more complex because the reproduction of structures is no longer understood as the only interest of societies. Broadly speaking, the reproduction and innovation of social and cultural structures is at stake. Fend (2006), for instance, distinguishes four functions of institutionalised education, which are paradigmatic for structural-functional theories of schooling:

1. Cultural reproduction.
2. Qualification.
3. Allocation.
4. Integration (Fend, 2006, pp. 32f., own translation).

For Fend, cultural reproduction is a process of enculturation of the children into their symbolic environment. Qualification refers originally to economic competitiveness, includes the transmission of knowledge and skills necessary for "concrete" work. Integration aims at social coherence by forming social identities and preparing for political participation. Note that Fend, as does Parsons, refers to "allocation" and not to "selection", since he regards the legitimate distribution of pupils to future professions as a social function of institutionalised schooling, and not the disqualification and exclusion from the desired careers. Fend (2006, p. 44, own translation) is not arguing cynically when he concludes:

> For the society, the core function of the education system is allocation. For the individual, schools pre-structure educational and professional paths and are thus among the institutions, which facilitate the development of differential biographies. For the adolescent, the school system is the most important instrument of personal planning for the future.
>
> Therefore, we should not relinquish the function of the school system from participation in allocation. The teardown of systematic exclusion has already brought about opportunities for life to many children from the underprivileged strata which they would not have been offered without.

To sum up, educational theory is sufficiently clear about the fact that education systems, and particularly the institution of schooling, do not function only for the benefit of each individual student. It would be naïve to deny the school's social function of allocation—actually the public opinion seems to be quite aware of it, as we will see below. Allocation is only one function of schooling, albeit an important one, among others. This fact is independent of curricular decisions. Even if mathematics would only be accessible for a minority, and something else would be *mise-en-scène* for the remaining majority, the currency of success and failure would be maintained. Where Pais borrows the concepts "universal Lie" and "'repressed' truth" from Žižek, the effect is excessively drastic.

> The "universal Lie" is no more than the slogan "mathematics for all," the "repressed truth" being the crude reality of those who year after year continue to fail in school mathematics. The systematics failure of people in school mathematics points towards the system's antagonistic character: the condition of impossibility of realizing the common goal ("mathemat-

> ics for all") is simultaneously its condition of possibility. That is, what makes schooling such an efficient modern practice is precisely its capability of excluding people by means of promotion. (Pais, 2012, p. 58)

A kind of metaphorical antagonism constitutive for the conception of lies and truths is transferred to the functions of schooling. This is problematic because first of all, as Fend has pointed to (above), allocation and qualification are only two of the functions of institutionalised education. Second, the truth that failure is a systemic characteristic of every school system involved in allocation is hardly repressed. It is not even made invisible or implicit. I think Pais (2012, p. 72) is wrong when asserting:

> In order for school to be the most important ideological apparatus, to function as a credit system, it is not productive for it to be presented as an exclusionary institution. That would cause criticism from the whole of society, and would be unbearable from an educational or political point of view. In order to perform well in the role of a credit system, schools need to be presented as inclusionary and emancipatory places, places where phenomena such as exclusion and failure are seen not as necessary parts of the same system which purports to be trying to abolish them, but as contingent problems, malfunctions of an otherwise good system.

In 2010, the government of Hamburg (Germany) changed the legislation of primary and secondary education. Before, primary education was organised in form of a 4-year school for all children. After those 4 years, the pupils had been tracked into different types of schools of which only the highest track (Gymnasium) permitted access to university studies. The new legislation extended the common time of all children in the primary school by 2 years, and provided only two different secondary school types both permitting access to future university studies. The new legislation tried to change the public school towards a more inclusive and equitable system of education. Surprisingly, or not, the residents of Hamburg organised resistance and founded an "Initiative: 'We want to learn'—an association for a better education in Hamburg". The resistance legally transformed into a political referendum, in which the population of Hamburg was entitled to decide about the new legislation by vote. In the end, 58 % of the voters followed the Initiative and the government had to withdraw the new legislation. The majority preferred to maintain the openly allocative tracking system. The criticism of the majority was directed towards the tendency to weaken the role of school as a credit system and to inclusion. Apparently, if concerned about the allocation function of schooling and its consequences on the individual, the majority of Hamburg's inhabitants believed that the traditional rules of allocation worked well, or to their own benefit.

The exclusive character of schooling seems to be widely acknowledged. On the same page as above, Pais questions the material character of exclusion by referring to Baldino and Cabral (2006):

> Baldino and Cabral (2006) create a parody concerning where one can find exclusion in school. The authors suppose that we enter an elementary school and ask the staff where the so-called "exclusion" is happening. Who will be able to answer such a question? Where to locate exclusion in schools? It seems as if exclusion has no "materiality," no precise site where it is happening. (Pais, 2012, p. 72)

Again, where I live and work, teachers would willingly answer the question: They would point to the high number of students in front of them, they would point to the lack of qualified support, and they would point to the social background of many of the students. All these issues act as obstacles for inclusiveness in education.

Finally, I would like to comment on Pais's (2012, p. 79f.) suggestion for a different approach to exclusion:

> What if the exclusion associated with school mathematics is not a particular negativity, a vicissitude of a "good" system, but, on the contrary, represents a glimpse of what the school system really is: a credit system with the main goal of social selection by means of deciding who is capable and who is disposable?

As Fend (2006; see above) has argued, allocation actually can be conceived as the core function of the school system. However, he argues against using the term "selection". The concept of selection refers to a binary decision: after the selection process, an object is "in" or "out". Allocation, on the contrary, denotes that the school system distributes the students to different career paths, on which a different deepness of mathematical knowledge, or a different kind of mathematical knowledge, is functional. Actually, only very few students will later work as professional mathematicians—and it is of course a task of the school system to determine and to stimulate these few. Many more students will use mathematics, in varying degrees, as a sort of grammar: in physics, chemistry, engineering, and also in economic science. And even more students will require knowledge and skills of more basic mathematical operation as well as the ability to understand (and sometimes to communicate) mathematically coded information. In part, mathematical knowledge transmitted by schooling is indeed a requirement for professionalism (I will discuss this issue below). Nevertheless, justifications for mathematics being a core subject in general education are drawing on the particular importance of technological development for the wealth and prosperity of a country or the world population. However, the dichotomous distinction between the mathematically "capable" and the "disposable" as a result of schooling produces an oversimplified picture of the tensions between the different functions of the school.

Mathematics and Society

For discussions of the social importance of mathematics, Pais (2012, p. 65f.) suggests that it is more appropriate to focus on values than on knowledge:

> I suggest that the reasons why people need [mathematics] are not related with mathematical knowledge or competences, but with the school valorisation that mathematics gives to people. People need school mathematics not because they will use it directly in democratic participation (as knowledge or competence), but to continue having success in school, undertake a university course and find a stable job, so that they become "normal" social beings. I argue that the importance of mathematics must be discussed not in the field of knowledge but in the field of value.

In order to support this argument, Pais refers to Dowling's (1998) deconstruction of a "myth of participation" and to studies of mathematics at the workplace. Indeed, Dowling's analysis of schoolbook texts seems to be a good reference. What I would like to discuss is the role of mathematical knowledge and skills at the workplace. For that issue, Pais (2012, p. 67) draws on studies by Riall and Burghes (2000) and by Hudson (2008):

> Another useful resource is the study conducted by Riall and Burghes (2000), who gathered together employees from a wide range of industry, commerce, and the public sector. Their intention was to evaluate the extent to which these people use mathematics in their professions. They conclude that "almost the entire population of the study said that they had had to learn at school some maths that they had never then used again" (p. 110). In the voice of one of the workers, who stands for the general opinion: "I think that a lot of maths that is taught is not used in later life. I've forgotten most of what I had to learn and I never use it" (p. 104). Also Hudson (2008) found that the people of his study did not transfer what they learnt in school mathematics to their daily work activities. Rather, they developed their mathematical skills in the workplace.

There are two important observations in this statement. First, that much of what is learnt of mathematics in school is never again used in adult life. Second, in case that the workplace requires mathematical skills, these are learnt on the job. The first observation might hold for mathematics as much as it holds for other school subjects. Few people use facts about historical power relations, knowledge of chemical reactions, biological classification, interpretation of French literature classics, construction of scales in music, or grammatical rules of mother tongue etc., in their life. Does this automatically mean that teaching these matters is irrelevant? Quite the contrary, it is an indication that the debate about the appropriate curriculum should never end. The contents of the (mathematics) curriculum, that is, what the institution tries to transmit to the students, need to be the object of permanent reflection: What is so generally important that it should be taught to the new generation, given the fact that most of it will never be applied in later life? As schooling is compulsory in most countries, it seems to be a duty of those involved in the construction and reflection of curricula to not stop scrutinising the legitimacy and the justification of these curricula. One of the main issues, then, is questions such as: *What kind of "mathematics for all"?* Or: *The same "mathematics for all"?* From a functional point of view, the responsibility of mathematics education research is, first of all, to reflect on the production and impact of, as well as to inform about possible alternatives to, curriculum-at-work.

The second observation—the one of workplace mathematical skills being acquired at the workplace—is suggestive, but it might profit from some clarification. To a certain extent, and studies such as Hudson (2008) have made it very clear, the mathematical skills necessary to operate at the workplace are indeed acquired at the workplace. How could it be different? As workplace mathematics is a kind of applied mathematics that is oriented to the necessities and characteristics of the workplace, thus subordinating mathematical logic and structure to the requirements of production and service, and as these requirements are constantly changing, the relationship of school mathematics and workplace mathematics is principally loose. It might well be the case that workers and professionals acquire sufficient mathematical skills on the job in order to comply with what is only minimally expected of them. As Hoyles, Noss, Kent, and Bakker (2010) studying techno-mathematical literacies in manufacturing and financial service demonstrate, the employees in these setting show enormous difficulties in understanding mathematical information that is presented to them in technological forms, whether it be as graphs in a computer system or as explanatory texts of financial products. A major problem seems to be that the technologies entrap the employees to forget about the mathematical relationships that operate below the surface of graphs and tables. What employees need to develop, but rarely do, are Techno-mathematical Literacies (TmL):

> The requirement for TmL does not apply to all employees; only those who engage in situations where there is a need both to use technically expressed information or to communicate such information, either to fellow employees or to customers/suppliers outside the company. This we have argued is a requirement frequently of intermediate-level jobs, which are typified by shop-floor managers in manufacturing and customer sales/service agents, such as in financial services. [...] Fortunately, we suggest that more effective training of mathematical skills that gives voice to employee skills is a feasible option. (Hoyles et al., 2010, p. 184)

Hoyles et al. designed training courses with the explicit aim of advancing the employees' techno-mathematical literacies. On the basis of technology-based objects (graphs, tables, and spreadsheets, etc.,) they make the mathematics behind these objects visible and thus an object of the employees' attention. But it is only in the exceptional case of such carefully designed training courses that the employees acquire, or refresh, the mathematical knowledge required to operate with technologised mathematical information.

Do medical doctors develop their mathematical skills at the workplace? Gigerenzer, Gaissmaier, Kurz-Milcke, Schwartz, and Woloshin (2008) report on a study with gynaecologists' understanding of positive mammograms. They confronted the gynaecologists with the following situation (p. 55):

> Assume you conduct breast cancer screening using mammography in a certain region. You know the following information about the women in this region:
>
> - The probability that a woman has breast cancer is 1 % (prevalence)
> - If a woman has breast cancer, the probability that she tests positive is 90 % (sensitivity)
> - If a woman does not have breast cancer, the probability that she nevertheless tests positive is 9 % (false-positive rate)
>
> A women tests positive. She wants to know from you whether that means that she has breast cancer for sure, or what the chances are. What is the best answer?
>
> A. The probability that she has breast cancer is about 81 %.
> B. Out of 10 women with a positive mammogram, about 9 have breast cancer.
> C. Out of 10 women with a positive mammogram, about 1 has breast cancer.
> D. The probability that she has breast cancer is about 1 %.

The gynaecologists could calculate the best answer or they could recall what they should have known anyhow. The results show that only 21 % of the doctors chose C as the best estimate, with 19 % underestimating the probability about an order of magnitude (D) and 60% grossly overestimating the probability, about an order of magnitude (A and B). Gigerenzer et al. conclude:

> This study illustrates a fundamental problem in health care: Many physicians do not know the probabilities that a person has a disease given a positive screening test—that is, the positive predictive value. Nor are they able to estimate it from the relevant health statistics when those are framed in terms of conditional probabilities, even when this test is in their own area of speciality. (p. 56)

And this is not only the case with gynaecologists and mammography, as Gigerenzer and colleagues show in further studies in the context of the use of contraceptive pills, survival and mortality rates; smoking and cancer; and down syndrome tests. Of course, test situations are different from real practice; but health care is a terrain with many restrictions for research because of ethical matters. Gigerenzer et al. suggest, "if you want to find out yourself if this is the case, ask your doctor" (p. 56). I did, and unfortunately my personal experience confirmed Gigerenzer et al. conclusion. Apparently, even in contexts in which false appraisals or recommendations might lead to severe psychical and physical stress, the required mathematical skills to prevent misestimation are not acquired at the workplace.

In order to broaden the discussion, let me add that the role of mathematics in terms of access to socially powerful positions seems to be another more complex issue than often claimed. First, it might be asked what the powerful positions are and who a member of this social elite is. Second, the way access is regulated to these positions follows different traditions depending on each country. As I will outline, these two differentiations indicate that general conclusions about the role of mathematics for access to social positions of power are problematic.

Research on elites and power in Europe often differentiates between the elites in the fields of economy, politics, justice and administration, the most influential being the economy, followed by politics, and finally, justice and administration (Hartmann, 2007, 2010). In countries, where the recruitment of the elites is mainly a matter of social class distinction, the role of success in school mathematics for access to powerful social positions is different from its role in those countries, where the social class hierarchy matters less. Hartmann (2007) contrasts the social recruitment of elite groups in various European countries (Table 5.1):

Table 5.1 Recruitment of social elites (synthesised from Hartmann (2007), p. 220 and p. 222)

	Economic elite (CEO/PDG etc., of largest companies)			Political elite (Cabinet of the year 2006)		
	Upper classes (%)	Bourgeoisie (%)	Lower middle class/working class (%)	Upper classes (%)	Bour geoisie (%)	Lower middle class/working class (%)
Switzerland	29	21	50	14	14	72
France	57	30	13	62	13	25
Germany	52	33	15	12	50	38
Italy	52	16	32	10	35	55

The recruitment of the economic and the political elite in France is at one pole of the European spectrum. In France, both economy and politics are dominated by upper class elites. In Germany and Italy, class hierarchies are reflected much more clearly in the economic than in the political sector. In Switzerland, both the economic and the political sector integrate many lower middle class and working class people as members of the elite group. As Bourdieu has pointed out (e.g. 1989), where the elites reproduce the social class structure, with the upper classes dominating the economic and political elites, the selection process is less oriented to merit than to habitus.

As Hartmann shows in addition, the different elite-sectors are differently integrated in the various countries. It is, for instance, not unusual for the elite in France to change from one elite sector to the other. In Germany, in contrast, this still is a rare case and the public reacts allergically when a leading politician switches over to the economic sector after finishing his political mandate. In countries, in which the elites regularly change between the different elite sectors, the elites are more homogenous and socially enclosed and less "affected" by social mobility.

The reputation and prestige of mathematics varies within the different European countries. This fact is reflected in the career paths of the members of the elite. For instance, whereas studies of law predominate the academic formation of the German elites, the École Polytechnique is the preferred institution for the French economic elite (Hartmann, 2007). Furthermore, entrance to the Classes Préparatoires, which pave the way to studies in the prestigious Grandes Écoles, is rather dependent on a mathematics/science focus in the Baccalaureat (Hartmann, 1996). However, this is not to say that brilliant performance in school mathematics paves the way into the French economic elite. As discussed above, knowledge is clearly subordinated to habitus. To sum up, if we talk about really powerful social positions, mathematics does not seem to play a major role. Of course, particularly in France, mathematical knowledge and skills are a necessary requirement to access a place in the treasured Grandes Écoles, in which the future elite receives formation. However, excellent performance in mathematics is not a sufficient requirement to access because other mechanisms of allocation dominate.

Is Mathematics Education Research Naïve?

Questions such as "What counts as research in the field of mathematics education?" will never receive a final response, of course. What is part of the body of scholarly knowledge in mathematics education? Those "results", "evidences", "reflections", "theories" and the like that have been accepted for publication in the scientific journals of the field? These are tricky questions. There are some researchers who advocate conceiving of the research field as clearly centred on the teaching and learning of mathematics (in schools), others define a "nucleus" of the research field (again, the teaching and learning of mathematics), and still others point to the importance of multidisciplinarity and transdisciplinarity. In relation to the issue of equity in mathematics education, the first view of mathematics education as a rather small and clearly curtailed research field seems to be inapt, as Pais (2012, p. 50) emphasises:

> Although studies dealing with equity in mathematics education acknowledge the social and political dimensions of the problem, I shall argue that such studies insist on addressing the problem of inequity as if it could be understood and solved within mathematics education. It is as if we admit that the problem has an economical and political nature, going way beyond the classroom, but, since we are mathematics educators, we must investigate it in the classroom.

This approach—which consists in reducing a political problem to a didactical one, thus possible to be solved through the development and implementation of better stratagems to teach and learn mathematics—cannot be said to have produced the desired results, namely the commonly shared desire of "mathematics for all". (p. 50)

It is indeed important to include the social and political dimensions of the issue of equity not only in the depiction of a research problem, but also in the methodology used to generate insight. However, isn't there a substantial number of researchers who are doing exactly this: developing methodologies that build on the social, political and/or economical dimensions of institutionalised mathematical instruction? Pais (2012, p. 51) extends the critique:

> Mathematics education research has not been appreciative of research that is not immediately concerned with action, in the sense of providing solutions or strategies for improving the teaching and learning of mathematics.

Perception seems to relate, among others, to the position from which the issue is perceived. It is thus subjective, and it is difficult to argue against. My own perception is different: There is a dominant publicly shared and publicly reinforced rhetoric. It says: The aim of research in mathematics education is to improve the teaching and learning of mathematics. Researchers know this and many behave accordingly when they meet the public. Within the research community however, general insight and theories seem to outvalue a more direct relation to action and improvement of teaching and learning. Look at the announcements of the recent and past Felix Klein and Hans Freudenthal awardees on the ICMI webpage. Few of them are praised for their attempts to improve teaching and learning. All of them are praised for their seminal theoretical developments. With the Emma Castelnuovo award ICMI is only recently taking the importance of research for practical instructional action into account.

Apparently Pais (2012, p. 77) is arguing more radically than I am, when classifying access to resources as a non-economical and non-political issue.

> Gates and Zevenbergen (2009, p. 165) identify a common basis for such measures: "What might we all agree on then as fundamentals of a socially just mathematics education? Perhaps we can list: access to the curriculum; access to resources and good teachers; conditions to learn; and feeling valued." The first thing that is evident here is the complete absence of a political conceptualisation of equity. What is recognized as an economical and political problem ends up being addressed in a technical fashion: better ways to teach and learn mathematics for all students.

Finally, Pais borrows G. Biesta's notion of the "learnification" of education. This tendency is fuelled when "theory" is interpreted as "tool for action":

> On the one hand, if we take a look at two of the recent articles on the role of theory in mathematics education research (Cobb, 2007; Silver & Herbst, 2007), we can notice how "theory" is perceived as providing "tools for action," where action is normally the practice of school mathematics, thus reducing research to a matter of providing the solutions for the problems of practice. (2012, p. 52)

As I have discussed elsewhere (Gellert, 2010), what Cobb (2007) and Silver and Herbst (2007) explicate in the *Second Handbook of Research on Mathematics Teaching and Learning* (ed. F. Lester) is indeed a particular, and particularly problematic, conception of theories. Without going into detail, here it should not be forgotten that the handbook mentioned is a project of the US National Council of Teachers of Mathematics (NCTM), and assumedly, reflects a highly pragmatic approach to theory. Pragmatism has a long tradition in US (educational) philosophy, but it is not a standard, let alone dominant, perspective everywhere.

Conclusion

In summary, the claim "mathematics for all" aims at universality of mathematical education. What do I like to convey when focusing on "universality" here? In the introduction to the German edition of Butler, Laclau and Žižek (2000), Posselt (2013) synthesises universality not as a ground, but as a horizon, as a practice of cultural translation and as a negative condition of (political) articulation. The concept of "horizon" (sensu Laclau) alludes to a structural contingency as the arena for hegemonic articulation of particular content. "Mathematics for all" occupies this empty space and is thus an indication of the hegemonic power of mathematics in the field of education. However, universality can only be achieved, as Butler points to, if it permanently articulates the cultural particulars by processes of cultural translation. "Mathematics for all" seems to be a sufficiently open programme/slogan to permit articulation with particulars as diverse as, for instance, the curriculum and physical access to mathematics education. With Žižek, we can question "which secret privileging" had to be carried out so that the empty space could emerge that "mathematics for all" is claiming to fill.

Though engaging dialectically, rather than concordantly, with universality, Butler et al., agree that any emancipatory practice necessarily refers to a dimension of it. Moreover, universality seems to be unavoidable since, according to Žižek (in Butler et al., 2000, p. 315), "each particular position, in order to articulate itself, involves (implicit or explicit) assertion of *its own mode of universality*." By transferring this statement from its original context of emancipatory political projects to the politics of mathematics education and mathematics education research, we can see that research in mathematics education, in all its particular positions, refers necessarily to some kind of universality. To accept "mathematics for all" as such a universality leads us to the contingencies and the hegemonic forces that shaped, and still shape, this very concept. In that sense, Pais's critique is working towards more self-reflection of the field of research of mathematics education rather than against a mathematical education for all children. He points to the necessity of every emancipatory project to constantly scrutinise the validity of the reasons and the interests on which the project was constructed. I am looking very much forward to seeing how alternatives to "mathematics for all" emerge from these reflections.

Acknowledgments I want to thank Anna Llewellyn, Candia Morgan and the editors of this volume for their helpful comments and suggestions.

References

Baldino, R., & Cabral, T. C. (2006). Inclusion and diversity from Hegel-Lacan point of view: Do we desire our desire for change? *International Journal of Science and Mathematics Education, 4*(1), 19–43.

Bourdieu, P. (1989). *La noblesse d'état. Grandes écoles et esprit de corps*. Paris: Minuit.

Broomes, D. (1989). The mathematical demands of a rural economy. In C. Keitel, P. Damerow, A. Bishop, & P. Gerdes (Eds.), *Mathematics, education, and society* (pp. 19–21). Paris: UNESCO.

Butler, J., Laclau, E., & Žižek, S. (2000). *Contingency, hegemony, universality: Contemporary dialogues on the left*. London: Verso.

CIEAEM. (2000). *50 Years of CIEAEM: where we are and where we go. Manifesto 2000 for the year of mathematics*. Berlin: Freie Universität Berlin.

Cobb, P. (2007). Putting philosophy to work: Coping with multiple theoretical perspectives. In F. Lester (Ed.), *Second handbook of research on mathematics teaching and learning* (pp. 3–38). Charlotte: IAP.

Damerow, P., Dunkley, M. E., Nebres, B. F., & Werry, B. (Eds.). (1984a). *Mathematics for all: problems of cultural selectivity and unequal distribution of mathematical education and future perspectives on mathematics teaching for the majority*. Paris: UNESCO.

Damerow, P., Dunkley, M. E., Nebres, B. F., & Werry, B. (1984b). Introduction: report on the work of theme group 1 "Mathematics for All" at ICME 5. In P. Damerow, M. E. Dunkley, B. F. Nebres, & B. Werry (Eds.), *Mathematics for all: Problems of cultural selectivity and unequal distribution of mathematical education and future perspectives on mathematics teaching for the majority* (pp. 1–10). Paris: UNESCO.

Damerow, P., & Westbury, I. (1984). Conclusions drawn from the experiences of the New Mathematics movement. In P. Damerow, M. E. Dunkley, B. F. Nebres, & B. Werry (Eds.), *Mathematics for all: Problems of cultural selectivity and unequal distribution of mathematical education and future perspectives on mathematics teaching for the majority* (pp. 22–25). Paris: UNESCO.

Damerow, P., & Westbury, I. (1985). Mathematics for all: Problems and implications. *Journal of Curriculum Studies, 17*(2), 175–184.

Davis, K., & Moore, W. (1945). Some principles of stratification. *American Sociological Review, 10*(2), 242–245.

De Lange, J. (1984). Mathematics for all is no mathematics at all. In P. Damerow, M. E. Dunkley, B. F. Nebres, & B. Werry (Eds.), *Mathematics for all: Problems of cultural selectivity and unequal distribution of mathematical education and future perspectives on mathematics teaching for the majority* (pp. 66–71). Paris: UNESCO.

Dowling, P. (1998). *The sociology of mathematics education: Mathematical myths/pedagogic text*. London: RoutledgeFalmer.

Fend, H. (2006). *Neue Theorie der Schule. Einführung in das Verstehen von Bildungssystemen*. Wiesbaden: VS.

Freudenthal, H. (1991). *Revisiting mathematics education: China lectures*. Dordrecht: Kluwer.

Gates, P., & Vistro-Yu, C. (2003). *Is* mathematics for all? In A. J. Bishop, M. A. Clements, C. Keitel, J. Kilpatrick, & F. K. S. Leung (Eds.), *Second international handbook of mathematics education* (pp. 31–73). Dordrecht: Kluwer.

Gates, P., & Zevenbergen, R. (2009). Foregrounding social justice in mathematics teacher education. *Journal of Mathematics Teacher Education, 12*(3), 161–170.

Gellert, U. (2006). Mathematik "in der Welt" und mathematische "Grundbildung." Zur Konsistenz des mathematikdidaktischen Rahmens von PISA. In T. Jahnke & W. Meyerhöfer (Eds.), *Pisa & Co. Kritik eines Programms* (pp. 277–291). Franzbecker: Hildesheim.
Gellert, U. (2010). Modalities of local integration of theories in mathematics education. In B. Sriraman & L. English (Eds.), *Theories of mathematics education: seeking new frontiers* (pp. 537–550). Berlin: Springer.
Gigerenzer, G., Gaissmaier, W., Kurz-Milcke, E., Schwartz, L. M., & Woloshin, S. (2008). Helping doctors and patients make sense of health statistics. *Psychological Science in the Public Interest, 8*(2), 53–96.
Gispert, H., & Schubring, G. (2011). Societal, structural, and conceptual changes in mathematics teaching: reform processes in France and Germany over the twentieth century and the international dynamics. *Science in Context, 24*(1), 73–106.
Hartmann, M. (1996). *Topmanager. Die Rekrutierung einer Elite*. Frankfurt: Campus.
Hartmann, M. (2007). *Eliten und Macht in Europa. Ein internationaler Vergleich*. Frankfurt: Campus.
Hartmann, M. (2010). Elites and power structure. In S. Immerfall & G. Therborn (Eds.), *Handbook of European societies: social transformations in the 21st century* (pp. 291–323). New York: Springer.
Howson, G., Keitel, C., & Kilpatrick, J. (1981). *Curriculum development in mathematics*. Cambridge: Cambridge University Press.
Hoyles, C., Noss, R., Kent, P., & Bakker, A. (2010). *Improving mathematics at work: The need for techno-mathematical literacies*. London: Routledge.
Hudson, B. (2008). Learning mathematically as social practice in a workplace setting. In A. Watson & P. Winbourne (Eds.), *New directions for situated cognition in mathematics education* (pp. 287–302). New York: Springer.
Jahnke, H. N. (1986). Origins of school mathematics in early nineteenth-century Germany. *Journal of Curriculum Studies, 18*(1), 85–94.
Jensen, U. K. (1984). An evolution towards mathematics for all in upper secondary education in Denmark. In P. Damerow, M. E. Dunkley, B. F. Nebres, & B. Werry (Eds.), *Mathematics for all: Problems of cultural selectivity and unequal distribution of mathematical education and future perspectives on mathematics teaching for the majority* (pp. 55–57). Paris: UNESCO.
Jones, P. S. (1970). *A history of mathematical education in the United States and Canada*. Washington: NCTM.
Kanté, S. B. (1989). Critical issues of mathematics education in the Ivory Coast. In C. Keitel, P. Damerow, A. Bishop, & P. Gerdes (Eds.), *Mathematics, education, and society* (pp. 78–79). Paris: UNESCO.
Keitel, C. (1987). What are the goals of mathematics for all? *Journal of Curriculum Studies, 19*(5), 393–407.
Keitel, C., Damerow, P., Bishop, A., & Gerdes, P. (Eds.). (1989). *Mathematics, education, and society*. Paris: UNESCO.
Kollosche, D. (2014). Mathematics and power: An alliance in the foundations of mathematics and its teaching. *ZDM: The International Journal on Mathematics Education, 46*(7), 1061–1072.
Lundin, S. (2012). Hating school, loving mathematics: On the ideological function of critique and reform in mathematics education. *Educational Studies in Mathematics, 80*(1–2), 73–85.
Naumann, J. (1989). Practical aspects of basic mathematics teaching in Senegalese villages. In C. Keitel, P. Damerow, A. Bishop, & P. Gerdes (Eds.), *Mathematics, education, and society* (pp. 79–81). Paris: UNESCO.
Pais, A. (2012). A critical approach to equity. In O. Skovsmose & B. Greer (Eds.), *Opening the cage: Critique and politics of mathematics education* (pp. 49–91). Rotterdam: Sense.
Parsons, T. (1959). The school class as a social system. *Harvard Educational Review, 29*(4), 297–318.
Posselt, G. (2013). Grundlinien einer Debatte. Einführung zur deutschen Ausgabe. In J. Butler, E. Laclau, & S. Žižek (Eds.), *Kontingenz, Hegemonie, Universalität. Aktuelle Dialoge zur Linken* (pp. VII–XXVI). Wien: Turia + Kant.

Riall, R., & Burghes, D. (2000). Mathematical needs of young employees. *Teaching Mathematics and Its Applications, 19*(3), 104–113.

Silver, E. A., & Herbst, P. (2007). Theory in mathematics education scholarship. In F. Lester (Ed.), *Second handbook of research on mathematics teaching and learning* (pp. 39–67). Charlotte: IAP.

Souviney, R. (1989). The Indigenous Mathematics Project: Mathematics instruction in Papua New Guinea. In C. Keitel, P. Damerow, A. Bishop, & P. Gerdes (Eds.), *Mathematics, education, and society* (pp. 106–109). Paris: UNESCO.

Stacey, K., Almuna, F., Caraballo, R. M., Chesné, J.-F., Garfunkel, S., Gooya, Z., et al. (2015). PISA's influence on thought and action in mathematics education. In K. Stacey & R. Turner (Eds.), *Assessing mathematical literacy: The PISA experience* (pp. 275–306). Cham: Springer.

Straehler-Pohl, H., & Pais, A. (2014). To participate or not to participate? That is not the question! In B. Ubuz, Ç. Haser, & M. A. Mariotti (Eds.), *Proceedings of the 8th Congress of the European Society for Research in Mathematics Education* (pp. 1794–1803). Ankara: Middle East Technical University.

Ullmann, P. (2008). *Mathematik, Moderne, Ideologie. Eine kritische Studie zur Legitimität und Praxis der modernen Mathematik*. Konstanz: UVK.

UNESCO. (1990). *World declaration on education for all*. Bangkok: UNESCO Regional Office for Education in Asia and the Pacific.

Chapter 6
Ethnomathematical Barters

Aldo Parra-Sanchez

Abstract This chapter identifies and criticizes one assumption of the ethnomathematical research field, regarding the ways in which the relationship between mathematics and culture is addressed. Many developments and theoretical conflicts within this field can be traced to that assumption, which has been widespread indistinctly by practitioners and critics of ethnomathematics. Looking for a new understanding of this field, an alternative approach is proposed, trying to respond to some theoretical critiques and prompting new horizons. This intended approach privileges non-colonialist interactions among stakeholders, recognizing their different interests, their different ways to conceptualize and their interdependence. It is discussed how interactions can be conducted to hybridize different kinds of knowledge, constituting political and epistemological endeavors. The essay concludes observing which types of problems would appear due to the new approach.

Now Ethnomathematics…

Since ethnomathematics emerged as a research field within mathematics education, a proper theorization and definition has been sought and almost every researcher has attempted to give his/her personal view regarding its definition and intend. Although this is common to new and growing fields of research, this diversity of methods and approaches might be seen as a sign of disorder, non-cohesion, or absence of a shared horizon. Most of the researchers identifying themselves with this field share a common conception of ethnomathematics as a research program in the history of ideas, that seeks to understand the "generation, organization, institutionalization and propagation of knowledge" (D'Ambrosio, 1993) throughout the history of humanity, in the contexts of different interest groups, communities, peoples, and nations. However, this conception is developed in different directions. Even D'Ambrosio, who introduced first such conception (D'Ambrosio, 1985), has been modifying parts of his seminal theoretical statements about ethnomathematics over the last 15 years.

A. Parra-Sanchez (✉)
UNESCO Center of PBL in Engineering and Science, Department of Planning and Development, Aalborg University, Aalborg, Denmark
e-mail: aldo@plan.aau.dk

H. Straehler-Pohl et al. (eds.), *The Disorder of Mathematics Education*,
DOI 10.1007/978-3-319-34006-7_6

He proposes to embrace a more holistic and transdisciplinary approach, not only regarding mathematics but knowledge in general (D'Ambrosio, 2001).

In the last 5 years, new attempts have been made to produce theories that allow room for such variety of comprehensions. Miarka and Bicudo (2012) studied the relationship between mathematics and ethnomathematics through a phenomenological explanation of academic fieldwork. Rohrer and Schubring (2013, p. 78) proposed an overarching conceptualization of ethnomathematics by claiming that this theory "needs to be regarded as an interdisciplinary discipline that covers theories from both the exact and social sciences."

Parallel to such theorizations, critiques of ethnomathematics—pointing to assumptions that research in this field make when using terms such as "culture" and "mathematics"—have been published. Rowlands and Carson (2002) warn about the uses of ethnomathematics in education, by stressing the fact that western mathematics encompasses and formalizes all previous cultural systems that humans had developed. If this particular critique were to be accepted, a study of previous systems would result in a throwback that education cannot allow. Other critiques question the effectiveness of ethnomathematics in achieving its own intended political goals (Pais, 2011, Vithal & Skovsmose, 1997). While the first of these critiques attempts to dismiss ethnomathematics as a whole, the other two invite to sharpen the core ideas within them. Nevertheless, all of them share an "external" positioning with respect to ethnomathematics, since the authors do not consider themselves as researchers in this area. Only a few "internal" critiques—this is to say, critiques made by scholars who consider themselves involved in ethnomathematics research—have been published. For example, Alangui (2010) warned about the very old fashioned concept of culture that is commonly used in research. Knijnik, Wanderer, Giongo, and Duarte (2012) also problematized how some assumptions about students' realities and the use of concrete materials for teaching became naturalized in ethnomathematical research.

…Has an "Intersection Approach"…

A common feature of inner trends and external critiques within the diversity of approaches and purposes, is the intention of addressing the existence/absence of shared objects between mathematics and culture (despite the diverse definitions of those terms). By considering the particular culture of a group as one set and mathematics as another set, ethnomathematics as a research field might relate to examine the *intersection* of these two sets. Such intersection can be called the ethnomathematics of that group or even the mathematics of that group. Whatever the chosen name is, and without considering neither the possible methodological procedures to perform this examination nor the theoretical considerations that would make impossible comparing those sets, the underlying assumption is such an intersection matters.

It is not difficult to show in the growth of ethnomathematics as a research field, how the intersection of mathematics and cultures has been considered as relevant.

For instance, the North American Study Group on Ethnomathematics sponsors a journal which on its online version states that: "the journal's contents examine the intersections between mathematics and culture in both western and non-western societies, and among both math professionals and non-professionals" (NASGEM).

It is common in research articles to find expressions such as "every culture has mathematics" (Selin & D'Ambrosio, 2001, p. xvii); "ethnomathematics seeks to revive mathematics *living* in different traditions and cultures, not by considering them to be exotic, but by including them in the new historiography of mathematics" (Rohrer & Schubring, 2013, p. 84, emphasis added); and even the dilemma pointed out by Bishop (1994, p. 15, emphasis added): "is there one mathematics *appearing in* different manifestations and symbolizations, or *are there* different mathematics being practiced which have certain similarities?". As we can see, these references share the common ground of assuming first a distinction between mathematics and culture and assuming then that there is *something* in the intersection of these two entities. Accordingly, this *intersection* becomes the main—though not the sole—object of study for ethnomathematics.

Even historically it seems to exist a continuity regarding the importance of such intersection. This can be already seen in early approaches, such as that of Ascher and Ascher (1986) who define ethnomathematics as mathematical practices of non-literate people; it continues with a reconceptualization by D'Ambrosio (2006, p. 1, emphasis added) who states that "ethnomathematics *is* the mathematics *practiced* by cultural groups" and it can be also seen in contemporary work like that of Furuto (2014, p. 122) who assumes that ethnomathematics is "the intersection of culture, historical traditions, sociocultural roots and mathematics."

Barton (1996) explicitly pointed to the intersectional approach in a visual way, using Venn-diagrams. When trying to analyze the comprehension of the field by three influential researchers, he drew a diagram for each of the approaches to exaggerate the differences existing among them (Fig. 6.1):

Through all those excerpts, it can be seen how the assumption regarding the importance of the intersection has been central to conceptualizations about ethnomathematics, providing a way of thinking and talking about the field as well as framing the debate. In the next section, I show the limitations of such approach and I point out some consequences of these limitations.

…Problem…

There are different ways in which critics and followers of ethnomathematics position their arguments within the intersection approach and according to their personal alignment with the ethnomathematics program. In this section, four possible types of positioning for researchers are considered.

The first possible situation is to simultaneously support the ethnomathematical research program and also claim that the aforementioned intersection is not empty. From this starting point, a researcher will try to show the presence of some mathematical notion, skill, or concept within cultural artifacts and practices. Roughly speaking,

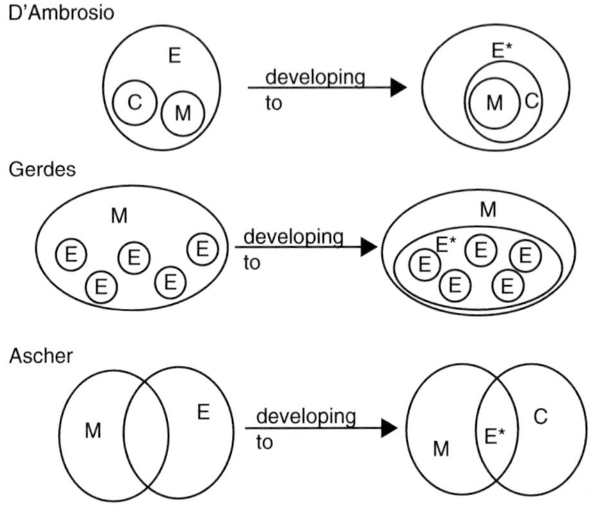

Fig. 6.1 Barton's diagram for ethnomathematics. (Barton, 1996, p. 213) (*E* ethnomathematics, *E** ethnomathematics as a research program, *M* mathematics, *C* culture)

the aim is about to "reveal the culture's hidden math." The ethnomathematician's goal becomes uncovering that presence and mathematical modelling appears as a natural complement in this endeavor. Therefore, there is a necessity of developing proper methodological tools to do that uncovering, as Ferreira (1994) and more recently Albertí Palmer (2007) have intended. If ethnomathematics is theorized using these arguments, the paradox of Millroy (1992) quickly arises as problematic. This paradox points out that it is not possible to find any knowledge other than the academic because researchers will be acting with their academically trained mathematical gaze. This is an echo of the anthropological "reflexivity problem" (Woolgar, 1988) resulting from considering ethnography as the proper methodology for ethnomathematics.

A second possibility is to be sympathetic with ethnomathematics as a research program, but to consider the intersection being empty, due to the nonexistence of the category "mathematics" in some cultures. This posture focuses on how knowledge is developed in different cultural groups by recognizing how it affects and is affected by educational discourses. A key question to this possibility was raised by Lizcano (2002, p. 1, own translation): "what can we see if, instead of looking at popular practices through 'mathematics,' we look at mathematics through popular practices?" This positioning puts in doubt the preeminence of mathematics as a superior knowledge over others. Conversely, efforts to embed holistic knowledge into the restricted boundaries of mathematical discipline are rejected. It seems to have a subtle essentialist view of culture as it leaves unexplained why a cultural group that does not have the category of mathematics would not be able to understand the existence of that category in the cultures of other groups, or at least to create an inner explanation about mathematics.

A third option presents similar reasons to consider the intersection as empty by definition, but differs from the second option as it does not follow ethnomathematics as a research program to expand the social understanding of mathematics as part of a culture. In this posture, it is common to use an argument of authority: cultural practices are not mathematics because they are not developed within a scholarly context. They have not been refined by one legitimate institution (universities, journals, the academic community of teachers, mathematicians, and so on), and because of that, they lack a "warranty" certificate. This attempt to create an essence of mathematical knowledge using the hegemony of one particular group was summarized by Rômulo Lins (2004, p. 99) with irony: "mathematics is the thing made by mathematicians when they said that they are doing mathematics." Although such a circular definition cannot be contested, its self-sufficiency is at the same time its big weakness as it implies an unacceptable omission: those institutions are cultural and historical. Lins reminds us how the professionalization of mathematics appears only in the nineteenth century and many of the literature recognized as mathematics before that time could barely satisfy current standards.

The last position does not follow ethnomathematics as a program neither, but is less radical; it considers the intersection as non-empty (i.e., it recognizes that mathematics can be present in several cultures). A hierarchical model drives this account of the development of the discipline, believing that the world has adopted conventions of mathematics, "because they have been sifted and tested and refined within the crucible of practical experience, which yields neither to passion nor to ideological persuasion" (Rowlands & Carson, 2002, p. 86). In such a model, mathematical knowledge has constantly been evolving in a universal process of improvement that transcends civilizations. If such an approach is accepted, any strong review of the history and epistemology of mathematics is impossible and accordingly ethnomathematics has nothing worthy to offer. Hence, the goal would be to fill in minor details of how the only possible rationality was achieved across space and time until now as an inevitable fate.

All these positions refer mainly to what is (or what has been) mathematical instead of what could be mathematical. I consider that the intersection problem conduces to a false dilemma, which is responsible for the critiques received and also for the growing "domestification" ethnomathematics has been the object to in the last decade as attested by Pais (2012).

…That Can Be Changed…

With this essay, I try to develop an alternative approach to theorize ethnomathematics that goes beyond the "intersection" problem by building on several research projects conducted in multilingual environments (Barton, 2008; Caicedo et al., 2009; Cauty, 2001; Meaney, Trinick, & Fairhall, 2011). These projects serve as an impulse to reflect upon the possibilities of developing dialogic processes within cultural groups around the concept of mathematics, its educational implications and its political uses. The political views of Knijnik et al. (2012) and Alangui's methodological contribution about "mutual interrogation" (Alangui, 2010) deserve to be considered because they give the interactions a central role, despite not working directly with linguistic issues.

The task is not to discover or find elements within the intersection of mathematics and culture, but to create links between them. When the Māori Language Commission proposed terms to be used within schools (Meaney et al., 2011) in a way that took care of sensible features of the Māori language and their cultural heritage, the research experience resulted in a collaborative and multilateral process that expanded boundaries in mathematics, culture, and language simultaneously. The new lexicon entailed new knowledge and new sorts of relations about that knowledge. Language provided a backdrop for interactions through translations and negotiations of meaning.

However, one can find the same type of movement in cases without that linguistic issue; Eglash (2000, p. 17) for instance, recognizes how Gerdes (2007) proposed new mathematical ideas inspired by cultural practices from Africa:

> Ethnomathematics of indigenous societies is not limited to direct translations of western forms, but rather can be open to any mathematical pattern discernable to the researcher. In fact, even that description might to be too restrictive: previous to Gerdes' study there was no western category of "recursively generated Eulerian paths"; it was only in the act of applying a western analysis to the Lusona that Gerdes (and the Tchokwe) created that hybrid.

The basic idea is to provide an interpretation of ethnomathematical research practice as intentional and deliberated processes. These processes generate connections between mathematics and culture in a non-essentialist understanding of both constructs.

This approach assumes a different role of the researcher: from that of one who looks for something hidden and preestablished to one who creates representations and meanings. With such consideration, researchers can be found on both sides, not only on the academic side. Practitioners and knowledge-holders become researchers as well, following their own agenda as proposed in post-colonial studies by Chambers (1996) and in ethnomathematics by Cauty (2001). Therefore, the intended connections produced are not unidirectional, as they do not create mathematical interpretations of cultural practices only, but they also provide culturally grounded explanations of mathematical practices. Another consequence is that the web of those multiple links creates a space for cultural encounter where unforeseen actions and situations can happen.

The features of this approach resemble the practice of *barter* in the manner that is done by some indigenous communities in South America (Townsend, 2012). In this type of barter, people from different villages participate in a meeting contributing in her/his own way with food, tools, and workforce. Participants bring food, tools, and elements that other participants do not produce. Sometimes these elements are either exchanged, or given as a gift. During the meeting, jokes and stories are told and people dance and sing songs. Barters are arranged to build a house or to help families through agricultural labor. Every person involved in the barter returns home with something new gained in the barter. Roughly speaking, when people are engaged in a barter, tasks that only can be done with a joint effort are accomplished and the interdependence of the agents is emphasized, benefiting everyone. A barter is therefore more than a mere exchange but an opportunity to share, create, and learn.

Ethnomathematical practice can assume the underlying principles of bartering as a way of addressing cultural encounters. These principles would put forward the relationship between academic researcher and communities far away from the realm of ethnography and demand unforeseen alternatives in each particular encounter.

…To an Interactional, Hybrid…

I explain the idea of creating connections that is central for bartering, using an example that Alexandre Pais (2013) proposed to illustrate his critique of ethnomathematics. He invites the reader to imagine a group of indigenous people observing students in a mathematics classroom where the topic of the day is the Pythagoras Theorem. After some time watching the students,

> They [the group of indigenous people] realize that what the students are doing while seated at tables with pens in their hands solving exercises on a sheet of paper *is actually* the construction of a house. Why does this sound absurd? Why is the direction of research always one of going to the local communities to recognize as mathematics what these people are doing? (Pais, 2013, p. 3, emphasis added)

As it is argued, it is irrelevant if this mathematical practice "is actually" one house or not. Certainly, it could be less problematic if the group says "this exercise *looks like* the way that we build a house" appealing to some *family resemblance* with mathematics (Knijnik, 2012). Nonetheless, the important aspect is the very act of the group claiming a connection between one system and the other. Pais found his story absurd because it underlies the colonial relationship on which classic ethnography relies. In such story, facts have no consequences and there are no interactions between people. Pais, like many of the followers and critics of ethnomathematics, does not conceive ethnomathematics as a form of barter.

Let us imagine a continuation of the story, a second part that decreases the colonial bias, by making relevant the diversity of voices and agents that are present in the situation, and involving those voices in a common goal. Let us imagine that one of the indigenous says that those equations on the chalkboard remind her of the building of a house.

> *Indigenous A*: This is a house
> *Student 1:* No, it is a theorem
> *Indigenous A*: Well, it seems a house to me
> *Student 2*: How so?
> *Student 3*: I do not understand
> *Indigenous B*: Of course, it is not the same, but when we construct, we put rows and pillars in a cross
> *Student 1*: Oh, perpendiculars?
> *Indigenous A*: Whatever, if you want to call it like that, it is ok. But we say "in cross." With that cross we put all thc tiles, caring that the water rain falls easy.
> *Student 3*: Teacher, that one would be the hypotenuse?
> *Teacher*: Not one, many of them, because there are many tiles in different directions
> *Student 2*: But Pythagoras has only one hypotenuse!
> -…

This endless hypothetical discussion starts involving different worldviews with explanations from multiple sources. This conversation sketches a process that requires collaboration among the agents. Such agents contrast, translate, criticize and appropriate ideas from the practice observed, constituting a barter of insights. This interaction is in itself an educational process that does not intend to arrive at a shared *happy end* by destroying differences in a common, unified knowledge.

This second part of the story does not sound absurd to me for a simple reason: it has already happened. There have been several experiences where the direction of research goes from the local communities to recognize cultural and political concepts and practices that are usually related with mathematics. Meaney et al. (2011) reported the challenging process of one Māori community in New Zealand trying to educate their children in a Māori-immersion school, highlighting how collaboration becomes central to confront the different challenges that arise in every stage. Gelsa Knijnik enquired how Brazilian farmers in a settlement discuss the different ways in which the land is measured, contrasting farmers' techniques with official ones used by banks and the state (Knijnik, 1996). Caicedo et al. (2009) registered the experience of an indigenous community in Colombia trying to appropriate mathematical knowledge for their political process of cultural resistance, applying their own idea of collective research along the way.

This new story is very close to the idea of mutual interrogation proposed by Alangui (2010) because the involved groups were engaged in processes that created new knowledge, instead of just more effectively and more equitably reproducing existing knowledge. Based on this image I would like to propose Fig. 6.2 as a first metaphor:

Fig. 6.2 "Square Heads 10" by Mana Neyestani

Fig. 6.3 Illustration by Aldo Parra-Sanchez, inspired in Square Heads

The effort to describe the other's practices in one's own terms demands a rearrangement of the own knowledge. Such reorganization implies the creation of *hybrid knowledge* which expands the "core" of the discipline and the culture. Those hybrids are particular, localized, and multiple at the same time. The conceptual movement from identification to creation demands that research projects in ethnomathematics become political processes of negotiation of meaning where differences are exposed and fostered, like in the imagined conversation of Fig. 6.2 or like in these other possible encounters presented in Fig. 6.3.

The situations proposed in Fig. 6.3 highlight the fact that the intended hybrid knowledge is by definition an attempt to communicate with others and requiring their response. It is the aftermath of a translation process that from the beginning knew the impossibility to achieve a final consensus. Indeed, the translation process was undertaken due to such impossibility. Success is not the result but a process of mutual adjustments that may never end.

…And Political…

Once the posture to consider interactions as barters is adopted, some of the critique developed on ethnomathematics can be employed as resource for the research field, in order to cast a new light on knowledge as a changeable social and historical practice. The self-referential definition of mathematics, provided by Lins (2004) entails an invitation to challenge authoritarian efforts, since what is assumed as mathematics cannot be predetermined. Every particular appropriation of a concept trough practice expands the concept's limits, transforming its meaning with the unavoidable presence of social life. By using a non-colonial perspective, one that understands power and knowledge as mutually constituted, it is no longer acceptable to be passive to the arguments of authority. Those appropriations of mathematical knowledge can be seen as acts of sovereignty and resistance against domination. For the case of indigenous communities Brayboy (2008, p. 342) asserts:

> Indigenous peoples engage in survivance through survival and resistance, and we are talking back. More than simply talking back, however, we are moving forward, claiming spaces and demanding acknowledgment of sovereignty that has existed since time immemorial. (…) If we as scholars are to consider the connection between Indigenous knowledge and sovereignty, then we must realize that our knowledge systems serve as a place of power and a source of continuance of our groups.

This idea of "talking back" reflects a bold trend among oppressed and marginalized communities understanding and assuming their agency in the struggles for power, not only at the concrete level of material needs, but also in an ideological dimension, breaking with the self-reinforcing cycle of hegemony-power-knowledge that Fasheh (1990, p. 24) explained:

> Hegemony does not simply provide knowledge; rather, it substitutes one kind of knowledge for another, in the context of a power relationship. Power in this sense, is almost defined by what is excluded. (…) To recognize my mother's activity as math was for me to recognize that education and knowledge are not only about facts but also about the inner logic of society, both within itself and in relation to outside forces (…) Hegemony is characterized not only by what it includes, but also by what it excludes: by what it renders marginal, deems inferior, and makes invisible.

This contention allows us to understand how political is the ethnomathematical activity involving and enhancing knowledge systems that have been dismissed by Eurocentric domination. As far as power and knowledge are imbricated, the struggle of marginalized groups and cultures to make their knowledge survive in time and space, emerges as a political action of resistance.

> Meaningful education, or community education, thus reclaims people's lives, their sense of self-worth, and their ways of thinking from the hegemonic structures, and facilitates their ability to articulate what they do and think about in order to provide a foundation for autonomous action. (Fasheh, 1990, p. 26)

It is fair to say that these ideas about the interaction between academic disciplines and nonacademic knowledge are not entirely new. Barton addressed something similar, when he conceived ethnomathematics as "a process of the social construction of knowledge at a cultural level" (Barton, 1996, p. 216) and claimed:

> Ethnomathematics does create a bridge between mathematics and the ideas (and concepts and practices) of other cultures. Part of an ethnomathematical study will elucidate why those other ideas are regarded as mathematical, and therefore why they might be of interest to mathematicians. Such a study creates the possibility both of mathematics providing a new perspective on the concepts or practices for those within the other culture, and of mathematicians gaining a new perspective on, (and possibly new material for), their own subject. (Barton, 1996)

…Approach…

I intend to add a second metaphor for these ideas, reworking the initial image of mathematics and culture as sets, and observing that it is possible to establish relations between the sets (see Fig. 6.4), instead of studying their intersection.

The first metaphor was that of an aesthetical interpretation, which highlights the presence of members (*knowledge-holders*) of different communities working together to produce, negotiate and interpret meanings. Complementary, this second metaphor of mathematical relation stresses the rise of those meanings, as new types of knowledge that do not belong to any established cultural (or mathematical) tradition.

These resultant meanings are plural and they do not unify, merge or melt worldviews; they perform a connection instead. Like any mathematical relation relating two domains without being part of them, those new meanings belong to a third domain and have another regime. Therefore, they cannot be treated within any essentialist approach because they are hybrids, anomalies, resulting in a process of interaction.

The mathematical diagram of Fig. 6.4 resulted from developing the analogy of culture and mathematics as sets: what could their elements be? How does ethnomathematics operate within that image? Following the intersection approach corresponding to Barton's Venn diagrams, there is only one possible intersection between two sets. Each member of a set is examined with the principle of excluded third: It must belong or must not belong to the other set. Alternatively, in the second metaphor there are multiple possible relations between sets; a relation is defined as a bundle of associations among the elements of two sets. One element in a set can be associated with (i.e., translated as) another element in the other set, associated with more than one element, or even associated with no element. If a connection seems "unsatisfactory," another relation is chosen, i.e., another bundle of associations is built. The relations are "customizable" while the intersection is not.

It is noticeable that the change from intersections to relations is not a small one at all. Such change entails an entirely different role for an ethnomathematics researcher. In the intersection approach, the researcher behaves as a detective looking for, uncovering, and trying to prove facts based on evidence. Researcher pursues a factual truth (timeless and univocal), which requires *proof*. In contrast, within the relational approach, the ethnomathematician acts like an artist: creating, proposing, and performing interactions; researcher tries to make sense through translations of meanings. The truth that the researcher aims at is a poetical one (ephemeral and polysemic) that deserves to be *experienced*.

This displacement does not intend to replace one image by another two, which simply describe better the same thing. It is rather an invitation to change the objects

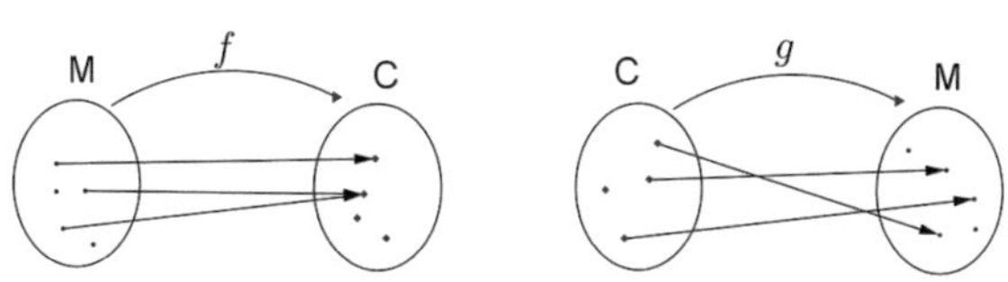

Fig. 6.4 *f* and *g* represent possible ethnomathematical research projects, comprised of multiple associations among objects of both sides (mathematics and culture); Therefore *f* and *g* are members of E* (*M* mathematics, *C* culture, *E** ethnomathematics as a research program)

of study and the ways in which practitioners can develop the field. For instance, a central role is proposed for the awareness and political intentionality of making conceptual connections. Consequently, a requirement to involve different voices gains preeminence to make possible the links between systems of knowledge. It is also critical to notice that collective processes of sense-making can be considered educational rather than merely curricular or school-bounded.

To conclude this section I want to make two comments about this new approach; First, I think it as an original idea, not because it is a novelty, but because it can be traced back to the origin of ethnomathematics. D'Ambrosio (1985, p. 47, emphasis added) stated in his breakthrough paper:

> We are collecting examples and data on the practices of culturally differentiated groups which *are identifiable* as mathematical practices, hence ethnomathematics, *and trying to link* these practices into a pattern of reasoning, a mode of thought.

Second, I emphasize the importance of grounding conceptual images in empirical research. Theoretical standpoints shape and condition the empirical research, by establishing what is thinkable, what deserves to be studied and what procedures of study can be applied. Theories are particular arrangements of concepts, agents, relations, hierarchies, and taxonomies. By using analogies, metaphors, diagrams, pictures, and other conceptual images, we can make explicit our understanding of those arrangements, and move through them to produce new insights.[1] Hence, the current effort to conceptualize ethnomathematics using diagrams and drawings is more than a flirtation with images.

At the same time, the theoretical reflection through images can have concrete consequences. For instance, if the intersection approach is assumed, and cultural practices are researched with the only purpose of finding examples of cultural differences, then ethnomathematics research would be reduced to collecting exotic cultural practices. If researchers, on the other hand, problematize the encounter of cultural practices with academic practices through a relational image, then ethnomathematics can be conceived as a barter, a kick-starter of new understandings and new types of hybrid knowledge that will defy hegemonic forms of power.

However, it is important to acknowledge that any conceptual image implies—by its nature—a particular delimitation of the problem. Such delimitation is both a reduction of complexity as well as a possibility to generate fruitful insights that cannot be achieved otherwise. Thus, I have presented several images, hoping that their overlap can minimize the reduction of complexity. The rationale of this strategy is explained further at the end of this essay.

[1] For instance, the term most used to describe the theoretical interplay is precisely the image of field, as a place where objects of a theory coexist and can be gathered in particular ways. Accordingly, researchers declare which is their position in the field and how they see the objects, when they state their personal understandings of the theory.

...Which Is Also Problematic!

The aim of ethnomathematics to expand the social understanding of concepts like mathematics or knowledge becomes clearer in this approach. It states that the object of study of ethnomathematics is no longer the intersection of mathematics with culture, but the multiple connections that people can build among them. Academic researchers, local communities and other stakeholders, attempt to appropriate and hybridize knowledge from various sources in their particular ways and for different purposes. Instead of previous and preestablished elements to be uncovered, we might consider the multiple and unexpected possibilities to be developed.

If ethnomathematics uses this approach to (dis)solve the "reflexivity problem" posed by the Millroy paradox, then a "symmetry problem" arises. Those negotiations and new meanings imply an active participation of different stakeholders in long-term processes. How can this participation be guaranteed? How can cultural groups be interested in establishing such processes? All the examples provided here were embedded in broad political projects of organized communities, which started earlier than these particular ethnomathematics research projects. These communities understand the relationship between mathematics and culture as a field crossed and impacted by politic and economic forces, where issues of identity, heritage and survival with dignity are at stake. Accordingly, these communities could conceive research in ethnomathematics as a relevant and strategic part of their political struggle for self-determination. However, not all groups have articulated their demands or concerns through an organizational strategy. How can dialogic processes be undertaken by non-organized groups? How can dialogue be established in the constrained space of a school system?

More considerations can be raised about the interactions. How much time does a process of dialogue require? It seems that it is not compatible with the increasingly short times imposed by universities for academic research, suggesting that interaction might need additional scenarios to be undertaken. What types of instances could they be? This challenge meets another natural tension, if dealing with dialogue and interaction, issues of mutuality and co-responsibility become central. Therefore, it is natural to ask how can "the others" do research on their own terms. Particularly, how can ethnomathematics be engaged in a political/epistemological level with other systems of knowledge, and in a way that respects self-determination and sovereignty? Although a first guess may be to follow the path of the mutual interrogation proposed by Alangui (2010), some issues remain regarding the risk to fail in a sort of tokenism. For instance, by prompting local communities to formulate questions and developments through a mimicry of traditional academic research not expressing their sincere concerns. New types of research results and new validation procedures emerge when the agency and insights of the local communities are a constitutive part of the theoretical tools deployed in research, and not simply the data to be collected.

Many of these questions cannot be answered directly or require an analysis that goes beyond the scope of this text. Nevertheless, they configure a promising landscape for the approach as long as they emphasize a condition of inherent uncertainty for every piece of ethnomathematical research. The vision of ethnomathematics as a process of barter could restore the seminal impulse to reveal the historical and cultural grounding of mathematics and could also reinforce its critical position in the relationship among power and mathematical knowledge.

Ethnomathematics should not only observe the past but also look towards the future. The field should not be concerned only with a better understanding of current western knowledge, through the study of how others cultural groups build their knowledge. Rather, ethnomathematics can also engage those other groups in the change of the accepted body of knowledge. That "broader vision of knowledge" claimed by D'Ambrosio (2012) cannot be static, but dynamic. I think this is the central contribution of the barter that I am fostering.

This chapter brings forward an old idea that considers mathematics as central to ethnomathematics. Certainly, I could have deployed more or less the same argument by substituting "mathematics" for "western/academic knowledge." However, I preferred it this way because by including mathematics in comparisons, links and images, those connections turn to be heretical. If mathematics are left out of the focus, the field of ethnomathematics subsumes in a general discussion, losing its strongest feature: the problematization of mathematics education and mathematics epistemology.

In the recurrent discussion about the role of mathematics in ethnomathematics, this approach rejects dichotomies. It assumes that ethnomathematics must and can reject the accusation to superficially empower people because their culture would be "one step up, closer to the divine conventions of mathematics," as some critiques suggest, i.e., Pais (2013) or Rowlands and Carson (2002).

On the contrary, by assuming that mathematics is not above other knowledge, and that it has its specificities like any other knowledge, it is not a problem if we refer to some of its objects in yet unfamiliar ways to gain new insights. Paradoxically, to forbid or to avoid the use of mathematics embodies a new way to enthrone the discipline by reinforcing its supposed untouchable character. It is important to realize that ethnomathematics does not wish to break or discard mathematics; it wants to break its sacredness. Accordingly, within this approach, mathematics is demystified when bartering reveals it as being mundane, just as any other affair.

I attempt such demystification not only with the arguments deployed so far but also with the way in which these arguments were deployed. I decided to play with metaphors throughout this text, intending that the overlap of images constitute a hybrid. A connection can be seen as a translation, which is like an aesthetical interpretation of a painting, resembling a mathematical relation that works like a bunch of arrows, which suggests a connection, etc. I drew all these family resemblances on purpose. By combining a variety of analogies in a network with multiple agents and voices (see Fig. 6.5), ethnomathematics could contribute to change mathematics at an epistemological level.

In this chapter I argued:

> Now ethnomathematics has an "intersection approach" problem that can be changed to an interactional, hybrid and political approach, which is also problematic!

This content is a way to invite interplay with the dynamic condition of culture and mathematics, instead of merely watching it. As mathematics are shaped by our efforts, the main problem is not to perceive the difference, but what to do with it. How can we live with, and through, the difference?

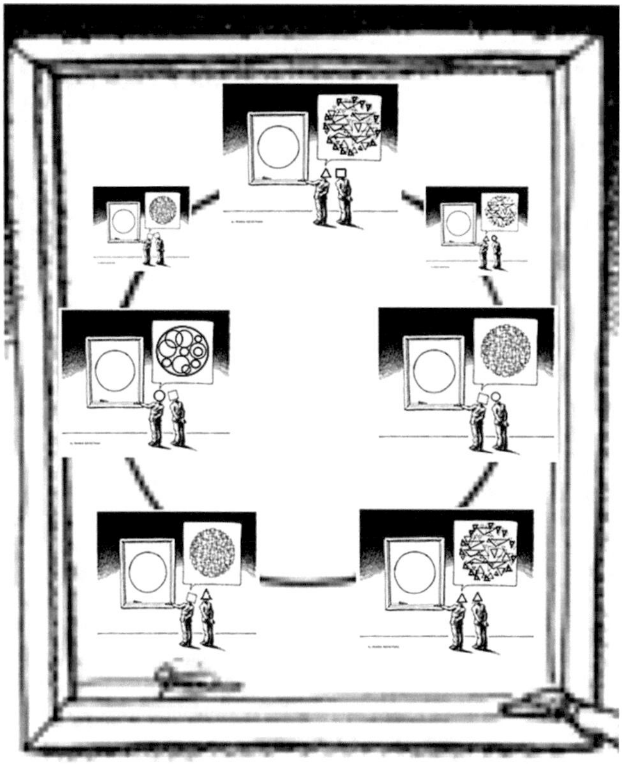

Fig. 6.5 A multiplicity of intentioned interpretations of a practice becomes the practice. Illustration by Aldo Parra-Sanchez, inspired in Square Heads

Acknowledgments A brief and early version of this chapter was published in the Proceeding of the Eighth International Mathematics Education and Society Conference, in 2015. The research leading to this chapter was funded by Departamento Administrativo de Ciencias, Tecnología e Innovación (COLCIENCIAS) and Aalborg University in Denmark. I would like to thank the members of the Science and Mathematics Education Research Group (SMERG) at Aalborg University for their comments on previous drafts of this chapter. This research is also part of the NordForsk Center of Excellence "JustEd."

References

Alangui, W. (2010). *Stone walls and water flows: Interrogating cultural practice and mathematics.* Auckland, New Zealand: University of Auckland.

Albertí Palmer, M. (2007). *Interpretación matemática situada de una práctica artesanal.* Barcelona: Universitat Autónoma de Barcelona.

Ascher, M., & Ascher, R. (1986). Ethnomathematics. *History of Science, 24*, 125–144.

Barton, B. (1996). Making sense of ethnomathematics: Ethnomathematics is making sense. *Educational Studies in Mathematics, 31*(1–2), 201–233.

Barton, B. (2008). *The language of mathematics: Telling mathematical tales (Mathematics Education Library).* New York: Springer.

Bishop, A. J. (1994). Cultural conflicts in mathematics education: Developing a research agenda. *For the Learning of Mathematics, 14*(2), 15–18.

Brayboy, B. M. J. (2008). Yakkity yak" and "talking back": An examination of sites of survivance in Indigenous knowledge. In M. Villegas, S. Neugebauer, & K. Venegas (Eds.), *Indigenous knowledge and education: Sites of struggle, strength, and survivance* (pp. 339–346). Cambridge, MA: Harvard Education Press.

Caicedo, N., Guegia, G., Parra, A., Guegia, A., Guegia, C., Calambas, L., et al. (2009). *Matemáticas en el mundo Nasa*. Bogotá, Colombia: CIIIT.

Cauty, A. (2001). Matemática y Lenguajes: Como seguir siendo amerindio y aprender la matemáticas que necesitara? In A. L. G. Zapata (Ed.), *Pluriculturalidad y aprendizaje de la matemática en américa latina: Experiencias y desafíos* (pp. 49–87). Madrid: Ediciones Morata.

Chambers, I. (1996). Signs of silence, lines of listening. In I. Chambers & L. Curti (Eds.), *The post-colonial question: Common skies, divided horizons* (pp. 47–62). London: Routledge.

D'Ambrosio, U. (1985). Ethnomathematics and its place in the history and pedagogy of mathematics. *For the Learning of Mathematics, 5*(1), 44–48.

D'Ambrosio, U. (1993). Etnomatemática: Um Programa. *Educação Matemática em Revista: Revista da Sociedade Brasileira de Educação Matemática–SBEM, 1*(1), 5–11.

D'Ambrosio, U. (2001). *Etnomatemática: elo entre as tradições e a modernidade*. Belo Horizonte: Autêntica Editora.

D'Ambrosio, U. (2006). *Ethnomathematics. Link between traditions and modernity*. Rotterdam: Sense Publishers.

D'Ambrosio, U. (2012). The Program Ethnomathematics: the theoretical basis and dynamics of cultural encounters *Cosmopolis. A Journal of Cosmopolitics/Revue de cosmopolitique, 3–4*, 13–41.

Eglash, R. (2000). Anthropological perspectives on ethnomathematics. In H. Selin & U. D'Ambrosio (Eds.), *Mathematics Across Cultures* (pp. 13–22). Dordrecht: Springer.

Fasheh, M. (1990). Community education: To reclaim and transform what has been made invisible. *Harvard Educational Review, 60*(1), 19–35.

Ferreira, E. S. (1994). A importância do conhecimento etnomatemático indígena na escola dos não-índios. *Em Aberto, 14*(62), 89–95.

Furuto, L. H. L. (2014). Pacific ethnomathematics: Pedagogy and practices in mathematics education. *Teaching Mathematics and Its Applications, 33*(2), 110–121. doi:10.1093/teamat/hru009.

Gerdes, P. (2007). *Etnomatemática—Reflexões sobre Matemática e Diversidade Cultura*. Ribeirão: Edições Húmus.

Knijnik, G. (1996). *Exclusão e resistência: educação matemática e legitimidade cultural*. Porto Alegre, Brazil: Artes Médicas.

Knijnik, G. (2012). Differentially positioned language games: Ethnomathematics from a philosophical perspective. *Educational Studies in Mathematics, 80*(1–2), 87–100. doi:10.1007/s10649-012-9396-8.

Knijnik, G., Wanderer, F., Giongo, I., & Duarte, C. G. (2012). *Etnomatemática em movimento*. Belo Horizonte: Autêntica.

Lins, R. C. (2004). Matemática, monstros, significados e educação matemática. In M. A. V. Bicudo & M. C. O. Borba (Eds.), *Educação matemática: pesquisa em movimento* (pp. 92–120). São Paulo: Cortez.

Lizcano, E. (2002). Las matemáticas de la tribu europea: un estudio de caso. In *II International Congress on Ethnomathematics, Ouro Preto, Brasil* (pp. 5–8).

Meaney, T., Trinick, T., & Fairhall, U. (2011). *Collaborating to meet language challenges in indigenous mathematics classrooms (Mathematics Education Library* (Vol. 52)). New York: Springer.

Miarka, R., & Bicudo, M. A. V. (2012). Matemática e/na/ou Etnomatemática? *Revista Latinoamericana de Etnomatemática: Perspectivas Socioculturales de la Educación Matemática, 5*(1), 149–158.

Millroy, W. L. (1992). An ethnographic study of the mathematical ideas of a group of carpenters. *Journal for Research in Mathematics Education. Monograph, 5*. doi: 10.2307/749904.

NASGEM. *Journal of Mathematics and Culture*. Retrieved October 23, from http://nasgem.rpi.edu/pl/journal-mathematics-culture.

Pais, A. (2011). Criticisms and contradictions of ethnomathematics. *Educational Studies in Mathematics, 76*(2), 209–230. doi:10.1007/s10649-010-9289-7.

Pais, A. (2012). A investigação em etnomatemática e os limites da cultura. *Reflexão e Ação, 20*(2), 32–48.

Pais, A. (2013). Ethnomathematics and the limits of culture. *For the Learning of Mathematics, 33*(3), 2–6.

Rohrer, A. V., & Schubring, G. (2013). The interdisciplinarity of ethnomathematics: Challenges of ethnomathematics to mathematics and its education. *Revista Latinoamericana de Etnomatemática: Perspectivas Socioculturales de la Educación Matemática, 6*(3), 78–87.

Rowlands, S., & Carson, R. (2002). Where would formal, academic mathematics stand in a curriculum informed by ethnomathematics? A critical review of ethnomathematics. *Educational Studies in Mathematics, 50*(1), 79–102. doi:10.1023/A:1020532926983.

Selin, H., & D'Ambrosio, U. (2001). *Mathematics across cultures: The history of non-Western mathematics* (Vol. 2). Dordrecht: Springer.

Townsend, S. (2012). Minga: The communal work tradition of Bolivia. *Americas, 64*(3), 14–17.

Vithal, R., & Skovsmose, O. (1997). The end of innocence: A critique of "ethnomathematics". *Educational Studies in Mathematics, 34*(2), 131–157. doi:10.1023/A:1002971922833.

Woolgar, S. (1988). *Knowledge and reflexivity: New frontiers in the sociology of knowledge*. London: Sage.

Chapter 7
Installing "Good Mathematics Teaching": Hegemonic Strategies and Alliances of Researchers

Eva Jablonka and Christer Bergsten

Abstract We discuss some examples of direct or indirect involvement of mathematics education researchers in teacher evaluation and curriculum design; and point to hegemonic strategies of persuading sponsors and policy makers how to install "good mathematics teaching". We illustrate how particular research approaches stabilise "good mathematics teaching" by structuring the meaning around interpretations of learning outcomes in the form of measurements, which are taken as symptoms of a range of social phenomena. Students' scores on mathematics tests are interpreted as indicators of their potential to become skilled "knowledge workers", citizens and consumers; teachers' and schools' effectiveness in producing gain scores as indicators of the quality of mathematics teaching for which they can be made accountable; and improvements in national measures as symptoms of innovative capacity that predicts relative competitive advantage. Our concern is the alliances researchers might seek in capitalising on the privileged status of mathematics that relies on the reiteration of those imaginations, in particular in contexts where funding of research favours "findings" that emerge from studies that identify "what works".

Evidence by Means of Mathematicoscience

In some countries attacks on research in education have been followed by the creation of government commissions, institutions, and funding agencies to promote an agenda for increasing "impact" (Levin, 2004). This is to be achieved by educational research that develops curriculum technologies that "work".[1] The discourse

[1] For instance, the UK government has recently recruited proponents of randomised controlled trials for promoting their use in evaluating education and public policy, with the intention to identify

E. Jablonka (✉)
Department of Education and Professional Studies, King's College London, London, UK
e-mail: eva.jablonka@kcl.ac.uk

C. Bergsten
Department of Mathematics, Linköping University, Linköping, Sweden

H. Straehler-Pohl et al. (eds.), *The Disorder of Mathematics Education*,
DOI 10.1007/978-3-319-34006-7_7

constructs "evidence-based" practice as opposed to "slick marketing, misleading demonstrations, word of mouth, tradition, and politics" (Slavin, 2008, p. 124). This reading of "evidence" is affirmative as it ought to produce knowledge in the form of sets of rules for practice, and excludes critical readings. Hence the negativity here is not qualitative research, but analysis. Even though it appears that small-scale action research is also excluded, as often are other meanings of learning outcomes than test scores. This then allows the production of meta-studies based on statistical measures of effect sizes (e.g. Hattie, 2009).

This conception of evidence produces room for new alliances between research and policy. Governments can recruit what looks like a purified expertise of education researchers who produce "scientific evidence" in order to demonstrate accountability, and researchers can promise gains to be achieved in limited amounts of time to make their research worth financing. The "evidence-based knowledge" is not only relevant for rationalising and legitimising political decisions. As it produces sets of rules for pedagogic practice it also lends itself to commercialisation in the form of curriculum packages, textbooks and teaching materials.

At the supranational level, the production of knowledge for education policy has been driven by the OECD (for an insider perspective, see Schuller, 2005). The comparative mathematics assessments conducted by the OECD provide a particular form of "evidence" about mathematics education, which constitutes what counts as a basis for public arguments and political decisions about education (for the latter see Mangez & Hilgers, 2012, who provide an analysis in terms of Bourdieu's fields). This evidence relies on the production of quantitative measurements for student learning.

The examples above might be understood as constituting a hegemonic discourse that Dowling (2009, p. 142) termed "mathematicoscience", a public (educational) discourse to which international comparative studies contribute (such as the IEA's Third International Science and Mathematics Study or the OECD's PISA[2]). This "mathematicoscience" excludes the subjective by establishing a legitimate relation to the empirical (modelled by science) together with a legitimate form of logical argumentation (modelled by mathematics). Public accountability for decisions calls for the elimination of the subjective to avoid the impression of contingency, even though "[n]o one seriously believes that policies are developed, implemented or evaluated by reference to research evidence alone, in some kind of aseptic rationalist bubble" (Schuller, 2005, p. 173).

interventions with large effects for low cost. One priority is "to increase understanding of 'what works' in education" (DfE, 2014). A decade earlier, the U.S. National Research Council issued a report on "Scientific Research in Education" (Shavelson & Towne, 2002) and has set up a clearinghouse for "what works" in education (http://ies.ed.gov/ncee/wwc/).

[2]As to the PISA, Jablonka (2015) observed that mathematics is attributed a privileged position over science in the relation to the empirical and is expanded into a new version of "mathematicoscience". Mathematics is presented as directly formalising the empirical, as "mathematical concepts, procedures, facts and tools to describe, explain and predict phenomena" (OECD-PISA, 2014, p. 17, our emphasis), whilst the science test is only about drawing "evidence-based conclusions about science-related issues" (OECD-PISA, 2014, p. 28, our emphasis).

Student Scores as Indicators

Supranational organisations not only produce mathematicoscience as a particular public (educational) discourse, but also provide arguments for the relevance of mathematics education in developing a nation's "human capital" or "Knowledge Capital" (OECD, 2015; PIAAC Numeracy Expert Group, 2009). The assessments construct the importance of mathematics for the nation state with reference to new forms of labour, including "knowledge work" or "information work", in particular with respect to the innovative potential of information and communication technologies. For example, a recent report presents findings and figures from an analysis of the "collective cognitive skills" or "knowledge capital" of nations and economic growth (OECD, 2015, pp. 25–27). The collective skills are measured by aggregate scores on international mathematics and science tests. The conclusion is these are "by far the most important determinant of a country's economic growth" (p. 26).[3] Notably, the "citizen" who needs mathematics and who is invoked in the comparative achievement tests does not feature in this.

This and similar reports produce teleological explanations for assumed policy trajectories, underpinned by rationales provided by affiliated experts who demonstrate causality. From a structuralist position, Atzmüller (2004) argued that in the "Post-Fordist" context the state seeks to provide the institutional capacities, for which the highly-qualified "knowledge workers" and "symbol analysts" can be activated for a permanently innovative capacity of capitalism. Hence, the business function of innovation becomes the content of state politics. Moves might include the restriction of research funding to apparently more direct contributions to economy (Radder, 2010). The struggle, however, would then be over who decides upon innovation and who claims to produce or possess the means to do so.

Some modes of research in mathematics education can easily be established as more useful than others in this context, in particular if they claim or "prove" to produce direct impact on participation and students' scores by efficient means, for example, by developing curricula that "work". Researchers might also get involved in developing teacher or school effectiveness measures. Student test scores are then employed by a range of actors, not only as indicating a particular level of mathematical skills of individual students,[4] but serving a range of purposes: for public scrutiny of the service provided; for school evaluation or for performance related pay of teachers in relation to "gain scores"; for defining target outcomes; and for diagnosing improvement or decline

[3] Any concerns about reliability and validity, as well as the direction of any causal relationships, "can be satisfactorily answered once skills are correctly measured, and the basic growth relationships can support a detailed analysis of the economic implications of improving a nation's knowledge capital" (OECD, 2015, p. 26). The gains in GDP for all countries included in the study have been "estimated with an improved workforce over GDP with the existing workforce from 2015 until 2095" (p. 48).

[4] Wiliam (2010) points out that the distinction between norm-reference and criterion-reference is not a property of the test results but of the interpretations and inferences drawn.

in the quality of an education system in a country as a symptom of the innovative capacity that leads to relative competitive advantage.[5]

Together with the indicators, rankings and relations with other indicators, the imagination about the reality for which they are taken as symptoms, has to be *instituted*. This imagination would then have to be periodically restaged and updated in reports to ensure visibility. We interpret these as hegemonic articulations, in taking hegemony as a theory of decision taken in an undecidable terrain in line with Laclau and Mouffe (1985). Hegemonic articulation can be achieved by means of temporary fixation of meaning of contested signifiers. Particular key signifiers assume a universal structuring function in the discourse of "good mathematics teaching", while initially open to articulations from different political standpoints. This involves the production of what Laclau (2007) terms an "empty signifier" in a process in which these differences are erased. We are interested in differential positions of mathematics education researchers in terms of their alliances in persuading sponsors and policy makers how to install good mathematics teaching by articulations that stabilise one particular meaning of "good mathematics teaching".

Researchers' Alliances in the Shaping of "Good Mathematics Teaching"

Improving mathematics teaching appears as a unifying agenda of mathematics education researchers. Lerman (2006) suspects that "we all have a sense that there is such a thing as good teaching of mathematics and an appropriate description of the mathematics that should form the mathematical activities of school pupils" (p. 299).[6] If one accepts that there is a discourse about good mathematics teaching that might constitute the identity of mathematics educators, then a search for some temporary fixation of the meaning of "good mathematics teaching" seems appropriate.

In employing Laclau's (2007) notion of hegemony to investigate the discursive space in the terrain of mathematics education, the identity of a community of math-

---

[5] Policy studies have contributed with analyses of the workings of accountability mechanisms that increase bureaucratic control of schooling, which are based on steering through indicators, or "governance through data" (Ozga, 2009) in a range of modes and settings, including national examinations, literacy and numeracy tests, school inspection policies, evaluation of graduate programmes, and university rankings (Ball & Goodson, 2015, Perryman, Ball, Maguire, & Braun, 2011). Piattoeva (2015), refers to "elastic numbers" in her analysis of the use of the results of the Unified State Exam in Russia, the different readings of which lead to merging the official and the popular. While accountability suggests the efficiency of state institutions can be ensured through monitoring by members of society, the evaluation is in the end produced by officials.

[6] In educational research journals, one finds, for example, articles entitled *Synthesis of research on good teaching* (Porter & Brophy, 1988), *The good teacher and good teaching* (Murphy, Delli, & Edwards, 2004), *In search of the essence of a good teacher* (Korthagen, 2004).

ematics educators who aim at good mathematics teaching could only be brought about by a process of *purification* of such a thing as "good mathematics teaching", which is separated from some forms of teaching that are "not good". Simply listing instances of this negativity cannot establish the specificity of any of them as an incarnation of "bad mathematics teaching" and all other negativities implied by it. Instances of "bad mathematics teaching", which one typically finds mentioned, include: too much focus on increasing exam scores (teaching to the test); low national ranks in international tests; not making sufficient use of ICT; stressing procedural fluency to the detriment of understanding; not enough room for students' own thinking and argumentation; lack of connections with students' real life; and exclusion of students of a particular social category.

Different concrete manifestations of pursuing research that works towards overcoming any particular (or combinations) of these diverse shortcomings, would need to be established as equivalent for creating an identity of a group that works towards "good mathematics teaching". The equivalence between these differential (ambivalent) struggles cannot be signified in a direct way as they operate in pointing to an *absence* (of all types of good teaching, and the negative consequences of this absence). The longer the chain of equivalence, the less concrete in terms of one particular identity the "good teaching" will be. This process amounts to the production of an "empty signifier", which signifies a lack of a universality: "Attempts to represent the system as such, in opposition to the negative outside, involve a privileging of the similar or equivalent aspect of an element. The empty signifier is this element whose difference from the rest of the system is tendentially erased in order to represent the system as such" (Rebello, 2008, p. 9). This is the condition for a hegemonic operation, in which one particular identity may come to signify the universality of "good mathematics teaching".[7] One example of such identity, which has been widely distributed in US and European mathematics education literature and development projects is "Inquiry Based Mathematics Education" (see Dorier & Maaß, 2013).

In terms of Laclau's (2007) analysis, all of the concrete manifestations of pursuing a research agenda that works towards good mathematics teaching, are formally equally positioned as incarnating the struggle in overcoming a range of differential constraints. But in "reality" they are not: "Not any position in society, not any struggle is equally capable of transforming its own contents in a nodal point that becomes an empty signifier" (Laclau, 2007, p. 43). We have chosen three diverse examples of projects that seek to install good mathematics teaching, in which researchers are involved as supporters by means of providing commentary, as authors of series of connected studies in research programmes, and as leaders of research projects that aim at curriculum development.

[7] The most obvious example of a hegemonic struggle over good mathematics teaching is what has become known as the "math wars" in the USA.

Example 1: "Good Teaching" as "Effective Teaching": Teacher Knowledge in Relation to Student Outcomes

Scales based on ratings of a variety of aspects of teacher performance are increasingly used in the USA for a range of purposes. These include formative teacher assessment, evaluation of curriculum policy, and professional development (Hill et al., 2012). One such scale is the Mathematical Quality of Instruction (MQI) score that aims at "independent estimates of the mathematical quality and the pedagogical quality of instruction" (Learning Mathematics for Teaching Project, 2011, p. 27). The separation between mathematics and pedagogy, however, does not seem to be easy to sustain. It turns out that the "conceptualization is deeply disciplinary, but coordinates mathematical and pedagogical perspectives" (p. 31). MQI scores have been found to correlate with quantitative measurements of the construct of Mathematical Knowledge for Teaching (MKT), which then is the appropriate knowledge for being able to maintain a "high mathematical quality of instruction" (pp. 42–43).[8]

The MQI scales, each with several codes marked as present/absent and appropriate/inappropriate, comprise *Richness and development of the mathematics*, *Responding to students*, *Connecting classroom practice to mathematics*, *Language*, *Equity*, and *Presence of unmitigated mathematical errors*. This framework is an ad hoc construction of categories, as opposed to being informed by an analytical framework. The criteria for the code marking, which is done by observing video recordings of classrooms, are not discursively available. Hence they require expert rating. This amounts to creating what Burke, Jablonka and Olley (2014) termed an "originative mathematisation": The principles for the identification of appropriate instances for each category are not discursively available, neither is a theory for the construction of the scale. In cases when the expert who does the rating is not a researcher of the Learning Mathematics for Teaching Project team, introducing an apprentice into this marking practice is problematic: "We suspect that 'training' observers in the knowledge it takes to code these tapes is likely to be of limited use; instead, observers with strong knowledge in both these arenas must be found" (Learning Mathematics for Teaching Project, 2011, p. 31).[9] What the observers engage in is the recontextualisation of teaching from a practice with largely implicit criteria for the good teacher. The performance criteria could possibly be acquired through observing a master teacher, which would be one who gets full score on the MQI scale. As to the political agenda of the team, equity is incorporated into the criteria as "providing all students

[8] The construct of MKT (and similar formulations), conceptualised within the discourse of mathematicoscience, has an inbuilt potential of further differentiation into "sub-knowledges", which might lead to the attraction of additional research resources. Thus, in Herbst and Kosko (2012) an instrument to measure mathematical knowledge for teaching high school geometry is developed, and Subramaniam (2014) investigates prospective teachers' pedagogical knowledge for teaching the estimation of length measurements (see also Thanheiser and Browning, 2014).

[9] For quantifying the MQI score, in the study by Blazar (2015) "two certified and trained raters watched each lesson and scored teachers' instruction on 13 items for each seven-and-a-half minute segment on a scale from Low(1) to High (3)" (p. 19).

access to the mathematics" (Hill et al., 2008, p. 446) but has not yet been quantitatively graded in the form of levels, in contrast to the other categories.

Even though measures of instructional quality originate in the idea that students' scores on mathematics tests are an inappropriate measure of the quality of teaching and hence classroom teaching needs to be looked at, correlations with some measures of student outcomes are often incorporated in studies that use such measures, or indeed are used as providing an argument for their validity. In discussing the MQI measure, the authors write that "it is unclear whether and how well the various elements of this instrument correlate with student outcomes" (Learning Mathematics for Teaching Project, 2011, p. 44). In a recent attempt to implement such "validation" of the MQI instrument, along with other predictors for student achievement, Blazar (2015) took on the "challenge" to present evidence of "causal inferences" (p. 16) between MQI scores and measures of student achievement.

Another similar example is found in Germany, where a quantitative study with large random samples measured secondary mathematics teachers' content knowledge (CK, with a paper-and-pencil test), pedagogical content knowledge (PCK, tests administered in interviews), and instructional quality (see below) to be compared with students' achievement gains from Grade 9 to Grade 10 (PISA data and a standardised test), mental ability and social background data (Baumert et al., 2010). This project (COACTIV) set out to investigate the relation between content knowledge and pedagogical content knowledge and their relative impact on student outcomes. The conclusion is that "the COACTIV group has succeeded in distinguishing the CK and PCK of secondary mathematics teachers conceptually and empirically" (p. 166), and that it is pedagogical content knowledge that "makes the greatest contribution to explaining student progress" (p. 167), while "deficits in CK are to the detriment of PCK, limiting the scope for its development" (p. 167). In this case, the signifier extensively used in their writings is "high-quality instruction", which is measured by the dimensions *cognitively activating learning opportunities* (analysis of curricular and cognitive levels of tasks), *individual learning support* (student ratings) and *classroom management* (teacher and student perceptions), based on the assumption that "higher instructional quality should be reflected in higher student learning progress" (p. 146). The study includes constructs of socio-economic hierarchies in the form of regression models, and argues, based on mathematicoscientific evidence, for high-quality instruction in all tracks in German schools to be achieved by a focus on teachers' pedagogical content knowledge.[10]

Example 2: Value-Added Modelling

Quantitative measures of teaching are most prominently used in the USA. One example, in use since the early 1990s, is value-added modelling (VAM). In this statistical apparatus teaching *as such* is essentialised as a quantifiable teacher

[10] This does certainly not assist any argument against tracking.

attribute (their "value") that exists to a variable degree in teachers, who are producing gain scores in students, conceptualised in terms of profiles with differing quantifiable characteristics (such as gender, ethnicity, and language) that amount to differences in speed when their teachers steer them through a curriculum that leads to their achievement gains on standardised tests. The students' characteristics can be accounted for as noise in the model in order to get to the essence of the teacher "value". This is then their "unique" contribution. The outcome in turn contributes to funding in the "Race to the Top" (U.S. Department of Education, n.d.).

When looking at Table 3 in Wei, Hembry, Murphy and McBride (2012, p. 14ff), one sees that different more or less complex versions of such models amount to very different rankings of the same mathematics teachers. Hill, Kapitula and Umland (2011, p. 826), amongst others, offer some internal critique, but appear to sympathise with the form of the accountability procedure:

> Although we do recommend the use of value-added scores in combination with discriminating observation systems, evidence presented here suggests that value-added scores alone are not sufficient to identify teachers for reward, remediation, or removal.

These procedures establish accountability relations, which include both standards and standardised procedures for monitoring the standards that are defined and developed without involvement of the party made accountable (and becomes punished or rewarded). Researchers are obviously involved in developing the procedures for monitoring the standards and funded by those who purchase their research.

Interestingly, critique of this mathematicoscience also comes from mathematicians. For example, Ewing (2011), after describing VAM as an example of employing mathematics "as a rhetorical weapon" and being "heavily promoted with unbridled and uncritical enthusiasm by the press, by politicians, and even by (some) educational experts" (p. 667), concludes after an analysis of such models (referring to a high stake report promoting VAM):

> Why must we use value-added even with its imperfections? Aside from making the unsupported claim [...] that "it predicts more about what students will learn...than any other source of information," the only apparent reason for its superiority is that value-added is based on data. Here is mathematical intimidation in its purest form—in this case, in the hands of economists, sociologists, and education policy experts. (p. 672)

In a study led by an economist, VAM ratings had a similar relationship (contribution to variance) to gain scores in reading and mathematics as to changes in student height (Bitler, Corcoran, Domina, & Penner, 2014). In spite of these critiques by experts in quantitative methodology, the hegemonic constitution of "evidence" as mathematicoscience for rationalising reward, remediation or removal of teachers, and "good teaching" as gain scores is maintained because of VAM's alliance with accountability regimes.

Example 3: The Ultimate Curriculum: "The Biggest Bang for the Buck"

The increased amount of studies of *instructional effectiveness* of different teaching approaches by means of randomised controlled trials mark a comeback of experimentalism, as mentioned in the introduction. Experimental curriculum development studies occasionally include classroom observations in order to check the fidelity of the teachers' "dispending" of the intervention (the treatment), or to complement measurement of gain scores with scores from classroom observations (e.g. Clements, Sarama, Spitler, Lange, & Wolfe, 2011; Ross & Bruce, 2007). Classroom practice is only relevant in relation to the statistical regularity the black box produces as its student achievement outcomes. The role of the teacher is then to function as a technician of high fidelity curriculum delivery.

We look at an example from the USA (again), which included a large scale study that used a cluster randomised trial with data from 1305 pre-school children for the evaluation of a curriculum based on mathematics learning trajectories (Clements et al., 2011). Statistical analyses employing hierarchical linear modelling were used on both participant and school levels, based on instruments displaying high reliability measures, to investigate the effectiveness of the curriculum, which was "evidenced" by a substantial effect size: the children in the treatment group "outperformed those in the control group on the total mathematics test score, with an effect size of 0.72" (Clements et al., 2011, p. 153).

At the outset of the experiment, one concern was that "children from low-resource communities who are members of linguistic and ethnic minority groups demonstrate significantly lower levels of achievement than children from higher-resource, nonminority communities" (p. 128), making the case for providing these groups with high-quality educational experiences in school they do not receive at home. Indeed, no significant interactions were found between the treatment and socio-economic status, limited English proficiency, gender, individualised education plan status or ethnicity, apart from that African Americans "averaged higher gains than other children" in the treatment group but lower in the control group (p. 145).

In this project, development of teaching quality (in recruiting William James as authority) is constituted as based on "the science of learning and instruction", which "continues to lay down increasingly specific and useful guidelines" (p. 158). Curriculum is then a technology derived from this science involving teachers who will never have full access to it.[11] This approach is opposed to "focusing primarily on teachers' autonomously inventing individual curricula", or "idiosyncratic 'creativity' that does not build on extant science, and learning and instruction is less likely to serve either the profession or the classroom's students" (p. 158). In the randomised controlled trials part of the study of this curriculum, the teachers are

[11] In the project, teachers received 13 days of professional development and then got access to a website with examples for teaching activities and some explanations.

objects to the same exact science (of teaching) that measures "fidelity of implementation" and other variables, including "teacher personal attributes". The project is based in the assumption of relatively unskilled teachers, "especially those in early childhood, [who] have limited time and knowledge of mathematics and mathematics education research (Sarama, 2002; Sarama & DiBiase, 2004) required to plan, research, and write truly research-based curricula (as defined in Clements, 2007)" (Clements et al., 2011, p. 158). The production of the curriculum material is then perpetuating the de-skilling of teachers. While the approach is argued by its efficiency in expanding the pool of future "knowledge workers" through early intervention, the resulting curriculum technology amounts to a Taylorisation of the "knowledge work" of the teachers.

The importance of early-years intervention is argued from a "human capital perspective", as an area where "theoretically grounded and empirically grounded curriculum-based interventions in early childhood may constitute an efficacious and cost-effective route to raising achievement in low-resource communities" (pp. 128–129). While only expensive programmes designed for this level have proven to be effective, curriculum interventions tend to produce higher effect sizes and thus may "get the biggest bang for the buck" (p. 128). By developing a research based curriculum and proving its effectiveness by "evidence" provided by large-scale randomised controlled trials, researchers can enter the growing educational industry.

Reflections

Under the label "school effectiveness", the gauging of students, teachers, and schools has been travelling across ideologically diverse political regimes since at least the 1960s in the UK and the USA, and has been exported to other countries (e.g. Morley & Rassol, 1999). While early school effectiveness studies attempted to show that schools do make a difference, Rea and Weiner (1998) suggested that "university departments or centres advocating ESM [Effective Schools Movement] should be recognized formally as 'think-tanks' for policy makers, rather than independent research centres" (p. 23). More recent revivals emerged in the political contexts described in our introductory section. Researchers who develop quantitative measures of learning outcomes then become allies in a process of increasing bureaucratisation, while alternative modes of classroom research are at risk of being "written off for their alleged 'ideological posturing'" (Howe, 2004, p. 57).

The Mathematical Quality of Instruction scale mentions equity as a feature of good teaching that experts (only) can identify, but the category is omitted in the quantitative studies. In the value-added models, students feature as intake and social or cultural hierarchies are incorporated in terms of available categories that work for a quantification, which can be accounted for by statistics in order to get to the essence of the good teacher in the form of a measure that predicts good mathematics teaching. The German project that researched the relation between content knowledge and pedagogical content knowledge employs a similar strategy. In the teacher-proof early-years curriculum, equity for children from "low-resource communities who are members of linguistic and ethnic minority groups" is quantified in relation to what the majority students learn. Equity is then defined in relation to established social categories and its meaning becomes fixed in relation to gaps in student scores.

In the diverse examples we have discussed, with different chains of significations involved, the particular negativity to be overcome for realising good mathematics teaching is *unsatisfactory student learning outcomes*, the meaning of which becomes temporarily fixed in the form of quantitative measurements. For instance, a priori observational categories for the quality of mathematics teaching, converted into a scale for individual teachers, are linked to student scores for justification of the model. Student scores are used to differentiate between important types of teacher knowledge, and for proving the efficiency (in terms of time and money) of a curriculum scheme. Learning outcomes, in the form of student scores, then assume a central structuring function in the particular discursive space mathematics education researchers occupy related to the practice of mathematics teaching that we have investigated. These scores have come to signify a universality of "good mathematics teaching" embedded in a logic of outdoing. Students need higher scores for obtaining better positions in the "knowledge economy"; nations aim at improving national measures as symptoms of their innovative capacity that leads to relative competitive advantage; teachers and schools compete in increasing their value-added measures.

These learning outcomes then bind together a range of previously established meanings in different institutional terrains, articulated around nationalism, an imaginary efficiency that symbolically converts student gain scores into monetary value, accountability and equity. This is afforded by means of a particular conception of educational research. In our examples, researchers involved in the projects conduce to rationalising decisions and interpretations by means of mathematicoscience in line with policies that help constituting mathematicoscience as a dominant discourse in terms of both a privileged competence for the labour-force and a mode of producing "evidence" that assists in strengthening what Espeland and Stevens (2008) termed "new regimes of measurement". This analysis points to a self-stabilising system in constituting the meaning of "good mathematics teaching".

Gains in student scores as a nodal point for all interpretations of "good mathematics teaching" that produce truths about how it can be installed, however, do not (yet) constitute the identity of mathematics education research. There are certainly competing discourses that aim at producing alternative realities with more complex articulations. Unfixing the meaning of test scores signifying good mathematics teaching can be practised in day-to-day interactions with teachers, teacher education students, and in alliances between interest groups within mathematics education research.

Acknowledgments We would like to acknowledge the helpful critical comments to earlier versions of this paper by Nina Bohlmann, Tony Brown and the other participants at the DOME–Disorder of Mathematics Education meeting held at Freie Universität Berlin, 15–17 January 2015.

References

Atzmüller, R. (2004). Arbeit an der Veränderung—Überlegungen zur Staatstheorie im Postfordismus. *Grundrisse—Zeitschrift für Linke Theorie & Debatte, 12*. Retrieved November 22, 2015, from http://www.grundrisse.net/grundrisse12/12roland_atzmueller.htm.

Ball, S., & Goodson, I. (Eds.) (2015). Governing by numbers [Special Issue]. *Journal of Education Policy, 30*(3).

Baumert, J., Kunter, M., Blum, W., Brunner, M., …, Tsai, Y. -M. (2010). Teachers' mathematical knowledge, cognitive activation in the classroom, and student progress. *American Educational Research Journal, 47*(1), 133–180.

Bitler, M., Corcoran, S., Domina, T., & Penner, E. (2014). Teacher effects on student achievement and height: A cautionary tale. Paper presented at the Spring Conference of The Society for Research on Educational Effectiveness, March 6–8, 2014, Washington D.C. Retrieved November 22, 2015, from http://www.sree.org/conferences/2014s/program/downloads/abstracts/1137.pdf.

Blazar, D. (2015). Effective teaching in elementary mathematics: Identifying classroom practices that support student achievement. *Economics of Education Review, 48*, 16–29.

Burke, J., Jablonka, E., & Olley, C. (2014). Mathematical modelling: Providing valid description or lost in translation. In P. Barmby (Ed.), *Proceedings of the British Society for Research into Learning Mathematics 34.1* (pp. 31–46). Durham: BSRLM.

Clements, D. H., Sarama, J., Spitler, M. E., Lange, A. A., & Wolfe, C. B. (2011). Mathematics learned by young children in an intervention based on learning trajectories: A large-scale Cluster Randomized Trial. *Journal for Research in Mathematics Education, 42*(2), 127–166.

Department for Education. (2014, March). Assessment, curriculum and qualifications. Research priorities and questions. Retrieved November 22, 2015, from https://www.gov.uk/government/publications/assessment-curriculum-and-qualifications-research-priorities-and-questions.

Dorier, J., & Maaß, K. (2013). Inquiry based mathematics education. In S. Lerman (Ed.), *Encyclopedia of mathematics education*. Springer Reference (www.springerreference.com). Berlin, Heidelberg: Springer.

Dowling, P. (2009). *Sociology as method: Departures from the forensics of culture, text and knowledge*. Rotterdam: Sense Publishers.

Espeland, W. N., & Stevens, M. L. (2008). A sociology of quantification. *European Journal of Sociology, 49*(3), 401–436.

Ewing, J. (2011). Mathematical intimidation: Driven by the data. *Notices of the American Mathematical Society, 58*(5), 667–673.

Hattie, J. (2009). *Visible learning*. London: Routledge.

Herbst, P., & Kosko, K. (2012). Mathematical knowledge for teaching high school geometry. In L. R. Van Zoest, J.-J. Lo, & J. L. Kratky (Eds.), *Proceedings of the 34th Annual Meeting of the North American Chapter of the International Group for the Psychology of Mathematics Education* (pp. 438–444). Kalamazoo, MI: Western Michigan University.

Hill, H. C., Blunk, M., Charalambous, C., Lewis, J., Phelps, G. C., Sleep, L., et al. (2008). Mathematical knowledge for teaching and the mathematical quality of instruction: An exploratory study. *Cognition and Instruction, 26*, 430–511.

Hill, H. C., Charalambous, C. Y., Blazar, D., McGinn, D., Kraft, M. A., Beisiegel, M., et al. (2012). Validating arguments for observational instruments: Attending to multiple sources of variation. *Educational Assessment, 17*, 1–19.

Hill, H. C., Kapitula, L., & Umland, K. (2011). A validity argument approach to evaluating teacher value-added scores. *American Educational Research Journal, 48*(3), 794–831.

Howe, K. R. (2004). A critique of experimentalism. *Qualitative Inquiry, 10*(1), 42–61.

Jablonka, E. (2015). The evolvement of numeracy and mathematical literacy curricula and the construction of hierarchies of numerate or mathematically literate subjects. *ZDM Mathematics Education, 47*(4), 599–609.

Korthagen, F. A. J. (2004). In search of the essence of a good teacher: Towards a more holistic approach in teacher education. *Teaching and Teacher Education, 20*, 77–97.

Laclau, E. (2007). Why are empty signifiers important for political studies? In E. Laclau (Ed.), *Emancipation(s)* (pp. 37–49). London: Verso.

Laclau, E., & Mouffe, C. (1985/2014). *Hegemony and socialist strategy: Towards a radical democratic politics*. London: Verso.

Learning Mathematics for Teaching Project. (2011). Measuring the mathematical quality of instruction. *Journal of Mathematics Teacher Education, 14*, 25–47.

Lerman, S. (2006). Review of "New teacher identity and regulative government: The discursive formation of primary mathematics teacher education." Tony Brown & Olwen McNamara 2005. New York: Springer. *Journal of Mathematics Teacher Education, 9*, 299–305.

Levin, B. (2004). Making research matter more. *Education Policy Analysis Archives, 12*(56), 1–20.

Mangez, E., & Hilgers, M. (2012). The field of knowledge and the policy field in education: PISA and the production of knowledge for policy. *European Educational Research Journal, 11*(2), 189–205.

Morley, L., & Rassol, N. (1999). *School effectiveness. Fracturing the discourse*. London: Falmer Press.

Murphy, P. K., Delli, L. A. M., & Edwards, M. N. (2004). The good teacher and good teaching: Comparing beliefs of second-grade students, preservice teachers, and inservice teachers. *The Journal of Experimental Education, 72*(2), 69–92.

OECD. (2015). *Universal basic skills: What countries stand to gain*. Paris: OECD Publishing. Retrieved November 22, 2015, from http://dx.doi.org/10.1787/9789264234833-en.

OECD-PISA. (2014). PISA 2012 results: What students know and can do—student performance in mathematics, Reading and science (Vol. I, revised ed., February 2014). PISA, OECD Publishing. doi: 10.1787/9789264201118-en.

Ozga, J. (2009). Governing education through data in England: From regulation to self-evaluation. *Journal of Education Policy, 24*(2), 149–162.

Perryman, J., Ball, S., Maguire, M., & Braun, A. (2011). Life in the pressure cooker—school league tables and English and Mathematics teachers' responses to accountability in a results-driven era. *British Journal of Educational Studies, 59*(2), 179–195.

PIAAC Numeracy Expert Group. (2009). PIAAC Numeracy: A conceptual framework. OECD Education, Working Paper no. 35. Retrieved November 22, 2015, from http://www.oecd.org/officialdocuments.

Piattoeva, N. (2015). Elastic numbers: National examinations data as a technology of government. *Journal of Education Policy: Special Issue: Governing by Numbers, 30*(3), 316–334.

Porter, A. C., & Brophy, J. (1988). Synthesis of research on good teaching: Insights from the work of the Institute for Research on Teaching. *Educational Leadership (May 1988)*, 74–85.

Radder, H. (Ed.). (2010). *The commodification of academic research: Analyses assessments, alternatives*. Pittsburgh: University of Pittsburgh Press.

Rea, J., & Weiner, G. (1998). Cultures of blame and redemption—when empowerment becomes control: Practitioners' views of the effective schools movement. In R. Slee, S. Tomlinson, & G. Weiner (Eds.), *School Effectiveness for Whom? Challenges to the school effectiveness and school improvement movements* (pp. 21–32). London: Falmer Press.

Rebello, J. T. (2008). (Why) Do empty signifiers matter to political economy? Hegemony, non-commodity money and part one of *Capital*. New School—Umass Graduate Workshop in Economics, October 25–26, 2008.

Ross, J., & Bruce, C. (2007). Professional development effects on teacher efficacy: Results of randomized field trial. *The Journal of Educational Research, 101*(1), 50–60.

Schuller, T. (2005). Constructing international policy research: The role of CERI/OECD. *European Educational Research Journal, 4*(3), 170–180.

Shavelson, R., & Towne, L. (Eds.). (2002). *Scientific research in education*. Washington: National Academy Press.

Slavin, R. E. (2008). Evidence-based reform in education: What will it take? *European Educational Research Journal, 7*(1), 124–128.

Subramaniam, K. (2014). Prospective secondary mathematics teachers' pedagogical knowledge for teaching the estimation of length measurements. *Journal of Mathematics Teacher Education, 17*(2), 177–198.

Thanheiser, E., & Browning, C. (Eds.). (2014). The Mathematical Content Knowledge of elementary prospective teachers [Special issue]. *The Mathematics Enthusiast, 11*(2).
U.S. Department of Education. (n.d.). Programs: Race to the Top Fund. Retrieved November 22, 2015, from http://www2.ed.gov/programs/racetothetop/index.html.
Wei, H., Hembry, T., Murphy, D., & McBride, Y. (2012). Value-added models in the evaluation of teacher effectiveness: A comparison of models and outcomes. Research Report, Pearson Education. Retrieved November 22, 2015, from http://images.pearsonassessments.com/images/tmrs/ComparisonofValue-AddedModelsandOutcomes.pdf.
Wiliam, D. (2010). Standardized testing and school accountability. *Educational Psychologist, 45*(2), 107–122.

Chapter 8
Communicating Research in Mathematics Education: Theoretical and Ethical Problems

Candia Morgan

Abstract Connections between research, policy and practice are often problematic as politicians and practitioners bemoan the irrelevance of research, while researchers complain that their work is misunderstood. In this chapter, the movement of research from the field of research into other fields is understood through the lens of Bernstein's notion of recontextualisation. Examples are given from a recent project investigating changes in mathematics examinations in England, illustrating how research results may be incorporated into alternative discourses to support pre-existing positions and values. This raises questions for ethical researchers.

Introduction

Educational research is often criticised for failing to provide results that can be used to inform practice, and educational researchers are criticised for failing to communicate with practitioners and policy makers. Measures of the "impact" of research outside the academic domain now play an important role in the evaluation and funding of research in the UK and elsewhere in the world. Apart from the need to meet the expectations of our employers, partaking in the ritual of research identified by Lundin and Christensen (in this volume), there are strong ethical and political arguments to be made for educational researchers to pay serious attention to how their work relates to the experiences and practices of students and teachers. However, my concern in this chapter is not so much with whether research is communicated "effectively" and can inform "evidence-based practice" (Slavin, 2002) as with the nature of the communication, with the ways in which research outcomes are transformed as they move from the field of research into the fields of policy and practice and with the possible consequences of such transformation.

C. Morgan (✉)
Department of Curriculum Pedagogy and Assessment,
University College London Institute of Education, London, UK

Institute of Education, University College London, London, UK
e-mail: candia.morgan@ucl.ac.uk

H. Straehler-Pohl et al. (eds.), *The Disorder of Mathematics Education*,
DOI 10.1007/978-3-319-34006-7_8

This concern is motivated by my own recent experience, observing the ways in which researchers, teachers and other practitioners have responded to my attempts to communicate research outcomes and reflecting on how various groups may interpret these outcomes. As I wrote the previous sentence, I hesitated over the word *interpret*, wondering whether I wanted to write *misinterpret*. This hesitation is at the centre of the problem I address in this chapter. The outcomes of my research and my own interpretation of these outcomes are the product of my intellectual work and that of my co-workers.[1] In fact, every stage of the research process was shaped by our theoretical assumptions and decisions based on these assumptions, whether these were consciously acknowledged and openly stated or not: the choice of research focus; statement of research questions; choices of data, methods of data collection and analysis; identification and description of outcomes; choices about which outcomes to report; decisions about how to report and to whom. Can I therefore claim ownership of the outcomes and the right to say that my interpretation is the only correct interpretation? However strongly I may feel that I "own" the research and that my interpretation is correct, there are two reasons why the answer to this question has to be no.

The first reason is rooted in my understanding of the nature of communication, semiosis and discourse. Meaning does not reside in words, images or physical objects. It is produced in social practice—in the ways that words, images and objects are brought into being and used by people in order to perform functions within a specific practice. As outcomes of research are disseminated (by whatever means of communication) beyond the particular practice within which they were produced, the kinds of functions they may perform change. Moreover, as they are used, they come into relation with the range of semiotic resources already in use within the new practice, enabling the production of new kinds of uses—new meanings. The interpretation of research outcomes is thus always relative to the practice within which they are functioning. In order to think about this interpretation, I make use of Bernstein's notion of recontextualisation, as developed by Morgan et al. (2002), to analyse and understand how the outcomes and discourse of educational research are transformed as they move from the field of research into the field of policy and the field of practice. Bernstein (2000, p. 32) contends that, as a discourse (in this case the discourse of educational research) moves from one site to another, a space is created "in which ideology can play". This play of ideology transforms the original discourse into a new discourse. The process of transformation involves selective appropriation from the original discourse, refocusing that which is appropriated and relating it to elements drawn from other discourses. The way in which this occurs is ideological in that it varies according to the interests of the agents in the recontextualising field. We must therefore expect that when the outcomes of research are disseminated into the fields of policy and practice they will be transformed. A challenge for researchers, which I return to later in this chapter, is to anticipate the nature of this transformation and to "manage" it so that the uses made of the research

[1] The research discussed in this chapter was undertaken together with Anna Sfard and Sarah Tang. The views expressed here, however, are entirely my own responsibility.

as it moves into new fields are as consistent as possible with the ethical and political principles of the researchers, including ethical consideration of the relationships among the various research participants and other interested groups as discussed by Wagner (in this volume).

The second reason why I cannot claim a right of ownership of "my" research is an economic reason. I referred above to research outcomes as the product of intellectual work. As an employee of a university in the UK, I sell my labour to my employer and, indirectly, to the UK state, which funds much of the research activity of the university. Of course, direct analogies between material and intellectual labour are problematic. As researchers, we may not experience alienation from the products of our labour in the same way as the factory worker, who sees her product taken from her and sold for the profit of the capitalist. In Marxist economic terms we are "unproductive" workers in that we do not directly produce surplus value (Draper, 1978, p. 490). Nevertheless, it would be naïve to imagine that our employers and the state have no interest in the research we produce and in the uses to which it is put. A direct interest of the university employers is in the exchange of "quality" and "impact" for further research funding and recruitment of students. More indirectly, the state has an interest in research findings in so far as these serve to support their exercise of government.

Of course, not all the products of intellectual labour directly support the state and its governance. Indeed, the work of researchers represented in this volume and of many others concerned with sociopolitical aspects of education is often deeply critical of systems and policies of government. However, in a state in which researchers have the so-called academic freedom, the state's interest in research outcomes is served not by direct suppression of opposition but by two other strategies: dismissing it by construing it as irrelevant or inadequate; or recruiting it through a process of recontextualisation.[2] The first of these strategies may be seen, for example, in the valorisation of randomised controlled trials (RCTs) as a means of providing evidence of "what works". For example, the UK Secretary of State for Education stated in 2013:

> We need more hard evidence in the education debate. [...] Randomised controlled trials offer us the opportunity to establish which policies genuinely help children. I am delighted the DfE is embracing a more rigorous approach towards evidence.[3]

Forms of research that address aspects of education that are not susceptible to the kinds of measurement required by RCTs or that result in more contextualised and nuanced outcomes are construed in this statement as lacking rigour, providing inadequate "soft" evidence, not "genuinely" helping children. This valorisation of RCTs has even been materialised in some cases by restriction of funding for other forms

[2] Of course, in societies with separation of state, academic and commercial interests, critical research may well gain a voice through publication. This creates opportunities for development of alternative discourses drawing on such research in ways that may compete with official discourses.

[3] https://www.gov.uk/government/news/new-randomised-controlled-trials-will-drive-forward-evidence-based-research, Accessed 11 May 2015.

of research. For example, a major educational foundation in the UK commissions large-scale research investigating the effectiveness of educational innovations, with the condition that the research should involve RCT.

The second strategy, the recruitment of research to the interests of the state, is the main focus of this chapter. I will explore issues related to the recontextualisation of educational research through discussing examples taken from a recent research project. "The Evolution of the Discourse of School Mathematics"[4] (EDSM) addressed the question of how expectations about student participation in mathematical activity have changed over time in England. In this project we used high stakes examinations from the end of compulsory schooling—the General Certificate of Secondary Education (GCSE), taken by almost all students at age 16—as our lens onto the expectations of the curriculum, taking a discourse analytical approach to interrogate the nature of the mathematical activity expected of students. From the beginning we were conscious of a political dimension to this research. In England, as in many countries, the curriculum and examinations are regulated by the state and performance in assessments in mathematics are used as one of the key measures of the success (or failure) of the education system. In particular, we were aware that change in examinations over time is likely to be construed by many audiences in terms of a discourse about "standards"—a discourse that is found in policy debates and in public media (Sfard, 2009), which is used as a means of justifying changes to curriculum, assessment and teachers' working conditions.

In the proposal to the UK government funded Economic and Social Research Council we attempted to distance ourselves from this discourse, writing:

> We argue that the analysis of change produced by this approach will provide insight into how changes in curriculum and assessment may affect students' mathematical learning. We do not seek to address the complex question of comparability of standards over time, but to consider at a detailed level the ways in which students' mathematical learning may have changed.

The aim of the research was thus to produce a description. The description in itself may seek (or claim) to be free of value judgements, but of course descriptions may be used in a variety of ways and users (including the researchers themselves) ascribe their own values to the terms of the description. Indeed, it could be argued that the notion of value-free description is impossible, given that the writing and reading of any descriptive text necessarily take place within some social practice that ascribes its own values to the objects, processes and qualities present in the text. In the case of the EDSM project, the values of the researchers were not explicitly stated in the original proposal but may be seen to include implicit valuing of coherence of curriculum and assessment with some unarticulated notion of "quality" of mathematics and student mathematical experience. Vagueness in the presentation of researcher values may be seen as a gambit to enhance the possibility of gaining funding to undertake the research; at the same time it opens up greater opportunities for readers to interpolate their own values.

[4] Funded by the UK Economic and Social Research Council, reference: ES/1007911/1.

The ascription of unintended values to research outcomes is not a new problem and is not confined to the transformation of research descriptions as they move into fields outside the university. Paul Dowling, speaking to a group of research students, deliberately misquoted Marx; my recollection of his words goes something like: "sociological research does not try to change the world; the point is to describe it".[5] Dowling's research aim has always been to create languages of description (Dowling, 2009), yet, when applied to the empirical field, the resulting descriptions are inevitably interpreted, and used by many if not most readers, including readers within the mathematics education research field, in value-laden ways. The imputation of values to descriptions introduces a strong potential to motivate action and consequent changes to the world. For example, his most widely known work in mathematics education produced descriptions of textbooks designed for higher and lower attaining pupils (Dowling, 1998). Part of his description focused on the identities and trajectories projected for students in the two types of texts. Using the descriptive terms "apprenticed" and "dependent", Dowling concludes that the strategies used in the books for higher attaining students projected middle class identities (envisaging future professional employment and other middle class occupations) and "apprenticed" students into mathematical practices, positioning them as potentially able to engage as independent subjects in mathematical practices. On the other hand, the books for lower attaining students projected working class identities (envisaging future manual employment) and construed their positioning with respect to mathematics as "dependent", lacking access to the regulative principles of mathematical activity.[6] While for Dowling this language may function simply as description, in other discursive practices, including those common within mathematics education and mathematics education research, the terms "apprenticed" and "dependent" are value laden and are likely to be given connotations beyond those explicitly given by Dowling's definitions. The notion of access to higher levels of mathematics is also value laden: access is positively valued in contrast to restriction or denial; dependence contrasts with the valuing of independence as a goal of a liberal/progressive[7] mode of pedagogy; while mathematics in general and higher mathematics in particular are generally unquestioned positive goods (though see Pais & Valero, 2014). As users of Dowling's research interpolate such values, they transform the description into arguments for change. For example, Heggarty and Pepin's (2002) study of textbooks in three countries draws on Dowling's work as evidence of the stereotyping of different social groups in schools in England, incorporating this into a wider critique of current practice and culminating in recommendations for change, suggesting that textbooks in England should "begin to embrace

[5] Marx, Theses on Feuerbach: "Philosophers have hitherto only interpreted the world in various ways; the point is to change it."

[6] It is of course very possible that Paul Dowling would contest the account of his research that I am presenting here. I am using his research for a new purpose, relating it to my own theoretical perspective and making it serve the interests of the arguments presented in this chapter.

[7] "Liberal/progressive" in the sense used by Bernstein (2000) and Lerman and Tsatsaroni (1998).

the richer view of mathematics and its learning which takes account of children as makers of knowledge and not as receivers of that knowledge"(p. 588).

In the examples from the EDSM project that follow, I identify and discuss aspects of the description of mathematics examinations resulting from our research that have shown themselves to be susceptible to recontextualisation. In particular, I focus on how these lend themselves to recruitment by existing official policy discourses.

Examples from the EDSM Project

The approach taken by the EDSM project focused on the discourse of mathematics examinations, using a range of discursive and linguistic analytical tools. Previous studies of the language of mathematics examinations have tended to treat the language in which questions are written as independent of the mathematical activity demanded of students. Researchers have thus identified the effects of linguistic factors and other communicational aspects (such as graphic elements and the layout of text) as "obscuring" the mathematics (e.g. Fisher-Hoch et al., 1997; Pollitt, Hughes, Ahmed, Fisher-Hoch, & Bramley, 1998; Shorrocks-Taylor & Hargreaves, 1999)—making examination questions more or less difficult for students. In the EDSM project, we have taken a different theoretical perspective on the relationship between language and mathematical activity. Drawing on Halliday's social semiotics (Halliday, 1978) and Sfard's communicational theory (Sfard, 2008), we conceptualise the language in which mathematics is communicated as constitutive of the mathematics itself. Different forms of communication in an examination question entail differences in the mathematical activity expected of a student. Our investigation of discursive differences between examinations set in different years is thus not concerned only with changes in levels of difficulty (although this is one aspect of the research) but is more centrally concerned with the nature of the mathematics with which students are expected to engage. See Morgan and Sfard (2016) for an account of the theoretical and methodological framework of the project.

Case 1: Grammatical Complexity

One of the textual characteristics we investigated in the examination papers was grammatical complexity. In particular, as reported in Morgan et al. (2011), we looked at the extent of use of complex nominal groups, a characteristic component of scientific registers (Halliday, 1993), measuring their recursive depth.[8] Comparing

[8] Recursive depth is defined as the maximum number of decompositions that can be performed. Decomposition is possible when a unit of language (clause, phrase, or word) contains a unit of the same or higher rank. In the given examples, square brackets are used to enclose such "rank-shifted" units.

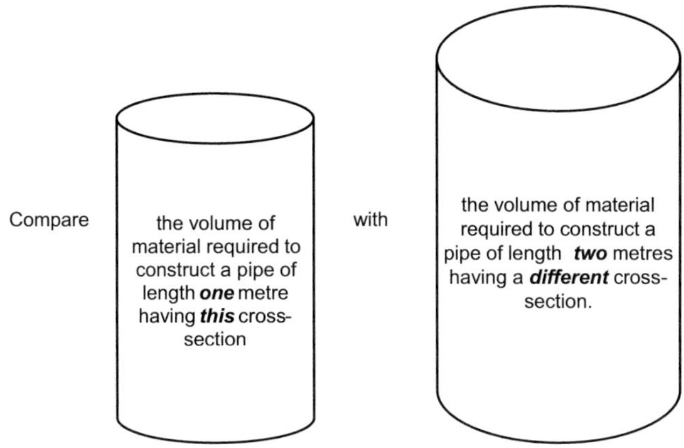

Fig. 8.1 Complex objects participating in a new process

examination papers from 1987 to 2011, the 1987 papers included deeply recursive nominal groups such as:

[the graph [of the curve $\lfloor y=5+3x-x^2$ [for $-2\leq x\leq 5$]]]]
[the volume [of material [required [to construct a pipe [of length [one metre]] [having this cross-section]]]]]

The most complex nominal groups found in the 2011 papers might be characterised as "everyday" objects:

[information [about the points [scored [by some students] [in a spelling competition]]]]

while the specialised mathematical objects in this year had a maximum recursive depth of 3:

[the expression [which is a factor [of $4n^2-1$]]]
[points [on the circumference [of a circle]]]

Halliday (1998) argues that the packing of information into complex nominal groups in scientific text is not arbitrary. Rather it functions to transform experience into knowledge, enabling the formation of precisely defined objects, which can act as participants in further processes. Figure 8.1 illustrates schematically how such information-packed objects may be inserted into an otherwise structurally simple instruction to "compare X with Y".

In our paper, we concluded:

> The dense nominal groups in the 1987 papers incorporate the results of several mathematical processes as qualities of a single object. A consequence of this is that the (apparently simple) instruction to "calculate" in fact demands analysis of the structure of the object to be calculated and consideration of the form of the answer. In 2011 it seems that the processes of analysis, approximation and consideration of units are separated from calculation and that mathematical objects, being generally less complex, contain less potential for further mathematical activity.

When presenting this paper at a mathematics education conference, we were faced by two opposing types of responses from those attending our session. The first (minority) response was to value the simplification of the syntax as it was assumed to make questions more readable and hence to make the mathematical task more accessible to students. This type of response is consistent with the principles applied by the examination boards, which have been concerned to avoid linguistic difficulties "obscuring" the mathematics. The second type of response (which on reflection was privileged by our own presentation) valued "the mathematics", regretting the loss of complexity as a sign of loss of demand for "higher" forms of mathematical activity or, in Dowling's terms, positioning students as dependent, denying them access to some of the principles of mathematical activity. Fellow researchers thus interpreted our description of syntactical features of examination questions in ways that were strongly coloured by their value systems. Participants in debate during the conference session drew on opposing positions within a discourse of access: either valuing access to opportunities to gain the material benefits of examination success or valuing access to the alternative intellectual benefits of specialised forms of participation in mathematics.

Case 2: Application of Mathematics in Context

Another discursive characteristic that we investigated in EDSM was the degree of mathematical specialisation in the discourse. This was investigated at the level of vocabulary but also at the level of whole questions, considering whether questions were located only within a specialised mathematical discourse or whether they also drew on everyday or other discourses in order to relate to non-mathematical contexts.

The utility of mathematics in the real world is a strong element of current curriculum discourse in England, offered as a justification for the privileged place of mathematics in the school curriculum. A rise in the presence of contextualised questions in examinations can be traced to the late 1980s, following the Cockcroft report (DES, 1982) and paralleling the introduction of the first version of the National Curriculum for England and Wales with its inclusion of "Using and Applying Mathcmatics" as a distinct attainment target (DES/WO, 1988). Since then, the presence of contextualised questions has fluctuated in response to various policy debates and curriculum changes. For example, concern expressed about levels of algebraic skills among more advanced students (Royal Society/JMC Working Group, 1997) and a report by the Qualification and Curriculum Authority noting the use of "trivial and distracting" contexts (QCA, 2006) can be associated with the increase in questions involving entirely specialised discourse from the late 1990s. The most recent development of the so-called "functional mathematics" was not represented in our data set but is likely to be reflected in an increase of some kinds of contextualisation.

The chart in Fig. 8.2 shows the variation in the proportion of examination questions involving only specialised mathematical discourse across the years in our sample. It also shows the proportions of contextualised questions necessitating

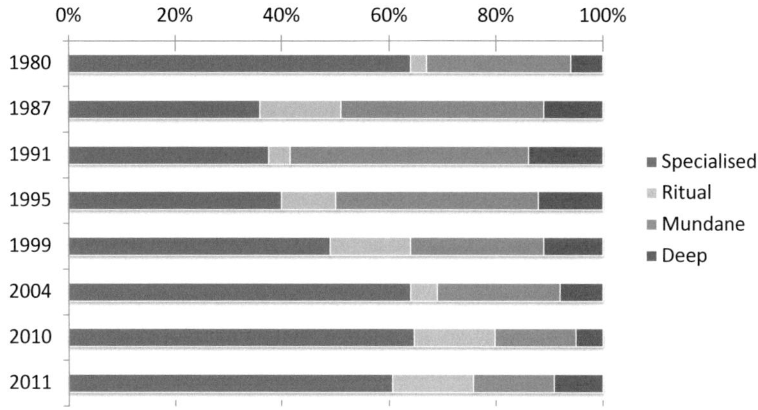

Fig. 8.2 Proportion of questions involving specialised mathematical discourse and different degrees of contextualisation

different degrees of engagement with the context itself, from "ritual" questions in which the relationship between context and mathematics is formulaic, with similar questions found extensively in textbooks and other classroom materials (e.g. questions involving the probability of drawing balls or sweets out of a bag or calculating the original price of an item on sale with a discount of 10 %), through "mundane" questions, which, while less familiar than the ritual questions, allow straightforward identification of the mathematical facts and procedures required to solve them, to "deep" questions in which students would need to analyse or draw on additional knowledge of the context in order to determine what mathematical techniques they should use.[9]

While the proportion of entirely specialised questions has returned nearly to pre-Cockcroft levels (although their distribution across the various areas of the curriculum has changed), the proportion of questions demanding any significant engagement with the context has remained small throughout and the level of engagement demanded in the most recent years in our sample is very low, with less than 25 % of questions in 2010 and 2011 involving anything more than ritual contextualisation.

At the beginning of this section, I set the scene for considering contextualisation by reference to utility and the value that is placed on application of mathematics in official curriculum discourse in England. Framing a response to our analysis within this official discourse would be likely to identify a failure of the examinations to support the curricular aim. However, there are alternative discourses that affect the ways in which groups with different interests, including critical sociopolitical perspectives, will respond. A variety of such responses may be found within the field of mathematics education, for example:

[9] The distinction between ritual, mundane and deep contextualization is adapted from Nabayanga (2002).

- Mathematics educators arguing from a critical or social activist perspective (e.g. Gutstein, 2006; Skovsmose, 1994) might also point to a lack of genuine opportunities to use mathematics to understand, critique or act in the social world (though their critique of the examinations might start from a different place).
- Others, drawing on the research of Cooper and Dunne (2000), might note that students from different social groups are likely to be differentially advantaged by questions demanding some level of engagement with context. Cooper and Dunne's conclusion that working class students are more likely to draw on aspects of their knowledge of the context that are not implicated in the expected mathematical solution might lead to the suggestion that questions providing less contextual information and demanding less engagement with the context may be more equitable.
- Discourses drawing on Piagetian theory of learning might see contextualisation as a means of relating abstract mathematics to concrete ideas and hence supporting students to make sense of the abstract. In this case, the depth of the contextualisation may not be considered so significant. The aim of contextualisation is to enable operation with the specialised discourse, not to engage with the real world context.
- As deeper engagement with context generally involves more extensive description of that context and hence more use of language, criticism from the point of view that "the language obscures the mathematics" applies more strongly. Less or simpler language entails less, simpler context. Hence questions involving less significant engagement with context allow more students to access the necessary mathematics to be successful in the examination.

The EDSM analytic framework has allowed us to produce a description of the way contextualisation of examination questions has varied over the time period studied. The description will, however, lead to different evaluations of the observed change, depending on the interests of the respondents and the discourses they draw upon.

So What Is the Problem?

However much we, as researchers, may seek to present the outcomes of our research as pure description it is clear that each audience or "user" of our research, whether from within the research community, or practitioners, or policy makers, will interpret it by drawing on other resources arising from their interpretations of other research, from policy discourse, from public media, from the discourses current in school practices. Their recontextualisation of our research will select from our reports, relate these selections to their own interests and to selections from other sources, and, most significantly, attach positive or negative values to each selection according to principles and sets of values that are located within the users' own practices. In particular, government agencies will select those aspects of the research that can be made to support the messages they wish to promote.

Of course, the premise that the outcomes of our research could or should be value-free description is open to challenge. I have argued elsewhere (Morgan, 2014) that the theoretical positions that underpin all our research decisions are related to our understanding of society and our political orientation. Moreover, many researchers, especially those with stronger awareness of sociopolitical aspects of education, adopt explicit sets of values and political positions, using these to shape their choices of research focus, methodologies, relationships to research participants, and the ways in which they communicate and disseminate their research. Nevertheless, however explicit we may be about the principles by which our research outcomes are generated, once they are in the public domain they are subject to processes of recontextualisation.

While we have no control over the principles and values employed by the users of our research, we can anticipate what some of these may be. As shown earlier in this chapter, from the beginning of our project we were aware of a dominant discourse of falling standards, current in policy and in public media. At the same time, we were aware of a discourse of access, also employed within policy discourse. Interestingly, the examination boards attempt to engage simultaneously with the values of both these discourses, manipulating questions in order to allow more students to give correct answers to at least some parts of the examination, while setting cut-off scores for grades in order to manipulate the proportions of students awarded each grade, thus ensuring the appearance of maintaining standards.

In an attempt to position ourselves outside these discourses, we identified the policy standards discourse as focusing primarily on numbers—the scores achieved by students and the numbers of students achieving high grades—and distanced ourselves from this focus on numbers by focusing instead on the nature of the mathematics demanded of students. We deliberately avoided using the expression "the quality of the mathematics"—but of course each of the various audiences of our research attaches their own values to particular kinds of mathematics, whether prioritising fluent and accurate reproduction of standard procedures, reasoning in ways associated with more specialised academic mathematics practices, or the use of mathematics to solve real non-mathematical problems or to engage critically with the social world.

Our project did not address the issue of access directly. The discourse of access tends to rest on assumptions that "mathematics for all" is an unquestionable principle and that assessment regimes should enable all students to demonstrate and gain credit for what they know and can do rather than be penalised for what they cannot do—an assessment principle enshrined in UK official policy rhetoric since the Cockcroft report (DES, 1982). The question of whether "mathematics for all" is a meaningful or desirable goal is dealt with elsewhere (see for example Pais in this volume, and Gellert in this volume). By focusing on an existing assessment regime, my concern in relation to access is not so much with the extent of students' opportunities to learn mathematics (although the nature of the mathematics they experience is a focus) as with the consequences of the assessment regime on students' current and future experiences and opportunities. The high-stakes of the GCSE regime for individual students, for teachers and for schools mean that the nature of the mathematics included in the examinations has a strong influence on what happens in mathematics classrooms. The extent of this influence is likely to vary between schools and classrooms in a way that is at least partially linked to the social background of the pupils. Private schools in England that serve the children of the elite may even opt out of the GCSE examination at age 16 completely, focusing instead on strategies for maintaining their disproportionate success in accessing the elite universities, including providing support for the challenging mathematics examinations required for entrance to such universities to an extent that is not widely available to students in state schools. While all state funded schools are more closely regulated, it is those that are officially deemed to be "inadequate"—overwhelming serving working class communities[10]—that are most likely to need to prioritise examination success at age 16 in order to preserve the continuing existence of the school and the reputation and jobs of the school management. I would argue therefore that our analysis of the nature of the mathematics of examination questions is likely to have its strongest influence on the classroom experience of working class children.

As examination success is strongly linked to future opportunities for further study and employment, it is necessary to consider the question of which groups of students may be advantaged or disadvantaged by particular ways of presenting examination questions. In particular, in considering how policy makers and practitioners may interpret and make use of our reports of the research, we must ask how resulting changes in policy and practice may affect students and teachers. Although we have attempted in a later phase of the project to identify how various characteristics of questions affect the ways students respond, the EDSM project did not attempt to look at whether or how such effects varied across groups of students. Nevertheless, we know from Cooper and Dunne's (2000) work that contextualisation of questions may disadvantage girls and students from working class backgrounds. It also seems likely that the use of complex syntax may disadvantage those with less fluency in the language of instruction or less access to academic forms of language. If policy makers interpret our research as evidence that the "quality" of the mathematics demanded needs to be raised, for example by increasing the grammatical complexity, this may lead to changes that exacerbate existing disadvantage. If our research is taken as evidence of falling standards, we will have contributed to a discourse that is used to regulate schools, teachers and students; and to devalue the achievements of all but an elite few.

Of course, the overall logic of my argument implies that educational research is always recontextualised as it is appropriated into the domains of policy and practice and that policy makers will always make use of research in ways that reflect their existing interests. Though this is a general conclusion, its consequences seem especially immediate and severe in the context of research related to qualifications and standards because of the effects of examination success and failure on the futures of students and teachers. In conceiving the EDSM project, we were clearly naïve in attempting to distance ourselves from the dominant standards discourse and presenting our research as being only about mathematics. Or possibly we were disingenuous, acting out an illusory replacement activity as argued by Lundin and

[10]The annual report for 2010–2011 by the government's school inspection agency Ofsted (2011) stated "this year the fifth of schools serving the most deprived pupils were four times more likely to be found inadequate than the fifth of schools serving the least deprived pupils. Seventy-one per cent of schools serving the least deprived pupils were judged to be good or outstanding this year compared with 48 % of schools serving the most deprived."

Christensen (in this volume). Does this mean that our research project was in itself misconceived, inevitably resulting in outcomes that may be recruited to support the dominant discourse and contribute to the maintenance of disadvantage? I argued earlier that our employers and the state either recontextualise our research to serve their own interests or dismiss and devalue it. However, there are other possible audiences for the products of our labour. As intellectual labourers we do not immediately give up all claims to our product in order that it shall be sold for the profit of our employer, but we may continue to use it for our own purposes as well. This allows an alternative, more optimistic conclusion. While recognising the likely fate of our research in the field of policy, we have the opportunity to offer our research outcomes to other groups—teachers, students, grass-roots community organisations, and opposition political activists—to use in ways that are shaped by the interests of these groups. This may involve stepping out of the role of researcher in order to engage in the social practices we seek to affect, being or becoming teachers, teacher educators, community members and political activists ourselves. In doing so, we can be agents in the recontextualisation of our research into a new field, relocating, selecting and transforming the "pure knowledge" of research into social action. We may attempt to shape the ways others interpret our research but must be realistic about the extent to which this is likely to be successful—and wary of the dangers of interpretations and consequent actions that may support policies and practices that we ourselves find inequitable or unethical.

References

Bernstein, B. (2000). *Pedagogy, symbolic control and identity: Theory, research and critique* (revised ed.). Lanham: Rowman and Littlefield.

Cooper, B., & Dunne, M. (2000). *Assessing children's mathematical knowledge: Social class, sex and problem solving*. Buckingham: Open University Press.

DES. (1982). *Mathematics Counts: Report of the committee of inquiry into the teaching of mathematics in schools under the chairmanship of Dr W. H. Cockcroft*. London: HMSO.

DES/WO. (1988). *Mathematics in the National Curriculum*. London: HMSO.

Dowling, P. (1998). *The sociology of mathematics education: Mathematical myths/pedagogic texts*. London: Falmer.

Dowling, P. (2009). *Sociology as method: Departures from the forensics of culture, text and knowledge*. Rotterdam: Sense Publishers.

Draper, H. (1978). *Karl Marx's theory of revolution, Volume II: The politics of social classes*. New York: Monthly Review Press.

Fisher-Hoch, H., Hughes, S., & Bramley, T. (1997). *What makes GCSE examination questions difficult? Outcomes of manipulating difficulty of GCSE questions*. Paper presented at the British Educational Research Association Annual Conference, York. Retrieved January 15, 2016, from http://www.cambridgeassessment.org.uk/Images/109646-what-makes-gcse-examination--questions-difficult-outcomes-of-manipulating-difficulty-of-gcse-questions.pdf.

Gutstein, E. (2006). *Reading and writing the world with mathematics: Towards a pedagogy for social justice*. New York, NY: Routledge.

Haggarty, L., & Pepin, B. (2002). An investigation of mathematics textbooks and their use in English, French and German classrooms: Who gets an opportunity to learn what? *British Educational Research Journal, 28*(4), 567–590.

Halliday, M. A. K. (1978). *Language as social semiotic: The social interpretation of language and meaning*. London: Edward Arnold.
Halliday, M. A. K. (1993). Some grammatical problems in scientific English. In M. A. K. Halliday & J. R. Martin (Eds.), *Writing science: Literacy and discursive power* (pp. 69–85). London: Falmer.
Halliday, M. A. K. (1998). Things and relations: Regrammaticising experience as technical knowledge. In J. R. Martin & R. Veel (Eds.), *Reading science: Critical and functional perspectives on discourse of science* (pp. 186–235). London: Routledge.
Lerman, S., & Tsatsaroni, A. (1998). Why children fail and what the field of mathematics education can do about it: The role of sociology. In P. Gates (Ed.), *Proceedings of the First International Mathematics Education and Society Conference* (pp. 26–33). Nottingham: Centre for the Study of Mathematics Education, Nottingham University.
Morgan, C. (2014). Understanding practices in mathematics education: Structure and text. *Educational Studies in Mathematics, 87*(2), 129–143. doi:10.1007/s10649-013-9482-6.
Morgan, C., & Sfard, A. (2016). Investigating changes in high-stakes examinations: A discursive approach. Research in Mathematics Education. doi:10.1080/14794802.2016.1176596
Morgan, C., Tang, S., & Sfard, A. (2011). Grammatical structure and mathematical activity: Comparing examination questions. *Proceedings of the British Society for Research into the Learning of Mathematics, 31*(3). Retrieved January 15, 2016, from http://www.bsrlm.org.uk/IPs/ip31-3/BSRLM-IP-31-3-20.pdf.
Morgan, C., Tsatsaroni, A., & Lerman, S. (2002). Mathematics teachers' positions and practices in discourses of assessment. *British Journal of Sociology of Education, 23*(3), 445–461.
Nyabanyaba, T. (2002). *Examining examination: The ordinary level (O level) mathematics examination in Lesotho and the impact of recent trends on Basotho students' epistemological access.* Unpublished PhD Thesis, University of the Witwatersrand, Johannesburg, South Africa.
Ofsted. (2011). The Annual Report of Her Majesty's Chief Inspector of Education, Children's Services and Skills 2010/11. London: HMSO. Retrieved January 15, 2016, from https://www.gov.uk/government/publications/hm-chief-inspector-of-education-childrens-services-and-skills-annual-report-2010-to-2011.
Pais, A., & Valero, P. (2014). Whither social theory? *Educational Studies in Mathematics, 87*(2), 241–248. doi:10.1007/s10649-014-9573-z.
Pollitt, A., Hughes, S., Ahmed, A., Fisher-Hoch, H., & Bramley, T. (1998). *The effects of structure on the demands in GCSE and A level questions*. Cambridge: OCR.
QCA. (2006). *Review of standards in mathematics: GCSE 1999–2004 and A level 1998–2004*. London: Qualifications and Curriculum Authority.
Royal Society/JMC Working Group. (1997). *Teaching and learning algebra pre-19*. London: Royal Society/Joint Mathematical Council.
Sfard, A. (2008). *Thinking as communicating: Human development, the growth of discourses, and mathematizing*. Cambridge: Cambridge University Press.
Sfard, A. (2009). Disabling numbers: On the secret charm of numberese and why it should be resisted. In L. Black, H. Mendick, & Y. Solomon (Eds.), *Mathematics relationships in education: Identities and participation* (pp. 9–18). New York: Routledge.
Shorrocks-Taylor, D., & Hargreaves, M. (1999). Making it clear: A review of language issues in testing with special reference to the National Curriculum mathematics tests at Key Stage 2. *Educational Research, 41*(2), 123–136. doi:10.1080/0013188990410201.
Skovsmose, O. (1994). *Towards a philosophy of critical mathematics education*. Dordrecht: Kluwer Academic Publishers.
Slavin, R. E. (2002). Evidence-based education policies: Transforming educational practice and research. *Educational Researcher, 31*(7), 15–21. doi:10.3102/0013189x031007015.

Chapter 9
Mathematics Teachers as Products and Agents: To Be and Not to Be. That's the Point!

Alex Montecino and Paola Valero

Abstract Studying mathematics teachers in the Political invites to understand how teachers' subjectivities emerge in the entanglement of the individual in discursive-material formations. We focus on the power effects of the expert discourses by international agencies such as OECD and UNESCO in the fabrication of the mathematics teacher's subjectivity. Deploying a Foucault-inspired discourse analysis on a series of documents produced by these agencies, we argue that nowadays cultural thesis about who the mathematics teacher should be are framed in a double bind of the teacher as a policy product and as a sales agent. Narratives about the mathematics teacher are made possible within a dispositive of control, which makes mathematics education and mathematics teachers the cornerstone for realizing current market-oriented, competitive, and globalized societies.

Introduction

In a conversation with a prestigious colleague, the topic of what it meant to adopt a political perspective to study mathematics teacher education came to the fore. Discussing the differences and similarities of mathematics teachers' work in different countries, the impact of international agencies such as OECD and its Program for International Student Assessment (PISA) became clearly a topic. In the colleague's view, PISA had not had a significant impact on the work of teachers because the ideas behind PISA had never made it to the classrooms. Teachers have difficulties in designing tasks that would realize the ideals of PISA in their everyday work with students. This kind of statement is an example of a truth that has emerged in the field of teaching and learning, as well as in mathematics education research:

A. Montecino (✉)
Faculty of Engineering and Science, Aalborg University, Aalborg, Denmark
e-mail: alexm@plan.aau.dk

P. Valero
Faculty of Engineering and Science, Aalborg University, Aalborg, Denmark

Faculty of Science, Stockholm University, Stockholm, Sweden

H. Straehler-Pohl et al. (eds.), *The Disorder of Mathematics Education*,
DOI 10.1007/978-3-319-34006-7_9

International comparative assessments contribute positively to increase people's mathematical competence. At the same time, mathematics educators ought to align in the realization of such good intentions, since better mathematical achievement of the kind that the assessments measure will lead to the improvement of living conditions at individual and national levels within a global economy. Mathematics teachers, in particular, ought to be committed to selling PISA's effective models.

We want to problematize such truth and consequently we use the statement of our colleague and its inherent rationality as a starting point in our analysis. We are not interested in asserting whether our colleague is right or wrong but in the way some ideas are accepted and become naturalized truths. Problematizing them implies, for example, recognizing the regime that the whole dispositive of PISA articulates. Such a task is beyond the scope of this chapter, and some other authors have started such endeavour (e.g. Kanes, Morgan, & Tsatsaroni, 2014). We focus on the discursive framing for the making of the mathematics teacher nowadays, and at this particular moment, we cannot ignore the force that the expert visions of international organizations display in such making. Thus, we intend to advance a research agenda of political studies on the mathematics teacher by displacing the analysis of teachers out of their minds, knowledge and beliefs, and out of classrooms, schools and teacher education. Accordingly, we pay attention to the *cultural theses* (Popkewitz, 2008) forming around who the mathematics teacher should be in the expert discourses of international agencies and research.

Gates and Jorgensen (2009) argue that while there is a myriad of political studies in mathematics education concerned with "social justice," little research has been done concerning the political dimensions with regard to teachers and their education. Two issues of the *International Journal of Mathematics Teacher Education* were dedicated to filling this gap in the literature. A political perspective on teachers and teacher education would ask "how teacher education plays a part in the furtherance of a practice which evidently works against the interests of many learners. Significantly, such socially unjust practices are not imposed upon teachers; they are enacted by them, and believed by them to be essential and natural" (Gates & Jorgensen, 2009, p. 164). The papers in the two issues include studies of practices in initial and in-service teacher education where an effort is made to challenge the implication of teachers themselves in the creation of inequalities. The political dimensions in this collection then seems to be connected to how teacher education can/cannot promote awareness for inclusive teaching and learning that would lead to more social justice.

Following Pais and Valero (2012) we would go a step further and argue that understanding teachers and their work in the Political requires a study of how teachers' subjectivities emerge in the inseparable entanglement of the individual and the discursive-material formations within which people and practices of teaching mathematics unfold. Brown and McNamara (2005) already conducted a study of primary mathematics teachers in England trying to understand the emergence of what counts as mathematical practices of teaching and teachers "shaped between the individual's grasp of the subject and the institutional definition of it" (Brown & McNamara, 2005, p. 2). They examined how curriculum and government policy impact on the teacher students' becoming teachers. Our approach diverges from Brown's and

McNamara's in that we direct our gaze away from concrete teachers and towards the expert discourses that nowadays seek to govern and to conduct teachers' professional life. Therefore, the attempt to delve into the current ways of reasoning and the encompassing cultural theses about who the mathematics teacher should be is an important effort to study mathematics teachers politically.

The "mathematics teacher" that we discuss here is not a concrete individual of flesh and bone. It is a discursive construction, where power is actualized in articulating ways of thinking about desired forms of being, and where the meaning of and expectations for the mathematics teacher is configured and negotiated. However, saying that the "mathematics teacher is not a concrete person" does not mean that we are simply talking about thin air formulations that have nothing to do with real people. On the contrary, these ways of reasoning frame possibilities of being and becoming. In particular, we examine the expert discourses about the teacher produced by international agencies such as OECD and UNESCO claiming to know how to fix the problems of education, particularly through the making of teachers as objects of policy. We also connect these to the expert discourses of research in mathematics teacher education existing in the literature.

With our analysis, we are seeking to show that the discursive frame for subjectivity of the mathematics teacher is configured in the tension between the mathematics teacher as a *policy product* and as a *sales agent*. The teacher is an object of policy (OECD, 2005) and, therefore, s/he is caught in a double bind: s/he is a product of governing technologies operating through policy, that respond to demands and requirements of society. At the same time, s/he is an agent for governing that has to sell effectively a highly valued knowledge—the mathematical knowledge—by conducting people's mathematical learning and achievement, for the betterment of the individual students and society. Furthermore, this double bind is made possible as an effect of power within the market-driven logic being performed through a dispositive of control, such as the expert knowledge systems, which international agencies and their comparative studies are part of. In other words, the double bind that frames mathematics teachers' subjectivity nowadays is closely connected to the expansion of particular capitalist understandings of education and teachers, where mathematics and mathematical competence, firstly, make particular sorts of people, and secondly, are key values to govern, control, and give value to people.

The chapter begins by positioning the mathematics teacher in contemporaneity and delineating the analytical strategy utilized in doing so. We present the network of discourses that configure and frame the mathematics teacher, which are putting in operation diverse lines of force for delimiting who the teacher is and must be. In the second section we trace and map circulating statements about teachers. We navigate, firstly, in the discourses fabricated by OECD and UNESCO about the teacher and the mathematics teacher as an object of policy, and secondly, in the discourses and ideas that circulate about students' mathematics achievement and how teachers produced it. We show how in these two types of discourses the mathematics teacher is framed as a social product and as a sales agent respectively. In the third section we open up the analysis of the documents by discussing the double bind of the making of the teacher in relation to the notion of societies of control. Finally, we discuss the contributions of this problematization to the political studies of mathematics teachers.

Researching Mathematics Teacher's Subjectivity in Discourse

In this study we operate with some concept tools for unpacking the discursive network framing mathematics teachers' subjectivity. These tools draw on Foucault's studies on discourse and subjectivity, and Popkewitz's cultural theses.

Enunciations are part of collective practices and the systematic and regular use to promote the conditions to configure statements that compose *discourses*. "We shall call discourse a group of statements in so far as they belong to the same discursive formation" (Foucault, 1972, p. 117) and, as Arribas-Ayllon and Walkerdine (2008) put it, discourses describe rules, divisions, and systems of a particular body of knowledge from specific spatiotemporal conditions. Furthermore, discourses "establish what kind of person one is entitled/obliged to 'be'" (MacLure, 2003, p. 176). Hence, discourse analysis is an analytical strategy that makes possible to trace the enunciations and statements that shape particular ideas about the mathematics teacher's subjectivity and his/her ways of thinking and being.

To trace statements is not a straightforward path to follow; some traces are lost and some seem unconnected. More than a clear line of argument, enunciations and statements are entangled in a discursive network where *cultural theses* about the mathematics teacher become visible. Popkewitz (2008, p. 5) argues that "to talk of cultural theses is to focus on how different sets of ideas, institutions and authority relations are connected to order the principles of conduct."

OECD and UNESCO are two institutions that in the last decades have gained prominence in enunciating what education around the world should be. Their documents encapsulate expert discourses that articulate ways of understanding and thinking about education, teachers and mathematics teachers. Popkewitz (2015, p. 1-2.) argues that the reports of international agencies:

> [p]rovide entrance to a style of thinking and acting that moves among different institutions and social actors, such as policy discourses and discussions among teachers' unions and public debates [...] The grey-zone area in which the reports operate, then, is more than mediating schemas between research and policy. They provide insight into the numbers as constituted in the international assessments as cultural practices about how to make judgments, to recognize types of objects, and draw conclusions in making manageable fields of existence that are never merely that of numbers.

Within the OECD and UNESCO documents, we can find that a large number of reports have focused on teachers. For example, OECD (2005, p. 220) asserts that teachers "[are] important not only for improving the knowledge base for teacher policy, but also as a way of introducing new information and ideas to schools." UNESCO (2015, p. 1) recognizes that "teachers are a critical education resource in every country." Therefore, these documents become an important source for examining the discursive framing of mathematics teacher's subjectivity in contemporaneity, seeking, firstly, to understand how the discourses in these documents generate systems of reason and cultural theses, which fabricate the desired mathematics teacher. For this study, the material analysed are the documents produced by OECD and UNESCO, such as: the documents by OECD *Teachers Matter: Attracting, Developing and Retaining Effective Teachers* (2005); *Mathematics Teaching and*

Learning Strategies in PISA (2010); *PISA 2012 Mathematics Framework* (2010); and *Equity, Excellence and Inclusiveness in Education: Policy Lessons From Around the World* (2014). The documents by UNESCO *The challenge of teacher shortage and quality: Have we succeeded in getting enough quality teachers into classrooms* (2015); *Evolution of policies on teacher deployment to disadvantaged areas* (2015); and *Challenges in basic mathematics education* (2012).

We deploy a Foucault-inspired discourse analysis (Arribas-Ayllon & Walkerdine, 2008, Jørgensen & Phillips, 2002). With this analysis, we seek to problematize truths that circulate in discourses and understand how these are established and configured. A discourse analysis helps to direct attention to questions of subjectivity, context and the socio-historical dimensions of discourse (Angermuller, 2014). From these ideas, the discourse analysis seeks to throw light on how in circulating discourses the current image of the mathematics teacher has been shaped. In other words, the discourse analysis that we deployed provides a way of thinking about how diverse cultural theses emerge in what is enunciated, conducting the mathematics teacher to particular ways of reasoning, thinking and being.

Tracing and Mapping Statements on the Teacher

To begin with, we need to consider that studies and reports developed by OECD and UNESCO about education are part of a field of expertise composed by diverse institutions, agencies, and users. This field is also of interest to a large number of people, institutions, and agencies. In these reports, mathematical knowledge and skills have taken relevance. Furthermore, international agencies give them moral attributes. For example, OECD (2014b, p. 6) asserts that:

> [f]oundation skills in mathematics have a major impact on individuals' life chances [...] poor mathematics skills severely limit people's access to better-paying and more-rewarding jobs; at the aggregate level, inequality in the distribution of mathematics skills across populations is closely related to how wealth is shared within nations. Beyond that, the survey shows that people with strong skills in mathematics are also more likely to volunteer, see themselves as actors in rather than as objects of political processes, and are even more likely to trust others. Fairness, integrity and inclusiveness in public policy thus also hinge on the skills of citizens.

Consequently, documents produced by international agencies set a logic and rationality in education where mathematical knowledge and skills are of great value for the development of a "good citizen." Teachers are the key element of a quality education system to produce high results for students, measured in terms of high scores in achievement tests. Therefore, it is asserted that mathematics teachers are important for society (OECD 2005, 2010c; UNESCO, 2009). Moreover, it is recognized that to think mathematically is a powerful mean to understand and control one's social and physical reality (OECD, 2010c). Additionally, UNESCO (2007, p. 6) states that "mathematics education is a key to increasing the post-school and citizenship opportunities of young people."

If mathematical competence becomes a desired qualification, then the mathematics teacher is considered a provider and developer of certain tools and skills to new generations, which should help people to undertake diverse tasks and problems of everyday life, and of their contexts (OECD, 2010c). Mathematical knowledge is essential for society and its development (Gellert, Hernández, & Chapman, 2013; OECD, 2010d). The mathematical knowledge gets a high value in society, and thus acquires a privileged position because it conduct students' ways of thinking and acting with this knowledge.

Within discussions about teaching and learning, educational achievement is related to factors beyond education; for example, OECD (2014a, p. 104) says that "[h]igher educational achievement benefits both individuals and society, not only financially, but in the well-being with which it is also associated, such as better health outcomes and more civically engaged societies." Educational achievement becomes the aim for the development of movements and efforts realized in diverse social spheres.

Moreover, when improving achievement is at stake, teachers are the only variable that policy can touch in significant ways to better students' achievement. As a result, what happens with teachers becomes a concern for several countries, policy makers, and social and school agents. For example, Schleicher (2012) states that school leaders reported a lack of qualified teachers, particularly mathematics and science teachers. A series of other issues acquire prominence: the need of good teacher training, the improvement of professional knowledge and skills that teachers have to develop, the increase in the effectiveness and competitiveness of teachers, and the implementation of policies to retain the best teachers, among others.

UNESCO (2015) recognizes that it becomes important to ensure that teachers are well trained, motivated and supported. Additionally, Schleicher (2012, p. 38) states that:

> [t]eachers need to be well-versed in the subjects they teach in order to be adept at using different methods and, if necessary, changing their approaches to optimize learning. This includes content-specific strategies and methods to teach specific content.

These reports are not alone in producing different statements of the sort. Mathematics education research literature also points out that the job of the mathematics teacher is a complex and demanding practice that requires a mixture of both theoretical and practical knowledge, skills, and deep understanding of children (White, Jaworski, Agudelo-Valderrama, & Gooya, 2013). A whole range of general and subject specific research resonates with the statements produced by international agencies.

Diverse statements about the mathematics teacher are formulated from an idealized and desired image of the teacher. However, at the same time, these statements shape an idealized and desired image of the teacher. For example, UNESCO (2007, p. 13) describe the *effective teacher*, which in turn embeds an image of the ideal teacher:

> [E]ffective teachers understand that the tasks and examples they select influence how students come to view, develop, use, and make sense of mathematics […] Effective teachers design learning experiences and tasks that are based on sound and significant mathematics;

they ensure that all students are given tasks that help them improve their understanding in the domain that is currently the focus.

The resonances between the multiple enunciations and statements produced by international agencies, research and other voices shape truths, which establish what is possible and desired. "Truth is a discursive construction, and different regimes of knowledge determine what is true and false" (Jørgensen & Phillips, 2002, p. 13). The reports—our focus here—create a new *grey-zone* (Lindblad, Pettersson, & Popkewitz, 2015) of authoritative expert knowledge located between policy and academic research, and thus the reports contribute to new truth regimes about teachers and their work. The resonances formed by a multiplicity of perceptions and understandings converge in shaping cultural theses about the mathematics teacher, configuring an ideal subject. The subject—the mathematics teacher—emerges through repetition and anticipation, and the subject is constituted in the given (Deleuze, 1991).

Hence, through the discourse analysis deployed, we seek to navigate through the discursive network, tracing and mapping the circulating enunciations and statements that constitute the framing of teachers' subjectivity nowadays. The analysis has two movements. First, we study how the mathematics teacher is shaped as a policy product. Second, we examine how the teacher is shaped as a sales agent.

The Mathematics Teacher as a Policy Product

Navigating through OECD and UNESCO documents it is possible to identify the articulation of a certain form of reasoning and arguing. Education is an important factor in the *social and economic development of countries* (OECD, 1989). For education to deliver the adequate formation of human capital, it is important to focus on the *quality of the education system.* A quality system will secure that as many students as possible acquire the needed competencies so that *students' achievement*, in general, can be high. Students' achievement is systematically monitored as a strategy to closely follow educational quality. The accumulation of extensive and detailed data about the quality of educational systems in many countries in the world reveals that there are factors of quality, which cannot be directly dealt with and easily influenced—what is called contextual factors. However, there are factors that governments can steer. The one key element is *the quality of teachers and their professional development.* It is within this type of reasoning that the four elements highlighted above—development, quality, achievement, and teachers—entangle in a discursive network framing the becoming of the mathematics teacher into a policy product.

A large number of reports and studies focus on the steering of education to produce effective students' learning and achievement. Diverse factors are recognized to have influence on *student's achievement, learning, and experience*. The reports state that contextual factors such as different abilities, attitudes and background that the students have and bring to school are "difficult for policy makers to influence, at least in the short-run" (OECD, 2005, p. 26).

But there seems to be taken as a fact that "*the quality of teachers and their teaching* are the most important factors in student outcomes that are open to policy influence" (OECD, 2005, p. 12, our emphasis). To conceive teachers as the targets of policy implies thinking that it is possible to design and fabricate teachers on the grounds of political ideas and agendas such as globalization and social progress. The teacher then becomes configured and controlled as a product for society to face the demands and needs of economic and political initiatives and interests.

This is connected with the emphasis in diverse documents for *the quality of the education system* and its relation with teacher's quality performance. For example,

> [a]ll countries are seeking to improve their schools, and to respond better to higher social and economic expectations [...] Teachers are central to school improvement efforts. (OECD, 2005, p. 19)
>
> PISA shows a clear link between student performance and teacher status, with students doing better in school systems that spend more on salaries to attract quality teachers. (Schleicher, 2014, p. 11)

UNESCO also recognizes a direct relation between the quality of the education system and the teacher. They warn that:

> [e]ducation quality can be jeopardized by hiring untrained teachers if they lack qualifications, preparation, motivation, appropriate working conditions and ongoing professional development. (UNESCO, 2015, p. 9)

Teachers are the key to the positive and sustainable development of education systems, constituting the principal challenge to quality mathematics education (UNESCO, 2012). Moreover, diverse investigations argue that teacher's quality is closely related to student's learning and his/her academic achievement (OECD, 2005); and that effective teachers help to close achievement gaps between advantaged and disadvantaged students:

> [e]ffective teachers are particularly important for disadvantaged schools and their students [...] Highly competent teachers can have large positive effects on student performance, strong enough to close achievement gaps between disadvantaged and advantaged students. [... Teachers] may help low performing students to catch up and improve. (OECD, 2012, p. 130)

Thirdly, the interest in strengthening the teaching profession has the purpose of striving for the *quality, effective teacher*. The teacher is considered as the means whereby it is possible to achieve the promise of improving the education system and to reach the desired quality level. In the search of an improved educational system, teachers become a priority issue for the society because "teachers are key to increasing educational quality" (Luschei & Chudgar, 2015, p. 3).

Achievement, quality, and teachers are meant to be geared towards high quality due to their aggregated significance for *social and economic development* in countries and between countries. Schleicher (2011, p. 45) states that the conditions for the teaching profession are important:

> [d]ata from Pisa show that high-performing education systems tend to prioritize the quality of teachers, including attractive compensation, over other inputs, most notably class size.

Such a statement indirectly states that low performing countries in PISA—which correspond with poor, developing countries—have problems providing good conditions for the profession. At the same time, teachers are being positioned as the key actors in bridging achievement gaps, which is also a socioeconomic gap. Hence, there emerge ideas such as the need for teachers to be given appropriate support and training for facing diversity in schools and classrooms:

> School education must therefore seek to overcome socio-economic inequalities throughout societies while at the same time utilise the benefits that diversity brings to schools and classrooms. A successful programme treats diversity as a source of potential growth rather than an inherent hindrance to student performance. One way to do this is to use teachers' strength and flexibility. Of course, for this to be effective, teachers need to be given appropriate support and training. (OECD, 2010a, p. 20)

In many countries, there is a high demand and need for qualified teachers. UNESCO (2007, p. 69) urges for the "need for better-trained mathematics teachers," that is, teachers trained with the highest standards of professional knowledge, skills, competence, and integrity; and teachers who must and can implement diverse initiatives to improve teaching. Such an effort is set as a priority in a context where "about half the countries report serious concerns about maintaining an adequate supply of good quality teachers, especially in high-demand subject areas" (OECD, 2005, p. 8).

Countries which have improved their performance in PISA have also set policies to improve their teaching staff (OECD, 2013b). Moreover, several high-performing countries took decided steps to raise the quality of the teaching profession—for instance, by inspiring people from other professions to give their talents to the teaching profession. Through marketing, for example, diverse recruitment campaigns can emphasize the fulfilling nature of teaching as a profession, and can attract candidates (OECD, 2014d). Such initiative is important because it is recognized that high performing countries, unlike other countries, recruit their candidates for initial teacher training from the top third of each cohort that graduates from their school system (OECD, 2010b). It is important to attract good candidates with potential for being a teacher as the raw material for the fabrication of the teacher. The recurrent idea concerning recruitment is that the better the candidates, the better the teachers. The teaching profession is thus being portrayed as "the option" for fulfilling and satisfying social demands and requirements. The satisfaction of societal needs and desires is secured through the configuration, use, and consumption of the object called "teacher."

However, good raw material is not enough. Teachers' continual professional development also promotes the social and economic development of a country. The retention of teachers is important: "Teacher policy needs to ensure that teachers work in an environment that encourages effective teachers to continue in teaching" (OECD, 2014a, p. 486). The instruments to secure recruitment go hand in hand with instruments to monitor the good quality of teachers' professional exercise and its improvement. For example, permanent evaluation of teachers, involvement in lifelong learning activities, and the monitoring of students' achievement are becoming control instruments for policy. Currently, it is needed to submit the teacher to constant testing with the aim of knowing if s/he is or is not competent. In other words, "quality control" becomes a constant measurement that the teacher must face. It is recognized that initial teacher training—whether good or bad—does not really matter, since it cannot prepare teachers to succeed in every challenge throughout a career (Schleicher, 2012, 2014). For example, in situations where there is "socio-economic heterogeneity in student populations, this heterogeneity is a major challenge for teachers and education systems" (OECD, 2014c, p. 36). Policies "should be implemented to ensure teachers have sufficient qualification and training" (UNESCO, 2015, p. 9).

The different lines that we tried to follow in the previous paragraphs cross, ensemble and intermesh. In the discursive network where the lines unfold there operates the mechanism where the teacher is controlled, produced and planned in function of what is desired by society. Moreover, the teacher is positioned in a market logic where supply and demand configure the teacher as a product that can be made and acquired by whoever has purchasing power. This favours a logic where the higher the purchasing power, the better teachers may be produced.

In this discursive network the mathematics teacher is configured as a political product, a product that results from policies. The market and society seek to satisfy the needs and desires that are established as urgent, through the making of the teacher. Moreover, the market—and its hunger for highly mathematically competent workforce—sets the attributes that the mathematics teacher must have, and thanks to globalization these attributes seem to been standardized. The mathematics teacher as a product of policy is subjected to the whims of the market, the development of policies, and the response to social demands.

The Mathematics Teacher as a Sales Agent

In the documents of international agencies there is a substantial concern about the development of mathematical knowledge in young people, and thus the teaching and learning of mathematics. This concern is expressed through attention on achievement of children, adults, teachers and even social achievement. Achievement is mainly measured through standardized tests, for example, PISA and Trends in International Mathematics and Science Study (TIMSS). Through the quantification of achievement, it is possible to know the level, competencies and expertise in mathematics that countries, groups or individuals have. It is argued that mathematics achievement is relevant since modern societies require a high level of mathematical competence for social development. Mathematics is referred to as the foundation of much of the scientific and technical activity that distinguishes advanced from less advanced societies, hence, "developing students' mathematical competence at a much higher level than is required for everyday communication is thus a goal of most school programs" (OECD, 2010c, p. 32).

In OECD and UNESCO documents mathematics achievement is set in a network of at least three lines: *social and economic differences*, *teaching and learning mathematics*, and *students' and teachers' performances*. Mathematics achievement is configured as a node, where the lines conducting the mathematics teacher in the discursive network converge. The mathematics teacher as a sales agent is a multifaceted salesman/woman. S/he must not only manage the mathematics teaching, but also, among others, the students, their motivation, experience, and expectations. The teacher must identify social needs and design offerings adjusted to the context according to the standards set for the product s/he is dealing with. S/he must be capable of promoting and selling a highly valued product—the mathematical knowledge and competence. Finally, the mathematics teacher becomes a defender for progress and success in society. The progress promised by this highly qualified agent is based on his/her capacity of developing of mathematical knowledge in students.

Firstly, within diverse documents of international agencies, it is possible to identify an interest in studying how *social and economic differences* in students and of countries constitute gaps in levels of performance. It is asserted that there is a direct relationship between students' social and economic background and their achievement, particularly in mathematics achievement. For example:

> On average, a more socio-economically advantaged student scores 39 points higher in mathematics than a less-advantaged student. This difference represents the equivalent of nearly 1 year of schooling. (OECD, 2013a, p. 17)
>
> On average across OECD countries, 13 % of students are top performers in mathematics (Level 5 or 6). At the same time, 23 % of students in OECD countries, and 32 % of students in all participating countries, are low performers in mathematics (i.e. they did not reach the baseline Level 2). […] Across OECD countries, 15 % of the difference in performance among students is explained by disparities in students' socio-economic status […] around 1 year of formal schooling—separate the mathematics performance of those students who are considered socio-economically advantaged and those whose socio-economic status is close to the OECD average. (OECD, 2014a, p. 189)
>
> Socio-economic measure is positively associated with mathematics performance in all countries. (OECD, 2010c, p. 76)

OECD and UNESCO studies have shown the differences that there are between countries and the differences that exist within each society or community that has been studied. These studies converge mainly on the concern to reduce the gap that emerges from social and economic differences. Here, the mathematical knowledge—its teaching and learning—has acquired a value in society, and the mathematics teacher is responsible for promoting the increased and improved acquisition of better mathematical knowledge. The mathematics teacher must guaranty that this knowledge comes to society; in other words, the teacher is positioned as the one who must sell and ensure massive consumption of this desired merchandise.

The results from studies that recognize social and economic differences generate antecedents for policy makers. OECD (2014a, p. 188) states:

> PISA results reveal what is possible in education […] The findings allow policy makers around the world to gauge the knowledge and skills of students in their own countries in comparison with those in other countries, set policy targets against measurable goals achieved by other education systems, and learn from policies and practices applied elsewhere.

Moreover, from successful experiences there emerge guidelines that set an ideal image of what is desired from the educational system, its functioning and participants. The OECD country reports clearly express this type of logic. For the case of Sweden, one of the countries whose performance in PISA tests declined dramatically in 2012, it is stated:

> The report makes extensive use of OECD's international knowledge base and of Swedish educational research, statistical information and policy documents. It identifies the main strengths and challenges of the school system and provides concrete recommendations and policy actions to serve as the foundation for a comprehensive school improvement reform to bring about system-wide change and strengthen the performance of all Swedish students [...] (OECD, 2015, p. 14)
>
> OECD Education Policy Reviews are tailored to the needs of the country and cover a wide range of topics and sub-sectors focused on education improvement. The reviews are based on in-depth analysis of strengths and weaknesses, using various available sources of data such as PISA and other internationally comparable statistics, research and a review visit to the country. They draw on policy lessons from benchmarking countries and economies, with expert analysis of the key aspects of education policy and practice examined [...] The methodology aims to provide analysis and recommendations for effective policy design and implementation. It focuses on supporting reform efforts by tailoring comparative analysis and recommendations to the specific country context, engaging and developing the capacity of key stakeholders throughout the process. (OECD, 2015, p. 15)

The recommendations of OECD to Sweden on how to create a "highway" to educational success (Lindblad et al., 2015, p. 137) highlighted the importance of designing targeted strategies for promoting better learning for all and for disadvantaged groups, and for raising the quality of teacher education and the teacher profession. The documents create clear images that are then sold around as effective solutions to fix the problems of education, by deploying different marketing tools for the consumption of the educational products of these agencies.

Secondly, in order to talk about mathematics achievement it is essential to focus on *teaching and learning mathematics*, since "it is clear that teaching and learning factors have a significant association with student performance in mathematics" (OECD, 2010c, p. 120). Furthermore, "there is a strong correlation between the teacher's knowledge of mathematics and successful classroom practice" (JMTE, 2014, p. 373). "The presence of qualified, well-motivated and supported teachers is vital for student learning. Effective teaching strongly influences what and how much students achieve in school" (UNESCO, 2015, p. 1).

Mathematical literacy is configured as an important issue in teaching and learning mathematics. It is relevant that people have mathematical skills and knowledge, but it is also important to know what can be done with these skills and knowledge. In OECD (2014a) it is enunciated that modern societies valorise individuals not for what they know, but for what they can do with what they know. Some OECD documents defend the relevance of mathematical literacy for society. For example:

> Mathematical literacy is an individual's capacity to formulate, employ, and interpret mathematics in a variety of contexts. It includes reasoning mathematically and using mathematical concepts, procedures, facts, and tools to describe, explain, and predict phenomena. It assists individuals to recognise the role that mathematics plays in the world and to make the wellfounded judgments and decisions needed by constructive, engaged and reflective citizens. (OECD, 2010d, p. 4)

The statements around this line contribute to set up the idea of the mathematics teacher as a sale agent promoting the education of a subject—the desired student (see also Andrade-Molina and Valero in this volume)—through the conduction of students' conduct and behaviour in learning mathematics and becoming mathematically literate. Hence, the teacher is portrayed as an agent for governing, subjecting and conducting children through mathematical learning.

Finally, it is possible to identify some policy initiatives around the teaching and learning focusing on the *improvement of teachers' and students' performance*. Students' mathematical achievement is recognized as "the educational outcome, student learning strategies and teaching strategies are its main predictors" (OECD, 2010c, p. 70). Teaching strategies and student learning strategies are characterized by OECD (2010c, p. 20) as:

> [t]eaching strategies refer to a broad range of processes, from the organisation of classrooms and resources to the moment-by-moment activities engaged in by teachers and students to facilitate learning. Student learning strategies refer to cognitive and meta-cognitive processes employed by students as they attempt to learn something new.

In OECD (2014a, p. 196), it is considered that:

> [t]op performers in mathematics are students who score at Level 5 or 6 on the PISA assessment. They can develop and work with models for complex situations, identifying constraints and specifying assumptions; select, compare, and evaluate appropriate problem-solving strategies for dealing with complex problems related to these models; work strategically using broad, well-developed thinking and reasoning skills, appropriate linked representations, symbolic and formal characterisations, and insight pertaining to these situations; and begin to reflect on their work and formulate and communicate their interpretations and reasoning.

High performance is of great interest for societies. "Student performance in mathematics is related to teachers' professional knowledge of mathematics, that is, their knowledge of mathematics per se, and the specialised knowledge of mathematics used in teaching" (UNESCO, 2012, p. 74). Regarding teacher's knowledge, research on mathematics teachers states that:

> [a]dditional research is needed to understand the relationship between […] knowledge and pedagogical moves. It is necessary, as part of the development of a robust theory of the knowledge teachers need, to understand how teachers' learning with understanding fits into teaching for understanding in their classrooms. (JMTE, 2015, p. 295)

The whole issue of the subject-matter qualifications of teachers is also highlighted as being of great importance. It is possible to identify some policy initiatives that have as aim to attract teachers in subjects such as mathematics, science and technology (Schleicher, 2011). The quality and effectiveness of teachers take relevance within social spheres:

> [t]eachers need to be capable of preparing students for a society and an economy in which they will be expected to be self-directed learners, able and motivated to keep learning over a lifetime. (OECD, 2005, p. 97)

Moreover, OECD (2014a, p. 18) recognizes that the professional development of teachers is politically relevant:

> [p]olicy levers and contexts typically have antecedents—factors that define or constrain policy […] For teachers and students in a school, for example, teacher qualifications are a given constraint while, at the level of the education system, professional development of teachers is a key policy lever.

Currently, students need to have an understanding of the fundamental concepts of mathematics. They need to be able to cope with a new situation or problem, recognizing the relevance of mathematics, identifying and using the relevant mathematical knowledge to solve the problem, and evaluating the solution in the original problem context (Schleicher, 2012). In addition, the mathematics teacher must be able to make sure that students will be able to do all these activities. The work of the teacher is more than teaching mathematics; rather the mathematics teacher must contribute to give value to mathematical knowledge for everyday life and for the future.

As a result, the teacher is a medium that extends and realizes the intentions of policy, for example, through the promotion and implementation of reform. Promoting reform is "considered by many to be a major responsibility of prospective teacher preparation" (JMTE, 2014, p. 295). S/he is made a sales agent that must favour a more equal and just society. The mathematics teacher is responsible for promulgating the desire for mathematics.

The Making of the Teacher Within a Dispositive of Control

Following the traces of enunciations and statements about the acclaimed and undoubted importance of teachers for building the future in documents of international agencies such as OECD and UNESCO, we entered an entangled discursive network where lines and forces cross. In our analysis a certain sense of repetition and circularity intended to grasp the folds and unfolds of the multiple stories told about who teachers are and who they should be. These stories are instantiations of power. Possible subjectivities become actualized in the discursive network (Jørgensen & Thomassen, 2015) unfolded by the prominent and increasingly decisive expert-knowledge of these agencies. It is precisely in the actualization of power in discourse and stories that possible cultural theses about the mathematics teacher emerge. Teachers' subjectivity is framed and entangled in a rationality of social progress, competitiveness, and globalization. The double bind of the mathematics teacher as a policy product and as a sales agent is made concrete in the demands and expectations of society and in the urge of making (mathematics) education work for the economy.

In our previous analysis the connection between these discourses and particular economic interests and agendas have been hinted at. In our conclusion we want to make such connection more explicit by opening up the political field of subjectification of which the network of discourses on the (mathematics) teacher is made possible. Deleuze's notion of *dispositive*—congruent with Foucault's notion of apparatus—helps us casting light on this issue. Foucault (1980, p. 195) wrote:

> I understand by the term "apparatus" a sort of—shall we say—formation which has as its major function at a given historical moment that of responding to an urgent need. The apparatus thus has a dominant strategic function. This may have been, for example, the assimilation of a floating population found to be burdensome for an essentially mercantilist economy: there was a strategic imperative acting here as the matrix for an apparatus which gradually undertook the control or subjection of madness, sexual illness and neurosis.

A dispositive—a "tangle, a multilinear ensemble" (Deleuze, 1992b, p. 159)—can be understood as a machine, which makes one see and speak (Deleuze, 1992b). A dispositive is immersed in the network of relations that can be established between "discourses, institutions, architectural forms, regulatory decisions, laws, administrative measures, scientific statements, philosophical, moral and philanthropic propositions—in short, the said as much as the unsaid" (Foucault, 1980, p. 194).

A dispositive making the teacher a policy object and a sales agent operates as a response to the urgent need expressed in different institutions and by expert knowledge producers, of securing a world order where education is meant to bring individual prosperity, collective competitiveness and international circulation of well-being, all following the patterns of those who are on the top of the rankings of wealth and progress. The making of mathematics teachers cannot longer be left to the whims of a single person's dream of teaching his/her favourite school subject, or of well-intentioned teacher education programs. Governments need to steer and secure the adjustment of a machinery to make the right agents of the desires of the state—which, by the way seem to conflate with the needs of the market.

The demands of the time are embedded in the different lines that we addressed in our analysis and in the ways they intersect: quality teachers and effective teachers are needed with the aim of closing gaps between students' achievement, professional development, and social and economic differences. For achieving this goal, teachers' professional development takes on particular relevance, since it secures compliance with what the ideal desire about who the mathematics teacher must be, and what s/he must do. The mathematics teacher needs to develop specific skills and knowledge, so that s/he can respond to central urgencies of society. But, how is it possible to ensure that the teacher has developed what is necessary or what is demanded? Here is where the continuous training and, specifically, standardized tests acquire importance. The highest mechanism of control in the education system is the use of standardized tests. These tests are setting a numerical language of control that marks access to information, and where people have become samples, data, or markets (Deleuze, 1992a). Hence, the test that measures students' performance and directly or indirectly teachers' quality and, as a whole, educational quality allows transforming education into controllable variables of a system attending a marketing logic.

> Marketing has become the center or the "soul" of the corporation [...] the operation of markets is now the instrument of social control and forms the impudent breed of our masters. Control is short-term and [...] continuous and without limit, while discipline was of long duration. (Deleuze, 1992a, p. 6)

The market sets supplies and demands around the mathematics teacher to satisfy social needs. There is always a demand determining what the teacher should know to satisfy the requirements of society, and these demands shape the double bind of the teacher: the teacher as a policy product is fabricated with the aims of meeting social demands and requirements, and as a sales agent is configured for conducting students towards the desires of society. The double bind increases the demands of professionally qualified teachers against the lack of them, which in turn installs a strong logic of competition. This logic implies that the teacher does not only need to compete in qualifications with other teachers to get a job; it is necessary to compete permanently with oneself for staying in the job, even in a situation when the person is highly needed. Secondly, and as a consequence of the previous, the focus is on the knowledge and skills of the mathematics teacher. These knowledge and skills must comply with special requirements of quality and expectations established by society. In research on mathematics teachers it is recognized that "[teachers need to] develop professional knowledge in support of their practice" (JMTE, 2014, p. 455). It is also pointed "that teachers' lack of content knowledge interfered with their judgements and that there [is] a mismatch between their perceptions of students' difficulties and the actual difficulties demonstrated by their students" (JMTE, 2014, p. 405). Hence, demands and social urgencies promote discourses and forces for establishing the idea of permanent training since it is recognized that initial teacher education is insufficient to satisfy new challenges that market sets. However, why did the need for permanent training and what is being sought with it emerge? A partial answer can be found in Deleuze (1992a, p. 4), who argues that:

> [i]n the societies of control one is never finished with anything—the corporation, the educational system, the armed services being metastable states coexisting in one and the same modulation, like a universal system of deformation.

Therefore, the idea of permanent training is a way of maintaining control of a never-ending process for the teacher. The idea of permanent training is operating as part of a dispositive by setting diverse forms of control, discourses, and forces. Consequently, the mathematics teacher is condemned to be incomplete and to have constant deficits to overcome, since society and the market will always be setting new requirements, demands, and urgencies that the teacher must face. The mathematics teacher will always be "a man in debt" (Deleuze, 1995, p. 181).

References

Angermuller, J. (2014). *Poststructuralist discourse analysis. Subjectivity in enunciative pragmatics*. Houndmills: Palgrave Macmillan.

Arribas-Ayllon, M., & Walkerdine, V. (2008). Foucauldian discourse analysis. In C. Willig & W. Stainton-Rogers (Eds.), *The SAGE handbook of qualitative research in psychology* (pp. 91–108). London: SAGE Publications.

Brown, T., & McNamara, O. (2005). *New teacher identity and regulative government: The discursive formation of primary mathematics teacher education*. New York: Springer.

Deleuze, G. (1991). *Empiricism and subjectivity: An essay on Hume's theory of human nature*. New York: Columbia University Press.

Deleuze, G. (1992a). Postscript on the Societies of Control. October, 59, 3-7. Retrieved from http://www.jstor.org/stable/778828.
Deleuze, G. (1992b). What is a Dispositif? In T. J. Armstrong (Ed.), *Michel Foucault Philosopher*, (pp. 159–168). Hemel Hempstead: Harvester Wheatsheaf.
Deleuze, G. (1995). *Negotiations 1972-1990*. New York: Columbia University Press.
Foucault, M. (1972). *Archaeology of knowledge*. London: Routledge.
Foucault, M. (1980). *Power/knowledge: Selected interviews and other writings 1972-1977*. New York: Pantheon Books.
Gates, P., & Jorgensen, R. (2009). Foregrounding social justice in mathematics teacher education. *Journal of Mathematics Teacher Education, 12*(3), 161–170. doi:10.1007/s10857-009-9105-4.
Gellert, U., Hernández, R., & Chapman, O. (2013). Research methods in mathematics teacher education. In M. A. Clements, A. J. Bishop, C. Keitel, J. Kilpatrick, & F. K. S. Leung (Eds.), *Third international handbook of mathematics education* (pp. 327–360). New York: Springer.
Jørgensen, K. M., & Thomassen, A. O. (2015). Maps of organizational learning in regional development projects: Stories, objects and places. *Tamara: Journal of Critical Postmodern Organization Science, 13*(3), 57–69.
Jørgensen, M., & Phillips, L. J. (2002). *Discourse analysis as theory and method*. London: SAGE Publications.
Journal of Mathematics Teacher Education [JMTE]. (2014). *Journal of Mathematics Teacher Education, 17*(1–6), 1–582.
Journal of Mathematics Teacher Education [JMTE]. (2015). *Journal of Mathematics Teacher Education, 18*(1–5), 1–499.
Kanes, C., Morgan, C., & Tsatsaroni, A. (2014). The PISA mathematics regime: Knowledge structures and practices of the self. *Educational Studies in Mathematics, 87*(2), 145–165. doi:10.1007/s10649-014-9542-6.
Lindblad, S., Pettersson, D., & Popkewitz, T. S. (2015). *International comparisons of school results: A systematic review of research on large scale assessments in education*. Stockholm: Swedish Research Council.
Luschei, T., & Chudgar, A. (2015). *Evolution of policies on teacher deployment to disadvantaged areas*. Background paper for EFA Global Monitoring Report 2015.
MacLure, M. (2003). *Discourse in educational and social research*. Buckingham: Open University.
OECD. (1989). *Education and the economy in a changing society*. Paris: OECD Publishing.
OECD. (2005). *Teachers matter: Attracting, developing and retaining effective teachers*. Paris: OECD Publishing.
OECD. (2010a). *Educating teachers for diversity. Meeting the challenge*. Paris: OECD Publishing.
OECD. (2010b). *Improving schools: Strategies for action in Mexico*. Paris: OECD Publishing.
OECD. (2010c). *Mathematics Teaching and Learning Strategies in PISA*. Paris: OECD Publishing.
OECD. (2010d). *PISA 2012 mathematics framework*. Paris: OECD Publications.
OECD. (2012). *Equity and quality in education: Supporting disadvantaged students and schools*. Paris: OECD Publishing.
OECD. (2013a). *PISA 2012 results: Excellence through equity (volume II): Giving every student the chance to succeed, PISA*. Paris: OECD Publishing.
OECD. (2013b). *PISA 2012 results: What makes schools successful (volume IV): Resources, policies and practices, PISA*. Paris: OECD Publishing.
OECD. (2014a). *Education at a glance 2014: OECD indicators*. Paris: OECD Publishing.
OECD. (2014b). *PISA 2012 results in focus. What 15-year-olds know and what they can do with what they know*. Paris: OECD Publishing.
OECD. (2014c). *PISA 2012 results: What students know and can do (volume I, revised edition, February 2014): Student performance in mathematics, reading and science, PISA*. Paris: OECD Publishing.
OECD. (2014d). *Teacher remuneration in Latvia: An OECD perspective*. Paris: OECD Publishing.
OECD. (2015). *Improving schools in Sweden: An OECD perspective*. Paris: OECD Publishing.

Pais, A., & Valero, P. (2012). Researching research: Mathematics education in the Political. *Educational Studies in Mathematics, 80*(1), 9–24. doi:10.1007/s10649-012-9399-5.
Popkewitz, T. S. (2008). *Cosmopolitanism and the age of school reform: Science, education, and making society by making the child.* New York: Routledge.
Popkewitz, T. S. (2015). *The cultural inscription of numbers: The grey-zone in international comparison of school result research and policy-making*. Main lecture prepared for "Education by the Number" Conference. University of Gothenburg. Retrieved June 8–9, 2015.
Schleicher, A. (2011). *Building a high-quality teaching profession: Lessons from around the world.* Paris: OECD Publishing.
Schleicher, A. (2014). *Equity, excellence and inclusiveness in education: Policy lessons from around the world.* Paris: OECD Publishing.
Schleicher, A. (Ed.). (2012). *Preparing teachers and developing school leaders for the 21st century: Lessons from around the world.* Paris: OECD Publishing.
UNESCO. (2007). *Education for all by 2015-will we make it?* Paris: UNESCO Publishing and Oxford University Press.
UNESCO. (2009). *Aportes para la enseñanza de la matemática. Segundo estudio regional comparativo y explicativo*. Chile: LLECE.
UNESCO. (2012). *Challenges in basic mathematics education.* Paris: UNESCO.
UNESCO. (2015). *The challenge of teacher shortage and quality: Have we succeeded in getting enough quality teachers into classrooms?* Education for All Global Monitoring Report. pp. 1–10.
White, A., Jaworski, B., Agudelo-Valderrama, C., & Gooya, Z. (2013). Teachers learning from teachers. In M. A. Clements, A. J. Bishop, C. Keitel, J. Kilpatrick, & F. K. S. Leung (Eds.), *Third international handbook of mathematics education* (pp. 393–430). New York: Springer.

Chapter 10
Technologies of (Re)production in Mathematics Education Research: Performances of Progress

Anna Llewellyn

Abstract In this chapter, I use Foucauldian theory to consider and critique the role of technologies of power, surveillance and governmentality within mathematics education research. I argue this deconstruction is pertinent within our current neo-liberal, market-driven education system, as both schools and universities are involved in overt and covert methods of governing. Moreover, this era is predicated on the fiction of the free, autonomous self, and discourses of "becoming". I examine this argument through the deconstruction of a key taken-for-granted truth of mathematics education research within the UK and other Western contexts, such as the USA and Australia—that it is heavily and uncritically invested in progress, progressive pedagogies and the "free" autonomous subject. I argue that this position relies on a "natural" mathematical child, who is posited as asocial, acultural and apolitical, where the focus is on what is to come, not what is already. Instead I suggest that mathematics education, and the mathematical child, are not natural but instead are social, cultural and political products. The natural, developmental, "free", child is (re)produced through both overt and covert surveillance, and monitoring from both schools and universities.

Introduction

Sometimes I wonder about my teaching. I sit down in a seminar, and I cannot help but facilitate it in a certain way. I push, I probe, I ask for explanations "why?". I ask students to expand their answers, or to clarify what they mean. I encourage responses. I encourage reflection. I encourage the expression of opinions and ideas. But I am not opening the dialogue freely; I am instead conditioning people to behave in a certain way, to become someone particular. My experience of teaching in schools and universities has taught me to manage situations, to control them, to push towards the

A. Llewellyn (✉)
School of Education, Durham University, Durham, UK
e-mail: a.e.llewellyn@durham.ac.uk

H. Straehler-Pohl et al. (eds.), *The Disorder of Mathematics Education*,
DOI 10.1007/978-3-319-34006-7_10

discussion I want to hear, or the mathematics I want to see. That does not always work of course and sometimes people do or say what they think is best. The nagging doubt I have is, "what if this is not the best way?" and "what if I am wrong?". Doubts about my manner of teaching, of course, make me a "good" reflective practitioner. However, we could alternatively state, that through reflection, I am a teacher, who conditions myself to develop in a predefined Eurocentric liberal manner of best practice. I promote a good Western education that is designed to offer emancipation, and autonomy. But am I? Or am I instead, supporting people into a preconditioned manner of thinking? Am I facilitating a modern Western straightjacket?

We cannot help, in some way, being a product of our experiences, and in turn, shaping the things that we encounter. For mathematics education, this means that I learnt, taught, and now research mathematics education in a way that is a product of my time and context, and as such, my views, and my work are framed by this; my discernments are not objective, or universal, but instead are woven within the culture of mathematics education practice and research. Within this, there will be certain things that are easier to say, and certain things are easier to hear (MacLure, 2003). For instance: it is easy to say that mathematics is important, it is less easy to question its relevance to the world. It is easy to listen to eminent professors, it is less easy to break free of the academic norms of practice and (re)production. As a result my work may (re)produce these positions; it formulates "truths" of mathematics education. With regards to the opening paragraph of this chapter, the freedom I offer(ed) my students and the freedom I proclaim(ed) to hold are an illusion. Instead, we are all part of regulatory discourses that permit acceptable norms of behaviour; we are all involved in technologies of governance and surveillance.

In this chapter I analyse one such regulatory discourse, that of progress and the 'progressive' child, arguing that mathematics education is not universal, in spite of dominant discourses that posit it as such. Instead, I suggest it is a particular social, cultural and political practice, in which mathematics education researchers are heavily embedded and invested. To construct this argument, I explore the (re)productive practices of the mathematics education community, including schools and universities; teachers and researchers. I examine the technologies of surveillance and governance that operate, overtly and covertly, perpetuating particular versions of practice and research.

First, I begin by discussing the current cultural context of neoliberalism within the UK. I contend that this current market-led approach to society and education forms the current cultural climate of mathematics education research, the "regime of truth … that is, the types of discourse which it accepts and makes function as true" (Foucault, 1980, p. 131); it makes and shapes what is possible. Then, I move on to explore in more detail my conceptual framework for analysis, that of the work of Foucault. By using Foucault, I can question such norms, and taken-for-granted truths, and the technologies that maintain us within a social system of (re)production. By examining neoliberalism, I am unpacking the present context within the UK, and other countries such as the USA and Australia.

Neoliberalism as a Cultural Context

The culture of education in the UK is currently caught up in a neoliberal performative agenda—a policy regime defined by "the progressive enlargement of the territory of the market" (du Gay, 1996, p. 56). Though neoliberalism is not constrained by national boundaries it "is affecting education in areas as diverse as Europe, the USA, South America and Australia (Grek, Lawna, Lingard, & Varjoc, 2009; Hultqvist & Dahlberg, 2001)" (Llewellyn & Mendick, 2011, p. 50):

> Neoliberalism is in the first instance a theory of political economic practices that proposes that human well-being can best be advanced by liberating individual entrepreneurial freedoms and skills within an institutional framework characterized by strong private property rights, free markets and free trade. The role of the state is to create and preserve an institutional framework appropriate to such practices. (Harvey, 2005, p. 2)

Hence "the most basic feature of neoliberalism is the systematic use of state power to impose (financial) market imperatives, in a domestic process that is replicated internationally by 'globalisation'" (Saad-Filho & Johnston, 2005, p. 3). This rise in neoliberalism has seen significant reform in education throughout the world (Apple, 2003), and it is this reform that provides "the appearance that the state has taken responsibility for improving society and, therefore, increases the state's legitimacy" (Hursh, 2007, p. 18). Significantly, the market has begun to drive general educational reform and policy (Ball, 1993; Ozga, Simola, Varjo, Segerholm, & Pitkanen, 2009; Whitty, Power, & Halpin, 1998), and terms such as "economy", "efficiency" and "measured outcomes" have entered both educational policy and discourses. Quality assurance and evaluation have become a method of governing (Grek et al., 2009) and "comparison for constant improvement against competition has come to be the standard by which public systems are judged" (Grek et al., 2009, p. 123). This has significant implications for anyone involved in education. For instance, it is imperative that official discourses, of governments, schools and universities, show that education is improving (see also Jablonka and Bergsten in this volume). As such, "reform" or "progress" has become central for institutions to illustrate their success and justify their presence. Within this, complex systems and processes are simplified into categories and data, for the purpose of "value" judgement (Ball, 2003), and *progress* takes on the form of a high status category. As such, education has largely become fabricated around these measurable statistics and can in effect be ascribed as "governing by numbers" (Rose, 1991). This is evident in the widespread normalised use of performance indicators, league tables and pupil data within schools. Similarly within universities, targets and impact measurements are both sought and valued. The neoliberal academic who wants a career invests in performances of measurement, such as citations, large grants, "impact" and higher-valued publications. They may also value students satisfaction data related awards. Not only is the production of knowledge constrained around these parameters, but the value of "knowledge production"/academia is perhaps superseded by these performance criteria.

The space to step outside this discourse is limited, as within both schools and university, surveillance is high as the public sector borrows management models of working from the private sector, not only of performativity, but of accountability (Ball, 2003). However, a key aspect of neoliberalism is that this monitoring is predicated upon the role of the "free", autonomous self, the "individual" that chooses their own empowered path (Rose, 1999a). For the modern person, this involves an investment in self-improvement via notions of the "productive self" (Rose, 1999a, p. 103), where the person perceives their life as an on-going project. Within this, subjects become self-monitoring and self-regulating (Ball, 1994, 2003, 2008). Hence, we are under the illusion that we govern ourselves (Rose, 1999a), although we are governed through discourses of autonomy and freedom.

An example of this is discussed in my opening paragraph. Whilst educating student-teachers, I mostly follow the accepted best practice of mathematics education research within the UK. I promote discussion, and questioning, of both the self and of the mathematics. In this, I allow my student-teachers the appearance of autonomy, and I encourage them to give their students a similar freedom. However, I monitor them, and I regulate them. I ask them "open" questions, yet moderate their answers. I classify their "progress" in relation to predetermined specified standards of practice. Hence, I am complicit in the production of the student-teachers, just as the student-teachers are complicit in the making of their mathematical students.

This preference for the student that exemplifies their mathematics through verbal engagement, is an accepted "truth" of dominant stands of mathematics education research within the UK. Moreover it this "active cognitive subject" (Valero, 2002, p. 542), that is "found" within "progressive" pedagogies who is preferred (Llewellyn, 2012; Lundin, 2012). This way of doing mathematics, (which is often presented as natural) is something which I deconstruct in this chapter and elsewhere (Llewellyn, 2012); this builds on the work of Walkerdine (1988, 1998a, 1998b).

I have been told that these ideas and my work are offensive; I have been patronised, and disdained, whilst giving presentations in UK universities that dare to question this norm. This monitoring and surveillance, and use of power, is an attempt to regulate my research and the production of knowledge within mathematics education; an order to bring myself in-line with the dominant voices. However, I "choose" to resist this; I position my writing, within an awkward place that does not offer "impact" or "what-works" solutions, but instead queries what we do, and what we pass off as "common-sense". Though this "field" is not removed from its own surveillance, or norms of practice. For instance, there are ways of presentation, in which I have to make my argument (such as this chapter). In addition, there are bodies that have more citational value than others; there are those that I must reference as key speakers. This argument holds for all of academic research: academic work in the social sciences posits itself upon citation and its hierarchies. The voices that acknowledge this as a technology of surveillance are rare.

Foucault and a Poststructural Framework

To deconstruct said practice, I utilise a poststructural, Foucauldian framework, one that takes language as ambiguous, constructive, and created through action (MacLure, 2003), and one that contests discourses of universal and timeless "truths". Instead, poststructuralism encourages us to ask, "what's going on just now? What's happening to us? What is this world, this period, this precise moment in which we are living? Or in other words: What are we?" (Foucault, 2003c, p. 133). It is these questions that allow us to examine how we are constituted and what we think we are.

For Foucault, there is no singular truth, nothing that accurately describes the world; Instead there are ways of making sense of the world. These ways of making meaning differ in their status and their power to have effects in the world. Some will acquire the status of truths while others will be dismissed:

> I am quite aware that I have never written anything but fictions. I'm not saying for all that this is outside truth. It seems to me the possibility exists to make fiction work in truth, to induce effects of truth with a discourse of fiction, and to make it so that the discourse of truth creates, "fabricates" something that does not yet exist, therefore "fictionalizes." (Foucault, 1989, p. 213)

Thus, Foucault and poststructuralism allows us to "think more about how we think" (Flax, 1987, p. 624), and how we act. Consequently, we may disrupt the proliferation of common sense assumptions, and the simplistic route to "best" practice. Instead, poststructuralism allows for the mess, and for the "daily struggle and muddle of education" (Donald, 1985, p. 242); in fact, it positively encourages it. This is vitally important in an education regime that tends towards generalisations through a "what works" (Oancea & Pring, 2008) movement. Furthermore, it is important in mathematics education, which is often built upon assumptions of reason, rationality and absolute truths. Indeed Foucault asserts reason and the rational subject to be the major arc of ongoing philosophical thought. He states (1984, p. 249):

> I think that the central issue of philosophy and critical thought since the eighteenth century has always been, still is, and will, I hope, remain the question: *What* is this Reason that we use? What are its historical effects? What are its limits, and what are its dangers? How can we exist as rational beings, fortunately committed to practicing a rationality that is unfortunately crisscrossed by intrinsic dangers?

Hence, Foucault allows us to question the reasoned mind and Western privilege (Moi, 1988). Moreover, it is an endeavour to pull "ourselves free of the web" (Walkerdine, 1998a, p. 15) of rules and power that keep us woven within common discourses.

In this chapter, I examine these rules through acknowledging and deconstructing governmentality, which is a means of legitimatising, systemising and regulating the use of power, but one that is discursively produced as for the good of both the system and the self; this is very different to an authoritarian method of governing (Foucault, 2003a). This is particularly pertinent within neoliberalism, as it is premised upon self-governance and reason, rather than a disciplinary power (Rose, 1999b). Both "liberalism and neoliberalism are seen as practices, reflexive modes of action, and special ways of rationalizing the governance" (Cotoi, 2011, p. 112). Thus, we are governed into the premise that we govern ourselves. Furthermore, this deconstruction is imperative within education, as education establishments are most often positioned as places of universal good, and are symbols of mastery of nature and society (Dale, 2001). Hence, power/knowledge can circulate uncontested; there being little questioning of the knowledge imparted and often little interrogation of the systems and strategies employed. However, "the school and the university both perform the function of a technology of power … They train people towards acceptable behaviour" (Walshaw, 2007, p. 102). Thus for a university, it is vitally important that as a critical institution, researchers, including myself, are aware of our own practice in the formation and maintenance of "truths". Moreover, within a neoliberal education system and society, it is important that we deconstruct the self and the narrative of governance:

> Studies of governmentality are not sociologies of rule. They are studies of a particular "stratum" of knowing and acting. Of the emergence of particular "regimes of truth" concerning the conduct of conduct, ways of speaking truth, persons authorized to speak truths, ways of enacting truths and the costs of so doing. (Rose, 1999b, p. 19)

In the rest of this chapter, I discuss how methods of governance and technologies of power, and surveillance from within mathematics education, work to maintain the privileging of progress and the progressive subject. Arguing that this heavy and uncritical investment in the "free" autonomous subject has particular appositeness within neoliberalism.

Progress and Progressivism as (Re)productions of Mathematics Education

(Re)productions of Progress

It is difficult to imagine a society that does not try to make things better; indeed "we have coined no political substitute for progressive understandings of where we have come from and where we are going" (Brown, 2001 p. 3). Since the modern era, we have become consumed and constrained by a teleological discourse of improvement. We are not easily content to reflect on where we are, but instead to look to where we have come from and where we are heading, "with the Faustian notion of becoming rather than being" being imperative (Popkewitz, 2008, p. 26).

The triumph of modernity, through rationality and reason has arguably seen societies become "enlightened" through the advancement of science (including mathematics) and the arts. Education is a fundamental part of this production; it is positioned as a way of correcting society's ills (Popkewitz, 1988, 2008). As a system of reason, education is both a measure of improvement (from both across and within countries), as well as a means of improvement, for both the self and society. This merges the individual and national, the former offering the promise of autonomy and emancipation that is prevalent within discourses of neoliberalism (Mendick, 2011).

For national progress, this is enhanced, as mathematics education is positioned as having responsibility for economic growth and progress. This maxim has international validity, such that "powerful supranational organizations, such as the OECD and the World Bank, view education primarily as a tool for improving economic performance" (Gilead, 2012, p. 113); indeed, according to the OECD, mathematical literacy is key to democratic citizenship (OECD, 2013). Again this is enhanced within, and thus governed by neoliberalism, where:

> [E]quity and enterprise, technological change and economic progress are tied together within the efforts, talents and qualities of individual people and the national collective—the "us" and the "we." (Ball, 2008, p. 17)

As such, economic progress is constructed around the promise of benefiting both the individual and the national.

This positioning permeates educational discourses such that progress is sought by pupils, teachers, schools, governments and often by mathematics education researchers. Pupils have to make progress in their lessons; teachers have to show that their pupils are making progress, and they, as professionals are making progress. Governments demonstrate that their policies result in progress for schools and societies, and many researchers demonstrate that their ideas result in progress; progress in mathematics education is an "anchoring narrative" (Mendick, 2011, pp. 50f.).

As mathematics education researchers, we may consider how readily we accept this position, especially as it is one that brings prestige to our subject, and one that legitimates grants and publications. If we accept it, as many in mathematics education research do (Lundin, 2012; Mendick, 2011) then we are complicit in governing a possible "truth" of mathematics, one that always already keeps mathematics as privileged. We may ask, where is the critique? And subsequently what is the role of academia? Furthermore, this investment may be "inevitable" in UK education, but does not necessarily hold for all university research, other genres of academic research, are broader in their outlook (Dale, 2001). Drawing on Dale (2001), Mendick states (2011, p. 50):

> As Dale points out, this is very different from researchers in sociology of religion or the family where the research agenda operates independently of their personal views on the social roles of religion or family. In contrast, in the sociology of education, education is treated less an object of study than as a resource.

Thus, particularly within the UK, educational researchers do not necessarily study education, but seek to improve it, and demonstrate that it can advance society. However, the evoking of progress within the academic discipline of mathematics education is ubiquitous—both "mathematics" and "education" being signs of development.

Within neoliberalism progress within education takes on a specific form of performativity. This concerns the improvement of the self, but moreover manifests through a measurable function that has the potential to curtail the field of mathematics education research further. This performativity, ensures that "from the 1990s, the official discourse tells the story of a sustained programme of improvement/reform of education in England, linked to the creation of objective "depersonalised" judgement and increasingly driven by apparently objective data." (Ozga et al., 2011, p. 114). Hence, education has to produce data that demonstrates improvement of schools, universities and governments.

For any educational research measurable data is problematic in that it encourages a specific way of knowing and creating truths. For instance, amongst other areas, universities in the UK are judged by the Research Excellence Framework (REF), where citation and "impact" are currently seen as paramount; this of course both enables and restricts knowledge production, along particular paths and within limited time frames. In addition, a certain type of research may be privileged; research that tells straightforward stories, and allows quantifiable, "objective results". This is shown in the recent preference in educational research for "evidence-based" practice and for the pursuit of figures that give uncomplicated answers to "improving" education (Hamilton & Corbett-Whittier, 2013). Oancea and Pring (2008) call this the "what works" movement, which they contend has come to dominate educational policy and practice. They give examples from the US and UK contexts to show how research such as Randomized Control Trials (RCTs) have been held up as the gold standard of education. With RCTs, researchers are looking to employ a cause and effect relation, hence treating students as commodities, and researchers as objectified scientists that design and report on experiments. Education is (re)imagined as a field akin to medicine, that can perform a diagnosis, provide the correct medicine, and the self and society are fixed.

As such, mathematics education research is caught up in and bound by this construction, such that the parameters that allow it also limit its production (Brown & Clarke, 2013). Some in mathematics education research may be concerned with impact, with short time-scales, and with advocating oneself unquestionably. Moreover, they may be concerned with the pursuit of statistical evidence to show "what works" and what is "best practice". A positivist may argue that this removes bias from any data collection, which may be preferable to doing research from certain already assumed positions, such as progressivism (discussed in the next section). However, any system, especially one that is based upon judgement and privileges impact, governs what is possible and what is not; it includes some and excludes others. Hence, it creates and limits our view of education in practice; in this instance, it suggests that studying and researching mathematics education can be removed from its context, and is universal and unproblematic. These research strategies make certain ontological assumptions. For example, they do:

> [N]ot fully account for knowledge of the world as "taken" by the person, rather than "given," or for research that aims to destabilise taken for granted concepts and frameworks rather than replace them with equally closed alternative systems. As such, it is a restrictive model. (Oancea & Pring, 2008, p. 22)

There are of course many studies worldwide that are aware of cultural significance when learning mathematics, particularly those in comparative research (for example, Clarke et al. (2006) and Stigler and Hiebert (2009)). However, my concern is for mathematics education within the UK where cultural relevance is absent, and where mathematical knowledge is often viewed as absolute.

Thus, through governance, and "technologies of surveillance" researchers can become complicit in (re)producing knowledge that fosters normalisation, a process that encourages a specific version of the "normal" that subsequently becomes taken-for-granted or "natural" (Foucault, 1977/1991, 1978/1998), as the many researchers add to the bank of knowledge that has already been said. We are not free to break from our own academic norms, practices and progress in spite of an illusion that we can. Instead, mathematics education research is governed and governs what is possible within itself.

An examples of a taken-for-granted "truth" that creates and constrains what is possible within mathematics education research is that mathematics is important to society, and furthermore mathematics is a means and measure of progress. Within this, mathematics becomes a gatekeeper for success (Gates, 2001), and a good grade at school mathematics is essential. Hence, if anyone were to suggest that mathematics education is not useful, which was the argument discussed recently by Pais (2013, see also in this volume) at the conference *Mathematics and Contemporary Theory*, they would struggle to be heard. Even within the liberal academic space of that conference, resistance to his argument stemmed from the principle that Pais is saying something that these people, who are invested in mathematics education and its utility, did not want to hear. Educational researchers have their own investments and own methods of governing, though we often do not articulate or acknowledge them.

(Re)productions of Progressivism

Progress and progressivism are not the same but they are both connected by the notion of reform, development and a better way. In addition, they are both "taken for granted truths" of mathematics education research, this "anchoring narrative" (Mendick, 2011, pp. 50f.), being facilitated by and hence governed through research in mathematics education.

Arguably, UK mathematics education research's promotion of progressivism is more overt, than it is of progress. Many researchers' assertion is that the best mathematics teaching involves active pupils who "understand" the mathematics. As alluded to in the opening paragraph of this chapter, it asks "why?" questions, and allows for discussion. It can be termed "progressive" or reform, open, non-traditional and student/child-centred, with foundations in educational reformers such as Rousseau (see Rousseau, 1763/1884), Dewey (see for example Dewey, 1902/1956, 1916) and/or development psychologists such as Piaget (see for example Piaget & Garcia, 1989, or Piaget & Inhelder, 1969/2000); it is commonly positioned against "traditional" pedagogies, also known as closed, teacher-led and subject-centred. These pedagogies are evident in the UK, but are also reflected in the "Math Wars" found in the USA, where proponents of reform and traditional mathematics education fiercely advocate each position and corresponding curricula.

They seem to hold a resilient position, in spite of several scholars warning against propagating these false binaries (including Alexander, 1994, 2010; Pring, 1989), and regardless of other practices around the world. My concern is the tendency of many to strongly advocate one position at the detriment to the other, neither of which are "real". These "discourses of dichotomy" (Alexander, 2010), are prevalent within UK education; indeed some educational research encourages comparative studies of apparent dualisms. However, this positioning limits possibilities (Alexander, 2010), as opinions are restricted to these binaries, discussion is focused around extremes, and the creation of new knowledge is considerably constrained (MacLure, 2003). Furthermore, people are positioned into feuding from mythical fixed locations. This tendency towards tunnel vision is common within political and educational discourses, where strong principles as statements of worth are required. The "good" politician knows their narrative, they know how to fix problems and society; the "good" educational researcher similarly knows their tale, they stay close to their specialised area of research as they defend and propagate it—that is their job. As already discussed, this is grievous within education establishments that are holders of knowledge and "truths", and that teach what is possible (Dale, 2001); this is governance most covert.

As I have discussed in Llewellyn (2012), and has been discussed elsewhere (e.g. Lundin, 2012), the preference for "progressive" education, the "active" learner and teaching for "understanding" is clear within dominant branches of mathematics education research within the UK; and other parts of the worlds such as the USA and Australia. For example, the UK's Association of Teachers of Mathematics (ATM) stance on this position is unbending.

Progressive education may draw from the past, however, it is implicated in the future. Specifically, progressive education is always already focused upon development, more so than the here and now. It is future orientated despite the focus on "free" play. This "cognitive model of the infant as problem solver mirrors that of the assembly worker, with research privileging those activities and products which will enhance performance" (Burman, 2008, p. 43); attention is given to what the child will become, rather than with what the child is (Burman, 2008). Whilst, there may be some aspects of progressive education that clash with performativity, it is progressive pedagogies focus on "becoming" that mirrors neoliberalism.

In addition, within progressivism the child is positioned as free, and knowledge is produced through experience. The progressive child is assumed to be predisposed to education and development, which is viewed as both natural and inevitable. Specifically "'the child' is deferred in relation to certain developmental accomplishments" (Walkerdine, 1997, p. 61), for example, Piagetian stages, where the child learns in "logical" developmental blocks. As such, it assumes that intrinsic motivation, inquisition, and curiosity are "real". Therefore, the concern is with a specific kind of development and anyone who does not follow this is deviant (Walkerdine, 1990). However, if the natural is not evident, the teacher's role is to produce this from "within" the student; "the behaviours [of the child] do not precede the practice precisely because their specificity is produced in these practices" (Walkerdine, 1990, p. 138). Hence, this version of mathematics and this mathematical child are not natural, but instead, are a production of discourses of progressive education, and developmental psychology (Burman, 1992, 2008; Burman & Parker, 1993; Henriques, Hollway, Urwin, Venn, & Walkerdine, 1998; Walkerdine, 1997, 1998a; Walkerdine & Lucey, 1989). Furthermore, it is these markers that have allowed for the standardization of children's development in schools, and although many progressive educators would loathe this, it is this normalisation that has facilitated an assessment and value culture that works within neoliberalism (Jenks, 1996). Furthermore, progressivism validates the fantasy of the autonomous self.

A simple cultural contrast to progressive education is provided by China, where the education system is built upon Confucian ways of being, and the culture of respect and hard work. Here, teaching for student involvement is not excluded; however, it is thought that this can be best achieved through mastery of skills first, via the repetition of mathematical methods and cognising through listening. This pedagogy is often constructed as a passive, deficit model by Western writers, although recent work in cultural and language studies questions this Othering and derisory, discursive positioning (Grimshaw, 2007). From the Chinese perspective, it is this practice that allows pupils to think and hence develop their own methods and autonomy (Jin & Cortazzi, 1998). This is possible within a Confucian based culture, where values of respect are paramount. However, we must be careful of adopting an essentialist position, where approaches to culture and learning are homogenised and viewed as fixed within a country. For instance, Chinese educators will be diverse in their practices. Moreover, China, through increased globalisation, is currently responding to the influence of Western approaches to learning (Grimshaw, 2007). Hence, whether following a discourse, or resisting it, objectifying and separating best pedagogy from the cultural regime of truth of the time and context is fallacious. As Radford states "classroom forms of knowledge production and human cooperation are not created in situ, on the spot. On the contrary, they are related to culturally and historically constituted political and economical forms of production of human existence" (Radford, 2014, p. 2).

In highlighting this example, I am not setting up these positions as real, indeed I would argue that mathematics education in China is not the polar opposite of the UK, the situation is more nuanced and complex. For instance: a child can be dictated to, but that does not stop them thinking; a child can have "freedom" but still feel constrained; "understanding" can occur at different stages for different people. Instead, I am arguing that what is seen as progress is produced by its context, and by our cultural, social and political norms; moreover, this progress is mostly conditioned via the norms of the "teacher", not of the child, regardless of the narrative of student-centred progressivism. Hence, in spite of globalisation, what is possible and preferable for one person in the UK, may not be the same as someone in China—and neither of these countries are homogenous in their production of people and privilege. My concern is that many, within mathematics education research within the UK, only accept a certain version of progress; students are only good enough if they perform in a certain "progressive" manner; this is the specific practice that is privileged and supported by mathematics education research. Through governance and technologies of surveillance, (such as articles, books, citation, teaching, impact, progress,) and most simply, norms of what is acceptable to say, some educators only allow one version of mathematics education—that of the progressive child. This "has effects of constraining human possibilities and marginalizing those who fall outside the 'nature'" (Pickett, 1996, p. 452)

Of course, Foucault rigorously critiqued the human sciences, his "initial critique of human sciences is that they, like philosophy, are premised on an impossible attempt to reconcile irreconcilable poles and posit a constituting subject" (Best & Kellner, 1991, p. 42). This is relevant in this case as "the presupposition of the individual as a unitary entity, a thinking, feeling machine which is self-directed as far as thought processes are concerned, is basic to a child-centred pedagogy and to developmental psychology" (Henriques et al., 1998, p. 102). Foucault states that "one has to dispense with the constituent subject, and to get rid of the subject itself, that's to say, to arrive at an analysis which can account for the constitution of the subject within a historical framework" (Foucault, 1980, p. 117). This critique is particularly pertinent as the human sciences are currently popular in both general culture and in educational discourses. This popularity encourages the fabrication of the subject around a "real" self to identify with, and aspire to; as such, the modern person can be drawn to essentialised discourses of the self that are aligned to psychological models. This fits well within neoliberalism where the autonomous, and entrepreneurial figure is central (Rose, 1999a). This involves striving to make lives meaningful and make sense of the self: to see their life as an ongoing project (Rose, 1999a). From a Foucauldian perspective, the modern person is "not the man [sic] who goes off to discover himself, his secrets and his hidden truth; he is the man who tries to invent himself" (Foucault, 2003d, p. 50). As Foucault would argue, we "govern (themselves and others) by the production of truth … the establishment of domains in which the practice of true and false can be made at once ordered and pertinent" (Foucault, 2003b, p. 252). Hence, we govern ourselves to accept fabricated fictions as the truth. Moreover, power is thought to be decentred and individuals are regulated through self-regulation. The role of the state, or the institution is to create a regime of truth that demands this individuality (Rose, 1999a, 1999b). It is not a dictatorship but a covert and subversive manner of governing that breeds a specific acceptable version of the normal. In schools, "students learn to monitor their own being, and they do this by practices of self-regulating, ever mindful of the gaze of others" (Walshaw, 2007, p. 131); similarly so do people in universities—including academics, and those in mathematics education research. We are all involved in governance, and surveillance of others, and the self.

Hence, through the idea of "becoming" the popularity of developmental psychology in mathematics education is reinforced by the rise of the individual within neoliberalism, and the governance of individuals through the promise and positioning of autonomy (Rose, 1999a, 1999b). As discussed, this governmentality, is governing by tactics as opposed to governing by law (Foucault, 2003a). The tactics constitute a system where there is "enforced obedience to rules that are presumed to be for the public good" (Walshaw, 2007, p. 102). In this case, mathematics education researchers govern and propagate their norms, that of "real" mathematics (Boaler, 2009) and of a "real" mathematical child, which relies upon an acultural, asocial and apolitical subject. This "modern conception of the student is largely based on the ideas of rational self-regulation, autonomy, and self-sufficiency. It assumes that the origin of meaning, knowledge, and intentionality is located *within*, and must come *from*, the individual ... [which is] a historical invention" (Radford, 2014, p. 4). My concern is that many in mathematics education do not acknowledge or critique the "free" subject as a cultural product; indeed they do not critique themselves and their role in the production of such norms. I propose it is the job of mathematics education researchers to contest such normalisation, rather than to propagate it. Particularly as education is often premised upon "truths" and reason, we must acknowledge our own governance.

Concluding Remarks

There are two main arguments to consider from this chapter. One concerns the privileging of progress, and progressivism, and the other concerns the manner in which these practices are proliferated as common sense norms through governance within mathematics education research. I have argued that within the sea of surveillance that many criticise within education, we often fail to recognise our own methods of governing; specifically that we are part of a production of norms and "truths". We cooperate in, and (re)produce technologies of power and surveillance, often without acknowledgement of our roles, or a critical stance concerning ourselves; for instance, citation is a dominant, yet uncontested method of governing. I am concerned that mathematics education researchers too often construct themselves as removed from the discourses that circulate and create positions or "problems" within mathematics. Our research is somehow above this, particularly with regards to progressivism—it is "natural"; and, just as the preferable way of doing mathematics is "natural", so is the preferable mathematical child. Using Foucault, and many who have drawn on his work (cited throughout), I argue there is no natural, but instead there are discourses that are cultural products of time and context. What is possible for someone within a mathematics classroom today in the UK is specific to cultural norms. What is possible for the mathematics education researcher is similarly so.

Of course, I too am not removed from this inward gaze. There are certain bodies I have cited and others I have ignored; as such, I am complicit in the (re)production of knowledge. However, I contend that I am not complicit in the (re)production of some norms of mathematics education. Specifically, in this chapter, I have attempted to disengage from trends in mathematics education, and its dominant discourses. Instead, I have queried the privileging of mathematics education, and of progress. These are arguments I know have offended fellow researchers. But we may ask ourselves—why is it that so? Why is it difficult to hear ideas that do not fit with your own view, or with the norm, particularly when it concerns progressivism, the promise of freedom, emancipation and development?

My concern, as always, is with that which is produced as "normal" and in this instance also as "natural"; this is particularly pertinent with the UK where our Western privileging seems to dismiss cultural specificities concerning the teaching and learning of mathematics. Furthermore, my concern is with that which is always already found within education, such as "progress". To counteract this, we should interrupt dominant discourses of mathematics education. For example, we must query what we mean by progress, and examine its cultural relevance, particularly within an era of self-improvement and self-governance such as neoliberalism. In addition, I suggest that the mathematics education research community "should" look for alternative ways around essentialising and/or normalising the mathematical child; particularly as asocial, acultural and apolitical; and as focused on what is to come, rather than what is present. More than this, we "should" examine ourselves in relation to what we do, what we promote and the technologies of power and surveillance that support this. It is the covert forms of governance, posing as freedom that are particularly concerning; particularly when they come through the always already esteemed academy of the university and education.

References

Alexander, R. J. (1994). *Innocence and experience: Reconstructing primary education*. Trentham: Stoke-on-Trent.

Alexander, R. J. (Ed.). (2010). *Children, their world, their education: Final report and recommendations of the Cambridge Primary Review*. Abingdon: Routledge.

Apple, M. W. (2003). *The state and the politics of knowledge*. New York: Routledge Falmer.

Ball, S. J. (1993). Education markets, choice and social class: The market as a class strategy in the UK and the USA. *British Journal of Sociology of Education, 14*(1), 3–19.

Ball, S. J. (1994). *Education reform: A critical and post-structural approach*. Buckingham, Philadelphia: Open University Press.

Ball, S. J. (2003). The teacher's soul and the terrors of performativity. *Journal of Education Policy, 18*(2), 215–228.

Ball, S. J. (2008). *The education debate*. Bristol: The Policy Press.

Best, S., & Kellner, D. (1991). *Postmodern theory: Critical interrogations*. Basingstoke: Communications and Culture.

Boaler, J. (2009). *The elephant in the classroom: Helping children learn and love maths*. London: Souvenir Press.

Brown, W. (2001). *Politics out of history*. Princeton, NJ, Oxford: Princeton University Press.

Brown, T., & Clarke, D. (2013). Institutional context for research in mathematics education. In M. A. Clements, A. J. Bishop, C. Keitel, J. Kilpatrick, & F. K. S. Leung (Eds.), *Third international handbook of mathematics education* (pp. 459–484). New York, London: Spinger.

Burman, E. (1992). Development psychology and the postmodern child. In J. Doherty, E. Graham, & M. Malek (Eds.), *Postmodernism and the social sciences*. New York: St Martin's Press.

Burman, E. (2008). *Deconstructing developmental psychology*. London: Routledge.

Burman, E., & Parker, I. (1993). *Discourse analytic research*. London, New York: Routledge.

Clarke, D., Keitel, C., & Shimizu, Y. (2006). *Mathematics classrooms in twelve countries: The insider's perspective* (Vol. 1). Rotterdam: Sense Publishers.

Cotoi, C. (2011). Neoliberalism: A Foucauldian perspective. *International Review of Social Research, 1*(2), 109–124.

Dale, R. (2001). Shaping the sociology of education over half-a-century. In J. Demaine (Ed.), *Sociology of education today* (pp. 5–29). Basingstoke: Palgrave.
Dewey, J. (1902/1956). *The child and the curriculum; and the school and society*. Chicago, London: University of Chicago Press.
Dewey, J. (1916). *Democracy and education: An introduction to the philosophy of education*. New York: Macmillan.
Donald, J. (1985). Beacons of the future: Schooling, subjection and subjectification. In V. Beechey & J. Donald (Eds.), *Subjectivity and social relations* (pp. 214–249). Milton Keyes: Open University Press.
du Gay, P. (1996). *Consumption and identity at work*. London: Sage.
Flax, J. (1987). Postmodernism and gender relations in feminist theory. *Signs, 12*(4), 621–643.
Foucault, M. (1977/1991). *Discipline and punishment: The birth of the prison*. London: Penguin.
Foucault, M. (1978/1998). *The history of sexuality (volume one): The will to knowledge*. London: Routledge.
Foucault, M. (1980). Truth and power. In C. Gordon (Ed.), *Power/knowledge: Selected interviews and other writings 1972-1977* (pp. 109–133). New York: Pantheon.
Foucault, M. (1984). Space, knowledge and power. In P. Rabinow (Ed.), *The Foucault reader* (pp. 239–256). London: Penguin.
Foucault, M. (1989). Power affects the body. In S. Lotringer (Ed.), *Foucault live: Collected interviews, 1961-1984* (pp. 207–213). New York: Semiotext(e).
Foucault, M. (2003a). Governmentality. In P. Rabinow & N. Rose (Eds.), *The essential Foucault: Selections from essential works of Foucault, 1954-1984* (pp. 229–245). New York, London: The New Press.
Foucault, M. (2003b). Questions of method. In P. Rabinow & N. Rose (Eds.), *The essential Foucault: Selections from essential works of Foucault, 1954-1984* (pp. 246–258). New York: The New Press.
Foucault, M. (2003c). The subject and power. In P. Rabinow & N. Rose (Eds.), *The essential Foucault: Selections from essential works of Foucault, 1954-1984* (pp. 126–144). New York, London: The New Press.
Foucault, M. (2003d). What is enlightenment? In P. Rabinow & N. Rose (Eds.), *The essential Foucault: Selections from essential works of Foucault, 1954-1984* (pp. 43–57). New York, London: The New Press.
Gates, P. (Ed.). (2001). *Issues in mathematics teaching*. London, New York: RoutledgeFalmer.
Gilead, T. (2012). Education and the logic of economic progress. *Journal of Philosophy of Education, 46*(1), 113–131.
Grek, S., Lawna, M., Lingard, B., & Varjoc, J. (2009). North by northwest: Quality assurance and evaluation processes in European education. *Journal of Education Policy, 24*(2), 121–133.
Grimshaw, T. (2007). Problematizing the construct of "the Chinese learner": Insights from ethnographic research. *Educational Studies, 33*(3), 299–311.
Hamilton, L., & Corbett-Whittier, C. (2013). *Using case study in education research*. London; Thousand Oaks CA; New Delhi; Singapore: Sage.
Harvey, D. (2005). *A brief history of neoliberalism*. Oxford: Oxford University Press.
Henriques, J., Hollway, W., Urwin, C., Venn, C., & Walkerdine, V. (Eds.). (1998). *Changing the subject*. London, New York: Routledge.
Hultqvist, K., & Dahlberg, G. (Eds.). (2001). *Governing the child in the new millennium*. London: RoutledgeFalmer.
Hursh, D. W. (2007). Marketing education: The rise of standardized testing, accountability, competition, and markets in public education. In E. W. Ross & R. Gibson (Eds.), *Neoliberalism and education reform*. Cresskill, NJ: Hampton Press.
Jenks, C. (1996). *Childhood*. London: Routledge.
Jin, L., & Cortazzi, M. (1998). The culture the learner brings: A bridge or a barrier? In M. Bryam & M. Fleming (Eds.), *Language learning in intercultural perspective: Approaches through drama and ethnography* (pp. 98–118). Cambridge: Cambridge University Press.

Llewellyn, A. (2012). Unpacking understanding: The (re)search for the Holy Grail of mathematics education. *Educational Studies in Mathematics, 81*(3), 385–399.
Llewellyn, A., & Mendick, H. (2011). Does every child count? Quality, equity and mathematics with/in neoliberalism. In B. Atweh, M. Graven, W. Secada, & P. Valero (Eds.), *Mapping equity and quality in mathematics education* (pp. 49–62). Dordrecht, New York: Springer.
Lundin, S. (2012). Hating school, loving mathematics: On the ideological function of critique and reform in mathematics education. *Educational Studies in Mathematics, 80*(1–2), 73–85.
MacLure, M. (2003). *Discourse in educational and social research*. Buckingham: Open University Press.
Mendick, H. (2011). *Is progress good for mathematics/education?* Paper presented at the British Society for Research into Learning Mathematics, London.
Moi, T. (1988). Feminism, postmodernism, and style: Recent feminist criticism in the United States. *Cultural Critique, 9*, 3–22.
Oancea, A., & Pring, R. (2008). The importance of being thorough: On systematic accumulations of "What Works" in education research. *Journal of Philosophy of Education, 42*, 15–39.
OECD. (2013). *PISA 2012 Assessment and Analytical Framework: Mathematics, Reading, Science and Problem Solving and Financial*. Literacy: OECD Publishing.
Ozga, J. (2009). Governing education through data in England: From regulation to self-evaluation. *Journal of Education Policy, 24*(2), 149–162.
Ozga, J., Simola, H., Varjo, J., Segerholm, C., & Pitkanen, H. (2011). Central-local relations of governance. In J. Ozga, P. Dahler-Larsen, C. Segerholm, & H. Simola (Eds.), *Fabricating quality in education. Data and governance in Europe* (pp. 107–126). London: Routledge.
Pais, A. (2013). *Conversation 2: Theory to analyze the macro dimension of classroom interactions*. Paper presented at the Mathematics, Education and Contemporary Theory 2.
Piaget, J., & Garcia, R. (1989). *Psychogenesis and the history of science*. New York: Columbia University Press.
Piaget, J., & Inhelder, B. (1969/2000). *The psychology of the child*. New York: Basic Books.
Pickett, B. (1996). Foucault and the politics of resistance. *Polity, 28*(4), 445–466.
Popkewitz, T. (1988). Educational reform: Rhetoric, ritual and social interest. *Educational Theory, 38*(1), 77–93.
Popkewitz, T. (2008). *Cosmopolitanism and the age of school reform: Science, education, and making society by making the child*. New York, Abingdon, Oxon: Routledge.
Pring, R. (1989). Subject-centred versus child-centred education—a false dualism. *Journal of Applied Philosophy, 6*(2), 181–194.
Radford, L. (2014). *On teachers and students: An ethical cultural-historical perspective*. Paper presented at the 38th Conference of the International Group for the Psychology of Mathematics Education and the 36th Conference of the North American Chapter of the Psychology of Mathematics Education, Vancouver, Canada.
Rose, N. (1991). Governing by numbers: Figuring out society. *Accounting Organisation and Society, 15*(7), 673–692.
Rose, N. (1999a). *Governing the soul* (2nd ed.). London: Free Association Books.
Rose, N. (1999b). *Powers of freedom: Reframing political thought*. Cambridge: Cambridge University Press.
Rousseau, J. (1763/1884). *Emile, or, concerning education: Extracts, containing the principle elements of pedagogy found in the first three books/by Jean-Jacques Rousseau; with an introduction and notes by Jules Steeg*. London, DC: Heath and Company.
Saad-Filho, A., & Johnston, D. (Eds.). (2005). *Neoliberalism: A critical reader*. London, Ann Arbor, MI: Pluto Press.
Stigler, J. W., & Hiebert, J. (2009). *The teaching gap: Best ideas from the world's teachers for improving education in the classroom*. New York: Simon and Schuster.
Valero, P. (2002). *The myth of the active learner: From cognitive to socio-political interpretations of students in mathematics classrooms*. Paper presented at the Mathematics Education and Society, Helsingør, Denmark.

Walkerdine, V. (1988). *The mastery of reason*. London and New York: Routledge.
Walkerdine, V. (1990). *Schoolgirl fictions*. London: Verso.
Walkerdine, V. (1997). Redefining the subject in situated cognition theory. In D. Kirshner & J. A. Whitson (Eds.), *Situated cognition: Social, semiotic, and psychological perspectives* (pp. 57–70). Mahwah, NJ: Lawrence Erlbaum.
Walkerdine, V. (1998a). *Counting girls out: Girls and mathematics* (2nd ed.). Abingdon: RoutledgeFalmer.
Walkerdine, V. (1998b). Developmental psychology and the child-centred pedagogy: The insertion of Piaget into early education. In J. Henriques, W. Hollway, C. Urwin, C. Venn, & V. Walkerdine (Eds.), *Changing the subject: Psychology, social regulation and subjectivity* (pp. 153–202). London, New York: Routledge.
Walkerdine, V., & Lucey, H. (1989). *Democracy in the Kitchen*. London: Virago.
Walshaw, M. (2007). *Working with Foucault in education*. Rotterdam: Sense Publishers.
Whitty, G., Power, S., & Halpin, D. (1998). *Devolution and choice in education: The school, the state and the market*. Philadelphia: Open University Press.

Part III
Disordering School Mathematics

Chapter 11
A Socio-critical Analysis of Students' Perceptions of Mathematics

David Kollosche

Abstract Rather than studying students' perceptions of mathematics from a beliefs or identity framework with the purpose of improving the learning of mathematics, this study develops a Foucauldian framework, which allows a socio-critical interpretation of students' perceptions, which are considered an indicator for their developing subjectivities. This allows me to discuss how diverging devotions to mathematics, suffering from mathematics as well as seeing personal relevance and challenges in mathematics connects to the institutional and societal functionality of mathematics education. Thereby, I also present data obtained in questionnaires from German ninth grade students.

Introduction

Mathematics education research is often assuming that mathematics education is primarily concerned with providing opportunities to all students to "learn" mathematics and develop mathematical "competences". Therefore, much research in mathematics education connects educational, psychological and mathematical theories in order to improve the learning of mathematics (Kilpatrick, 1992). This kind of research can be considered normative as it lays an ideological foundation of what mathematics education should be about. Within this endeavour, students' perceptions of mathematics have become of interest, on the one hand conceptualised as students' "beliefs" or "attitudes" about the nature and the learning of mathematics (Leder, Pehkonen, & Törner, 2002; Maaß & Schlöglmann, 2009), and on the other hand as part of students' mathematical "identities", which shape their belonging to communities of practice in the mathematics classroom (Grootenboer & Jorgensen, 2009; Sfard & Prusak, 2005). Although these approaches build on different theoretical backgrounds, both share the traditional assumption that mathematics education is primarily for the learning of mathematics and both approaches understand the

D. Kollosche (✉)
Humanwissenschaftliche Fakultät, Universität Potsdam, Potsdam, Germany
e-mail: david.kollosche@uni-potsdam.de

H. Straehler-Pohl et al. (eds.), *The Disorder of Mathematics Education*,
DOI 10.1007/978-3-319-34006-7_11

analysis and manipulation of beliefs, attitudes and identities as a contribution to the narrative of progress (cf. Llewellyn in this volume) in the teaching and learning of mathematics.

In contrast to the above-described conceptualisation of mathematics education, a growing branch of research is concerned with sociopolitical aspects of mathematics education (Valero, 2004), developing alternative perspectives on the mathematics classroom. Typically, authors from this branch strive to distance themselves from normative presumptions and attempt to provide descriptive analyses of the connections between mathematics education and the sociopolitical. As it has been shown that mathematical knowledge and competence is usually not transferred from school to other social domains but learnt in practice (Lave, 1988) and as it has been documented at least for the German case that adults remember hardly any contents, which they had learnt in secondary mathematics education (Maaß & Schlöglmann, 2000), mathematics education would be a gigantic failure, was its primary social function the learning of mathematical content knowledge. Consequently, it could be argued that there must be different social functions, which explain the persistence of such an enormously expensive institution as mathematics education—social functions, which may be at odds with the normative ideology of traditional mathematics education research (cf. Pais in this volume).

Different contributions have highlighted various dimensions in which mathematics education is connected to the sociopolitical formation of the individual and society. Mathematics education can be understood as a "gate-keeper" which decides over opportunities in further education and work (Stinson, 2004; Volmink, 1994). It has been shown that school mathematics systematically disadvantages students from low socio-economic backgrounds (Dowling, 1998) and from certain ethnicities despite their mathematical abilities (Gutiérrez, 1999; Martin, 2009; Stinson, 2013), thus reproducing existing distributions of power in society. Interestingly, these distributions of power are not only based on how mathematics education is organised, but on the ideology at work in the mathematics classroom (Straehler-Pohl & Pais, 2014). Stinson (2013) discusses the "white male math myth", an understanding of mathematics as an endeavour reserved for (Western and Asian) white males only. However, we can see ideology at work beyond the differences between socio-economic status, ethnicity and gender. In spite of its controversial philosophy, mathematics is often presented as an apolitical, undebatable, rational, omnirelevant and universally valid endeavour, thus installing mathematics as a tool of power throughout society and hindering students from questioning applications of mathematics or mathematical thinking (Dowling, 1998; Skovsmose, 2005; Ullmann, 2008).

From this perspective, mathematics education can be understood as an institution, which allows society to use mathematics for its organisation, preparing some students to participate in this form of organisation and preparing the rest to accept it:

> Could it be that mathematics education in fact acts as one of the pillars of the technological society by preparing well that minority of students who are to become "technicians," quite independent of the fact that a majority of students are left behind? Could it be that

> mathematics education operates as an efficient social apparatus for selection, precisely by leaving behind a large group of students as not being "suitable" for any further and expensive technological education? [...] Nonetheless, a large group of students might be left, and they will have learned a substantial lesson: that mathematics is not for them. To silence a group of people in this way might also serve a sociopolitical and economic function.
> (Skovsmose, 2005, p. 11)

Such form of critique is directed against the ideals and presumptions of mathematics education, for example, the presumption that mathematics education is good per se. It sets out to reveal what is hidden by these ideals and presumptions, for example, how mathematics education establishes mathematics as a tool of social power.

While the sociopolitical concerns above have been gained both from theoretical considerations and empirical observations, yet little research has documented if and in how far students report experiences that can severely influence their possibilities to engage with mathematics outside school, especially where mathematics is used to organise our society. However, students' perceptions of mathematics are of central interest as they can be considered manifestations of the socialisation processes the students underwent in the mathematics classroom. Therefore, the first research question of this chapter is: *How do students perceive mathematics?* Inseparably connected to this question is the difficult task to find a methodological and theoretical framework to document students' perceptions and to interpret them from a socio-critical perspective. Therefore, the second research question of this chapter is: *How can students' perceptions of mathematics be interpreted from a socio-critical perspective?* Together, both questions open up a wide field of study, and this chapter can only present a first grasp on the issue. Accordingly, it should be read as an explorative study.

Towards a Theoretical Framework

Much research on students' perception of mathematics has originated in the field of beliefs and affect concerning mathematics education (Leder et al., 2002; Maaß & Schlöglmann, 2009). Originally starting out to analyse students-held beliefs about the epistemology, teaching and learning of mathematics, this field of research has broadened its focus by also analysing students' attitudes and emotions towards mathematics, coining the term of "affect" (Goldin, 2002; Di Martino & Zan, 2011). For example, following an inductive approach, Pietro Di Martino and Rosetta Zan (2011) analysed 1662 essays from presumably Italian students from first to thirteenth grade in which they report their relationships with mathematics. They found that many students disliked following rules, the lack of emotions, the lack of individuality and a lack of sense-making. Maria de Lourdes Mata and colleagues (2012) conducted a qualitative study on the self-perceived competence, choice and value of mathematics among 1719 Portuguese fifth-to-twelfth graders. They showed that attitudes become less positive during the school career and that they correlate with self-perceived achievement, value, choice, competence and support in mathematics.

However, the studies focussing on beliefs and affects do not align with the sociological intention of this chapter (for a broad discussion of the shortcomings of beliefs and affects research cf. Skott, 2014). Firstly, they rely too heavily on quantitative methods, especially on Likert-scale questionnaires (e.g. Kislenko, Grevholm, & Lepik, 2007), which do not result in a description of the students' own voices but in a mere measurement of statements, which the researchers find most significant. Secondly, and more severely, research in this field hardly connects to sociological theories but stays psychologically oriented. Jürgen Maaß and Wolfang Schlöglmann (2009, p. vii) outline that "common to all research into affect is the idea that the categories of affect are based on mental systems", thereby excluding the sociopolitical a priori. Even qualitative research into beliefs and affect such as the study by Di Martino and Zan (2011) links its findings to motivational psychology but not to any theory of the social. Eventually, even such socially relevant findings, such as the perception of mathematics as "important but boring" (Kislenko et al., 2007), are not interpreted on a sociopolitical level.

An alternative approach to students' perception of mathematics is presented in the study of students' mathematical identities and their formation in communities of practice (Grootenboer & Jorgensen, 2009; Sfard & Prusak, 2005). This research perspective emphasises the social production of students' identities and regards these identities as decisive for the success in learning processes. While these contributions understand identity and learning rather as a social than as a cognitive phenomenon, they hardly address sociopolitical concerns. Especially, they do not distance themselves from the common narrative that mathematics education was primarily concerned with "learning" mathematics. Sfard and Prusak (2005) state explicitly that they develop their theory of identity to investigate and support learning processes. Grootenboer and Jorgensen (2009) report students' disengagement with mathematics, but do not wish to analyse it as a sociopolitical phenomenon but to find "a way out" by providing students with a professional understanding of agency as a working mathematician.

I propose to build on the work of Michel Foucault to find a theoretical framework, which allows to conceptualise and analyse students' perceptions of mathematics from a sociopolitical perspective (for a discussion of the use of Foucault for research in mathematics education see Walshaw, 2007; Kollosche, 2015). In his late concept of governmentality, Foucault (1982) argues that power should not be understood as a good, which a person or a group of people possesses, but as the control over techniques for the conduct of the self or others. Such techniques do not only comprise physical action, but also manners of feeling, thinking and speaking. By distinguishing the self and others, Foucault emphasises that people have power over themselves in that they can change their very existence. He calls the individual development of a technique for the conduct of the self an "ascesis". Foucault (1975/1979) is especially interested in what he calls "disciplinary techniques", that is, techniques for the conduct of others by means of their conduct of the self. For example, having students solve mathematical problems under the threat of bad marks is a teacher's technique for the conduct of others, while solving mathematical problems requires the development of techniques for the conduct of the self. Foucault argues that it is the need for developing individual techniques which accounts for the success and spread of disciplinary techniques throughout society in modern age. Indeed, it is also a core idea of any pedagogical action to expose students to demands, which provoke an ascesis. Teachers execute their power in the wish to improve their students' ability to calculate, to think logically or to apply mathematics. However, it is important to note that students cannot comply with these demands by simply imitating their teacher, but have to find individual techniques—techniques that may vary from student to student and result in many different ways of acting in and perceiving mathematics.

Apart from that, Foucault (1984, p. 334f.) uses a wide interpretation of the concept of knowledge, including beliefs, values, morals and presumptions. He then regards knowledge as inseparably linked to techniques of conduct (1979) and coins the concept of "power-knowledge" relations. On the one hand, knowledge may produce, improve and justify certain techniques of conduct. For example, mathematical considerations often inform social decisions, or mathematics education research produces knowledge to legitimate and improve the teaching of mathematics in schools. On the other hand, knowledge itself needs a basis of legitimisation, that is, techniques of conduct, which justify it as truth. For example, the knowledge of mathematics relies on logical, calculatory and other techniques for the conduct of the self and others, while the knowledge produced by mathematics education research justifies itself on the ground of educational, psychological, sociological, and other theories and methods. It is therefore impossible to separate knowledge from power. Indeed, knowledge requires power in order to become accepted, just as power needs knowledge in order to be executed.

This theory of the social has several implications for the concept of the individual. The individual finds herself exposed to the conduct by others, in the case of the school primarily by that of the teacher and fellow students. In order to cope with these external demands, the individual has to develop her own techniques for the conduct of the self. While these techniques can differ in the extent to which they allow a dignified survival in school—reaching from being a role model student to avoiding mathematics—these techniques can also differ in the way in which the ascesis is perceived by the individual: whether she finds it easy or hard to develop such techniques; whether she values or dislikes the techniques she creates; whether she fancies the existence the new techniques lead her towards. Knowledge, then, is not only a desired outcome of the pedagogical endeavour, but serves as a legitimising basis for both the teachers' techniques for the conduct of the students and the students' techniques for the conduct of the self. Knowledge is used to make sense of the ways in which the individual meets or avoids the demands; it is used to explain why it is reasonable to participate or not to participate. Consequently, students' perceptions of mathematics are no mere opinions on a socially impartial phenomenon, but an expression of the ascesis experienced in the mathematics classroom, constructing the mathematical individuality of each student.

Recent studies draw on Foucault's theory to analyse how mathematics is inseparably interwoven with the constitution of our society and how mathematics education is functional in constantly reproducing these connections. For example, Andrade Molina and Valero (in this volume) studies how geometry classes install a certain perception and understanding of "space" as a technology of the self. Elsewhere, I show that mathematics can be understood as a prototypical manifestation of the technology of logical and bureaucratic thinking, which are both used throughout society and introduced in the mathematics classroom (Kollosche, 2014). I argue these social functions of mathematics education are explicated neither in mathematics education research, nor in the mathematics classroom, but the mathematics classroom is organised as a disciplinary institution, which leads students either to adopt and reproduce the logical and bureaucratic thinking in order to be successful or to ignore and avoid mathematics in order to not be humiliated by constant failure. In both cases, the functionality of the mathematical power-knowledge relation is not threatened.

When analysing voices of students, the framework presented above may serve as an analytical lens. Understanding students' perceptions of mathematics as results of processes of ascesis, we may then ask, which techniques of the self students developed, what they developed them for, and what knowledge they use to make sense of their behaviour. Thus, the Foucauldian framework allows to build an analytical bridge between individual perceptions and sociopolitical phenomena.

Method

The data set consists of students' answers to an anonymous questionnaire developed with master students who had varying interests in this study. The questionnaire (Table 11.1) includes the wide range of 13 different open questions on the perception of mathematics and three questions on personal data (marks, age and gender). The purpose of the questionnaire was to raise data in the students' words for an exploratory analysis of students' perceptions of mathematics. Developing the questionnaire, we faced the problem of providing enough stimuli to gain a wide range of student answers without directing the students' attention to certain aspects more than necessary.

199 ninth-grade-students from nine different German secondary schools participated in the study. The data set is biased in the sense that the survey has been conducted nearly entirely in the North-East of Germany and addressed mostly students who aim at obtaining a certificate for higher education. Nevertheless, the data set proved to allow a widely focussed and differentiated analysis.

The analysis presented here followed several steps. First, thematic analysis was used to construct themes out of the data set. Second, the themes were analysed quantitatively, preparing the last step, where exemplary answers were analysed qualitatively in order to gain a deeper understanding of themes found and to open these for sociopolitical interpretations.

Table 11.1 English translation of the questionnaire used for the survey

Questionnaire	
1.	What is your favourite subject and which subject do you like least? Where would you position mathematics?
2.	Find at least three words that describe your mood and attitude towards mathematics!
3.	What animal comes to your mind regarding mathematics? Why does it fit well?
4.	What distinguishes mathematics from other subject? What do you like more or less in other subjects?
5.	What do you think of when you hear the word "mathematics"?
6.	What is easy in mathematics and what is hard?
7.	Some consider mathematics logical, others incomprehensible. What do you think?
8.	"Mathematics is not vivid enough." How do you evaluate this statement?
9.	Where does mathematics help in everyday life?
10.	Is it possible to learn mathematics on purpose or does one need talent? Explain!
11.	What do you like about mathematics and what repels you?
12.	How do you feel when you fail to understand something in maths?
13.	What was the mathematics mark on your last school report?
14.	How old are you?
15.	Are you male or female?

Thematic analysis as presented by Virginia Braun and Victoria Clarke (2006) is a qualitative method to reduce text data to a small set of well-reflected themes. In the first step of the analysis all answers were searched for data items referring to the perception of mathematics. Similar items were grouped, and groups expressing closely related perceptions were combined to themes. Contradictory and too differentiated items (e.g. "addition is easy, but fractions are hard") were not coded, whereas general answers were, even if exceptions were mentioned (e.g. "maths is easy, only fractions trouble me"). If ambiguous expressions were used (such as "confusion", which may be experienced as an excitement or as a burden), their meaning was assessed by the context. In this chapter, I only focus on those themes which occur in at least every tenth questionnaire of the complete set. Apart from that, all themes which focus on specific mathematical contents were excluded, for an analysis differentiated by curriculum contents, which may be interesting indeed, would over-expand the scope of this chapter.

Especially the composition of themes necessarily requires the researcher to decide on which groups express related perceptions. Especially, themes could often be merged further or conceptualised with more differentiation. In order to give a transparent account of the analysis, the themes constructed will be presented with exemplary items. Consequently, the thematic analysis followed an inductive approach. Text items were analysed literally without consideration of any latent meanings. In the extreme case, this might mean to interpret "sickness" as a somatic symptom instead of as a metaphor for refusal. However, I argue that this approach works best to avoid alienations of the students' perceptions through the analyst's lens. Eventually, even if the student did not indeed feel sick—what we cannot know—she might still have had her reasons to express her perception in such bodily terms.

Under the assumption that students' perceptions of mathematics differ by attitude, the data set was differentiated by self-assigned general attitude towards mathematics. General attitude was usually identified by the help of the initial question; only in unclear cases the answers to other questions were used for clarification. General attitude was grouped into three dimensions:

- Positive: The student expresses a positive relationship with mathematics.
- Neutral: The student expresses a contradictory or indifferent relationship with mathematics.
- Negative: The student expresses a negative relationship with mathematics.

Constructing Themes

Following the methodological approach outlined above, the associations expressed by the students in the questionnaires were grouped and combined to the following themes (Table 11.2):

Table 11.2 Themes originating from the thematic analysis with descriptions and examples

Name of theme(s)	
Description	Exemplary excerpts
Psychosomatic comfort vs. discomfort	
The students state that mathematics causes psychological or physical comfort or discomfort respectively	Exciting, fun, relaxing/stressful, frustrating, hopeless, fear, tired, headache
Easy vs. hard comprehension	
The students state that mathematics is easy or hard to understand	Comprehensible, easy/complicated, hard
Strong vs. little interest	
The students express strong or little interest in mathematics	Interesting/uninteresting, boring
High vs. low usefulness	
The students state that mathematics is useful or not useful for their current or future life	Important, meaningful/useless, senseless, superfluous, unnecessary
Challenging effort	
The students state that mathematics requires challenging efforts	Efforts, challenge, exertive, demanding, concentration, discipline
Logical dimension[a]	
The students state that mathematics is logical, that is, it follows a certain system of thought	Logic, logical
Evaluation	
The students state that they associate mathematics with exams or marking	Marks, bad marks, exams, tests

[a]In order to avoid any distortion of the data, this question was only applied to the questions 1–6

A comparison of the frequencies with which the themes occur in the questionnaire provided an insight into the themes most prevalent. However, these frequencies do not resemble any measure for the agreement or disagreement with certain statements. For example, not mentioning the term "logical" does not mean that a student does not consider mathematics logical—she just found other issues more important to tell. The mind map (Fig. 11.1) visualises the frequencies of the occurrences of themes. The text size indicates the frequency of each theme:

These findings already draw a picture of mathematics as a subject that is perceived challenging, logical, and useful; but simultaneously uninteresting, unpleasant and hard to understand. However, the data also show that students' perceptions are far from being coherent; they diverge into completely opposite ways of perceiving mathematics. Differentiating the groups by self-ascribed general attitude towards mathematics allows for a more detailed account. Fifty-seven students express a positive, 72 students a neutral and 70 students a negative general attitude towards mathematics. Given that classification, it is possible to compare the frequencies of the occurrences of certain themes for each group (Table 11.3):

This differentiation proved successful in providing more coherence within the data subsets. For example, "strong interest" was a frequent theme among students with a positive attitude, whereas it was not among students with a negative attitude. However, the groups with neutral attitudes towards mathematics are still incoherent. Possibly, there are various essentially different ways of having a neutral attitude towards mathematics. The following considerations focus on chosen points of interest, deepening the analysis by considering individual statements and opening them for a sociopolitical interpretation.

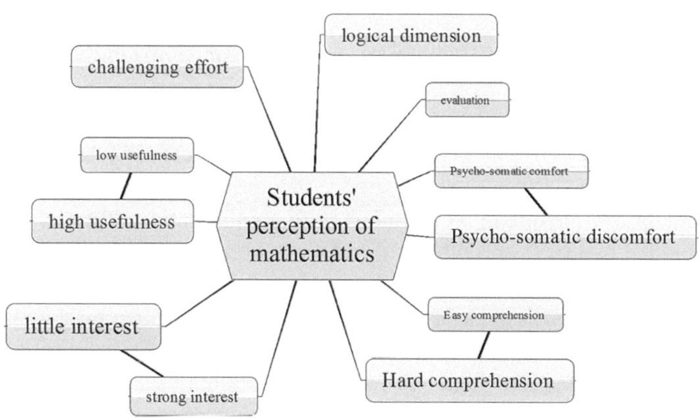

Fig. 11.1 Mind-map of themes with text size indicating frequencies among the whole set

Table 11.3 Relative occurrences of themes differentiated by general attitude towards mathematics

	+ (%)	± (%)	− (%)	Ø (%)
Psychosomatic comfort	33	6	0	12
Psychosomatic discomfort	4	29	61	33
Easy comprehension	37	7	0	13
Hard comprehension	2	31	61	33
Strong interest	42	22	0	20
Little interest	4	35	74	40
High usefulness	39	32	24	31
Low usefulness	7	18	21	16
Challenging effort	33	28	23	28
Logical dimension	47	24	20	29
Evaluation	7	11	20	13

Understanding the Themes

Mathematics as a Polarising Subject

Although many students state that they like mathematics, mathematics proves to be a *polarising subject*. While students with a positive general attitude towards mathematics frequently express psychosomatic comfort, easy comprehension and strong interest; students with a negative general attitude express psychosomatic discomfort, hard comprehension and little interest. For example, students with a positive general attitude towards mathematics state that mathematics was "interesting", "easy", "exciting", "fun" or that they would "relax" when performing mathematics. The extracts from students with a negative general attitude towards mathematics often provide more detailed information. With "boring" being the most prominent association with mathematics throughout the data set, many students and the vast majority of those with a negative attitude state that they have little interest in mathematics. Some students explain their lack of interest with mathematics being too "dry", "without emotions" or "unfriendly" and allowing "too little discussions".

Many students perceive mathematics as being "complicated" and "hard". They express psychological or physical symptoms such as "despair", "stress", "demotivation", "depression", "fear", "exhaustion", "headache" and "nausea". Apparently, feeling bad is a widespread association with mathematics: "Maths is the only subject where I panic. In the other subjects we aren't put that much pressure on. […] In my case, maths causes anxiety and headache." Interestingly, this burden is not presented as a pathology, as a yet-to-be-cured failure of teaching mathematics, but rather as an unavoidable characteristic of mathematics. For example, asked for what she thinks when she hears the word "mathematics" a successful student states in a factual manner that "desperate students" come to her mind.

Importance of Mathematics

Many students mentioned the *importance* of mathematics, which interestingly was rarely associated with its socially selective function. Probably, mathematics is not perceived as a device for social selection more than any other subject. Some students complained about "bad marks" or "math exams", but no student considered mathematics to be more selective a subject than any other school subject. However, one student remarked that she liked mathematics for its presumed fairness of marking: "In other subjects it does not please me that assessment is left to the discretion of the teacher. In maths it is not like this."

Nevertheless, there are some students who perceive mathematics as a gatekeeper to a happy future. Mathematics was able "to determine our future" and necessary "to learn a nice profession". One student point outs that mathematics could also be a gatekeeper to economic success: "I appreciate that mathematics is useful in life and that there are many professions for which you need mathematics and in which you earn much money."

Thus, while students do not perceive mathematics as a selective subject, they express an awareness of its allocating function in society.

Apart from that, the importance of mathematics was frequently associated with the utilitarian value of mathematics in contemporary everyday life and the expected future of the students. Even students who expressed a negative general attitude towards mathematics "appreciate mathematics as it can be applied everywhere". Other students argue more abstractly that they would need mathematics somewhere in the future: "When I hear 'maths,' I know that in the future it will become important and we will need it."

In spite of that, there is also a considerable group of students who regard mathematics as being "useless". They argue that "you do not need the bigger part of what you learn there", that mathematics was "needless and not important for my future life" and that most of what was taught would soon be forgotten. It is interesting to note that both the perception of mathematics as being useful and that of mathematics as being useless can be found in all groups of self-ascribed general attitudes towards mathematics. For example, the last two excerpts quoted above came from students who expressed a positive general attitude.

Challenge and Logic

In contrast to the mutually opposed themes discussed so far, the themes *challenge* and *logic* relate to the nature of mathematics and do not include a strong polarisation. The perception of mathematics as an intellectual challenge appears frequently and more independent of the students' self-ascribed general attitude towards mathematics. The students state that mathematics required a lot of "effort", "concentration" and "self-discipline". For example, a student associates mathematics with the following animal: "An alligator, as it is a dangerous animal. Seemingly invincibly it rises up in front of you, but with much effort you maybe can defeat it." While some students perceive this challenge as an unpleasant experience, other students appreciate this experience. For example, they state that mathematics was "a nice task to tackle" or that they liked "hard exercises where you have to consider skilfully to find a solution." Especially when mathematics is compared to other school subjects, this theme was associated with the idea of contemplation: "Mathematics differs from other subjects in that you have to contemplate a lot. Other subjects are often only learning by heart." Again and again, students associate mathematics with "exercising" and "understanding"; and reduce other subjects to memorising. Apparently, the students perceive mathematics as an intellectual endeavour of a special kind.

In every group of self-ascribed general attitude towards mathematics a considerable number of students explained before the corresponding question that they perceived mathematics as a logical subject. This perception may help to understand the challenge described by the students. Apart from explicitly stating that mathematics was "logical", many students allowed further insights. Some students generally stated that mathematics "stimulates cogitation", that it "keeps the mind fit" or that it "makes people smarter", forcing them "to switch on their brands and to contemplate a bit". Other students associated mathematics with certain epistemological traits. They often stated that mathematics knew only right and wrong and exactly one answer to each problem. They also explained that mathematics had an "ascending order", that it was "not leaving anything to chance" and had "a logical explanation for everything". Apart from that, students refer to the algorithmic dimension of mathematics by stating that you have to "apply formulas", follow "clear schemes" and that mathematical procedures have "hardly any exceptions, unlike vocabularies". While many students cherish the logical dimensions of mathematics, there are also some students who find it repellent. These students complain that "you have to do everything that accurately", many exercises are "only systematic and no fun" or "even a tiny mistake" results in failure. Some students complain that mathematics was too "dry" and did not involve any emotions: "I think that in maths there are no emotions. It is calculating. In stories in German tuition you can run riot and allow your fantasy free play."

Eventually, some students experienced logic as an epistemological obstacle. For example, a student wrote that "[s]ome can think well logically, others cannot" and that she did not belong to the first group. So, in summary, many students experience mathematics as a challenge and a logical endeavour, whereas they express divergent attitudes and opportunities to engage in these dimensions.

Interpreting the Results from a Socio-critical Perspective

In this section, the three topics discussed before are interpreted from a socio-critical perspective. Central to this perspective is the question of the existence, which mathematics education leads students to and its connections to individual and societal distributions of power.

Mathematics as a Polarising Subject

The first topic discussed was that of polarised perceptions of mathematics. The perceptions of mathematics differ severely and are closely connected to the students' general attitudes towards the subject. Nearly two thirds of the students with a negative general attitude towards mathematics state that they have little interest in the subject, find it hard to understand, and experience psychosomatic discomfort in the mathematics classroom. The reporting of hard comprehension indicates that a large group of students lack the techniques of the self to successfully cope with the disciplinary techniques exercised in the mathematics classroom. Showing little interest in the subject is an alternative form of ascesis that allows distancing oneself from permanent rejection. However, when a student fails, the disciplinary techniques exercised in the mathematics classroom appear not to allow a complete withdrawal, but keep their grip on her, often leading to various forms of psychosomatic discomfort. Therefore it can be argued on the basis of the data that contemporary school mathematics is an institution that exercises disciplinary techniques, which eventually lead to a dissociation of the big group of students who are not able or willing to understand mathematics, while the able and willing experience comfort, success and develop interest in the subject. The strong polarity in the documented perceptions of mathematics indicates that the perception of mathematics is often located very close to one of the two prototypes described above, while alternative techniques for the conduct of the self have not been found in the sample. Considering that a large group of students connects mathematics with hard comprehension, little interest in it and even psychosomatic discomfort, it seems legitimate to confirm Skovsmose's (2005) assumption that the substantial lesson for these students is that "mathematics is not for them", thus possibly excluding those students from an active role in the mathematical organisation of society. Mathematics can then be understood as a complex power-knowledge, which is erected in school and used to distribute power in our contemporary society. Although such a function of mathematics education may be very important for the functionality of our society, mathematics education systematically fails to include a large group of students into the mathematical discourse.

Importance of Mathematics

The second topic discussed was that of the perceived relevance of mathematics and its selective function. Interestingly, the findings contradict research on the issue on the first glance, as students hardly connect mathematics with a gate-keeping function for future opportunities. Indeed, some students mention grading and exams or state that they consider mathematics to be a more objective tool for selection than other subjects, but only a few students connect this to the determination of their future. However, many students state that mathematics would be useful for their contemporary or future life. These statements are general, mostly referring to the omnipresence or omnirelevance of the discipline or to a vaguely conceptualised personal future. Considering the contents dealt with in the grade 9 mathematics curriculum in Germany, it is apparent that these contents may be used in professional areas but not in private life, neither in that of the student nor in that of their peers or relatives. Accordingly, no single student qualifies the claim that mathematics was useful for them by discussing the relevance of a specific content from their classroom. Therefore, it can be assumed that the perception of mathematics as being useful for the students it not based on personal experience. The perceived usefulness of mathematics may then be interpreted as a dogmatic belief, as a power-knowledge relation in the Foucauldian sense, which is being fostered in mathematics education (Dowling, 1998; Lundin, 2012) and which might be effective in providing a technique for students and teachers, with which they can make sense of their own involvement in mathematics education, eventually reproducing that myth themselves. Considering the fact that German grade 9 contents are hardly useful in non-professional life (Heymann, 1996), it might be asked whether the belief in the mundane usefulness of mathematics is actually obscuring other functions of mathematics education, especially its function as a gatekeeper that also German ninth graders are subjected to. The technique of conduct to perceive mathematics as useful would then allow students to accept and live with the rank in the social order that the selective properties of school mathematics has imposed on them, hence subverting students' opposition.

Challenge and Logic

The last topic discussed was that of challenge and logic. Here, students report on interesting experiences which might be closely connected to the nature of mathematics and might thus shed light on further social functions of mathematics education. First, many students report that mathematics requires a lot of effort, and some of them refer more precisely to concentration, self-discipline and careful contemplation. Interestingly, many students contrast this experience with the experiences made in other subjects where learning could be realised by memorising. Interpreting the reported efforts when performing techniques of the self in the mathematics classroom as processes of ascesis it can be said that the students' techniques for the conduct of the self do not suffice, but that they are constantly experiencing ascesis, developing new techniques of the self to cope with the demands in the mathematics classroom. The differences the students state between mathematics and other subjects indicate that this ascesis is unique to mathematics and therefore has a unique function in the process of the students' construction of a mathematical individuality.

Logical thinking may be understood as a conduct of the self, which can be learnt in the mathematics classroom. However, the interpretation of this theme is difficult due to different understandings of the term "logical". On the one hand, some students connect the term to cogitation and smartness in general, using a rather universal and hardly differentiated understanding of it. Indeed, the belief that mathematics is logical may again be a dogmatic belief around mathematics, which is fostered in the classroom and used to legitimate the learning and the societal application of mathematics. Yet, on the other hand, some students share thoughts on the working mechanisms of the mathematics they experienced. They refer to the rigid antagonism of true and false, to the demand to logically explain statements and to the "clarity" and regularity of mathematical procedures. These perceptions of the nature of mathematics correspond well to the socio-critical analyses provided around mathematical thinking (Kollosche, 2014) and confirm that school mathematics is indeed connected to a specific form of thinking. Eventually, the ability to think logically is perceived as creating differences between students in the mathematics classroom. Students explicitly state that some students can think logically while others cannot, and they are aware of their own position within that field. In contrast to that, no student reported that she experienced mathematics as an opportunity to actually learn how to think logically. Therefore it can be argued that mathematics education is an institution which selects students by their ability to think logically while not providing visible opportunities to develop that kind of thinking.

To give a short summary, mathematics is perceived as a discipline, which is connected with a unique kind of thought whose learning is often experienced as a challenge and a burden, which is highly selective by in- and excluding students, which is accompanied by a dogmatic form of power-knowledge, and which legitimates mathematics as being useful for the students. The usefulness of learning mathematics is hardly associated with its content knowledge but with very general skills such as learning to think and with vague ideas of its importance for the future. This importance for the future of the student is sometimes connected to the role of mathematics as a gatekeeper in that it allows access to privileged education and professions. Ultimately, mathematics education is not perceived as an institution where meaningful contents are addressed in a fashion that allows a sovereign and supportive approach for all students, contradicting the dominant power-knowledge of mathematics as being important to be learnt by all students. Instead it can be argued that mathematics education privileges and sanctions students according to whether they develop and display techniques of conduct in line with the dogmatic power-knowledge of school mathematics. The analysis illuminates how certain ways of perceiving mathematics is a substantial ingredient of these techniques.

Looking Back and Forth

This explorative study focussed on the research questions how students perceive mathematics and how students' perceptions of mathematics can be interpreted from a socio-critical perspective. I argue that the use of Foucault's theory has proved successful in understanding students' perceptions of mathematics on a basis which is not a priori normative and not addressing negative perceptions as a pathology that could be "cured", but is open to describe and explain students' perceptions as manifestations of techniques of the self, which might be functional for the student even if they contradict the normative discourses of what learning mathematics should be about. For example, the finding that mathematics is perceived as "important but boring" (Kislenko et al., 2007) could not only be reproduced but be understood as a part of the working mechanisms of the institution of mathematics education. I suggest that this perspective allows to better understand the sociopolitical reality of the mathematics classroom and to challenge the normative convictions, which are often held in mathematics education research, and which might hinder a coherent comprehension of the forces at work.

While this study is only an exploration, it illustrates the relevance and productiveness of the chosen approach. The further development of methodology promises more detailed data, for example, by replacing questionnaires by interviews. Eventually, separate studies could focus on the issues touched upon in this study, for example, on the experienced comfort and discomfort in the mathematics classroom or on the unique challenges of mathematics education, leading to a deeper and better substantiated understanding of students' perceptions of mathematics; and the sociopolitical forces at work in the mathematics classroom.

Acknowledgments This study was realised with the help of the students of the master seminar on "Student Perceptions of Mathematics" at the Universität Potsdam, Germany, in 2014. I am very grateful for their support in collecting the data and gathering ideas for analysis and interpretation.

References

Braun, V., & Clarke, V. (2006). Using thematic analysis in psychology. *Qualitative Research in Psychology, 3*(2), 77–101. doi:10.1191/1478088706qp063oa.

de Lourdes Mata, M., Monteiro, V., & Peixoto, F. (2012). Attitudes towards mathematics: Effects of individual, motivational, and social support factors. *Child Development Research*. doi:10.1155/2012/876028.

Di Martino, P., & Zan, R. (2011). Attitude towards mathematics: A bridge between beliefs and emotions. *ZDM, 43*, 471–482.

Dowling, P. (1998). *The sociology of mathematics education: Mathematical myths/pedagogic texts*. London: Falmer.

Foucault, M. (1979). *Discipline and punish: The birth of the prison*. New York: Vintage. Original work published 1975.

Foucault, M. (1982). How is power exercised? In H. L. Dreyfus & P. Rabinow (Eds.), *Michel Foucault. Beyond structuralism and hermeneutics* (pp. 216–226). New York: Harvester.

Foucault, M. (1984). Preface to the history of sexuality, volume II. In P. Rabinow (Ed.), *The Foucault reader* (pp. 333–339). New York: Pantheon.

Goldin, G. A. (2002). Affect, meta-affect, and mathematical belief structures. In G. C. Leder, E. Pehkonen, & G. Törner (Eds.), *Beliefs. A hidden variable in mathematics education?* (pp. 59–72). Dordrecht: Kluwer.

Grootenboer, P., & Jorgensen, R. (2009). Towards a theory of identity and agency in coming to learn mathematics. *Eurasia Journal of Mathematics, Science and Technology Education, 5*(3), 255–266.

Gutiérrez, R. (1999). Advancing urban Latina/o youth in mathematics: Lessons from an effective high school mathematics department. *Urban Review, 31*(3), 263–281.
Heymann, H. W. (1996). *Allgemeinbildung und Mathematik. [General education and mathematics]*. Beltz: Weinheim.
Kilpatrick, J. (1992). A history of research in mathematics education. In D. A. Grouws (Ed.), *Handbook of research on mathematics teaching and learning* (pp. 3–38). New York: Macmillan.
Kislenko, K., Grevholm, B., & Lepik, M. (2007). Mathematics is important but boring: Students' beliefs and attitudes towards mathematics. In C. Bergsten, B. Grevholm, H. S. Måsøval, & F. Rønning (Eds.), *Relating practice and research in mathematics education. Proceedings of the Fourth Nordic Conference on Mathematics Education* (pp. 349–360). Trondheim: Tapir.
Kollosche, D. (2014). Mathematics and power: An alliance in the foundations of mathematics and its teaching. *ZDM Mathematics Education, 46*(7), 1061–1072. doi:10.1007/s11858-014-0584-0.
Kollosche, D. (2015). Criticising with Foucault: Towards a guiding framework for socio-political studies in mathematics education. *Educational Studies in Mathematics*. doi: 10.1007/s10649-015-9648-5.
Lave, J. (1988). *Cognition in practice: Mind, mathematics and culture in everyday life*. Cambridge: Cambridge University Press.
Leder, G. C., Pehkonen, E., & Törner, G. (Eds.). (2002). *Beliefs: A hidden variable in mathematics education?* Dordrecht: Kluwer.
Lundin, S. (2012). Hating school, loving mathematics: On the ideological function of critique and reform in mathematics education. *Educational Studies in Mathematics, 80*(1–2), 73–85. doi:10.1007/s10649-011-9366-6.
Maaß, J., & Schlöglmann, W. (2000). Erwachsene und Mathematik. [Adults and mathematics]. *Mathematica Didactica, 23*(2), 95–106.
Maaß, J., & Schlöglmann, W. (Eds.). (2009). *Beliefs and attitudes in mathematics education: New research results*. Rotterdam: Sense.
Martin, D. B. (2009). Researching race in mathematics education. *Teacher College Record, 111*, 295–338.
Sfard, A., & Prusak, A. (2005). Telling identities: In search of an analytic tool for investigating learning as a culturally shaped activity. *Educational Researcher, 34*(4), 14–22. doi:10.3102/0013189X034004014.
Skott, J. (2014). Towards a participatory approach to "beliefs" in mathematics education. In B. Pepin & B. Rösken-Winter (Eds.), *From beliefs to dynamic affect systems in mathematics education* (pp. 3–23). Berlin: Springer.
Skovsmose, O. (2005). *Travelling through education: Uncertainty, mathematics, responsibility*. Rotterdam: Sense.
Stinson, D. W. (2004). Mathematics as "gate-keeper" (?): Three theoretical perspectives that aim toward empowering all children with a key to the gate. *The Mathematics Educator, 14*(1), 8–18.
Stinson, D. W. (2013). Negotiating the "white male math myth": African American male students and success in school mathematics. *Journal for Research in Mathematics Education, 44*(1), 69–99. doi:10.5951/jresematheduc.44.1.0069.
Straehler-Pohl, H., & Pais, A. (2014). Learning to fail and learning from failure: Ideology at work in a mathematics classroom. *Pedagogy, Culture and Society, 22*(1), 79–96. doi:10.1080/14681366.2013.877207.
Ullmann, P. (2008). *Mathematik, Moderne, Ideologie: Eine kritische Studie zur Legitimität und Praxis der modernen Mathematik. [Mathematics, modernity, ideology: A critical study on the legitimacy and praxis of modern mathematics]*. Konstanz: UVK.
Valero, P. (2004). Socio-political perspectives on mathematics education. In P. Valero & R. Zevenbergen (Eds.), *Researching the socio-political dimensions of mathematics education. Issues of power in theory and methodology* (pp. 5–23). Boston: Kluwer.
Volmink, J. (1994). Mathematics by all. In S. Lerman (Ed.), *Cultural perspectives on the mathematics classroom* (pp. 51–67). Dordrecht: Kluwer.
Walshaw, M. (2007). *Working with Foucault in education*. Rotterdam: Sense.

Chapter 12
Interrupting Passivity: Attempts to Interrogate Political Agency in Palestinian School Mathematics

Jehad Alshwaikh and Hauke Straehler-Pohl

Abstract This chapter attempts to question the sociopolitical structure and conditions that Palestinian students live in and under which they are expected to learn mathematics and other subjects in schools. We describe these sociopolitical conditions and then focus on education and mathematics education in an attempt to illuminate the relationship between learning mathematics and the sociopolitical context that it "lives" in. This is achieved by using textbooks as an example of how ideology shapes education in Palestine generally, and mathematics education particularly. The analysis illustrates how learners of mathematics are constructed as passive subjects, separated from the sociopolitical conditions they live in. Based on this analysis, the chapter provides drafts for two hypothetical classroom activities that are designed as provocations for teacher education and seek to tease out the potentials for interrupting this passivity.

The Sociopolitical Context in Palestine in a Nutshell

Describing the situation in Palestine is hard—even for Palestinians who have spent most of their lifetime there. Palestinians have all sorts of governmental institutions (a president, a parliament, ministries, etc.) and at the same time live under military occupation—Palestine is a place full of ambiguities. Contradictions, such as this, massively complicate the project of presenting the internal Palestinian situation concerning education to an international academic audience—an audience that is used to read coherent stories that have been cleared from the contradictions emanating from the

J. Alshwaikh (✉)
Faculty of Education, Birzeit University, Ramallah, Palestine

Faculty of Humanities, University of the Witwatersrand, Johannesburg, South Africa
e-mail: jalshwaikh@birzeit.edu

H. Straehler-Pohl
Fachbereich Erziehungswissenschaft und Psychologie, Freie Universität Berlin, Berlin, Germany
e-mail: h.straehler-pohl@fu-berlin.de

H. Straehler-Pohl et al. (eds.), *The Disorder of Mathematics Education*,
DOI 10.1007/978-3-319-34006-7_12

immediateness of everyday life. In a similar way, these contradictions complicate the project of discussing attempts to facilitate educational change with an international academic audience. But we, Jehad who was born, raised and lives in Palestine and Hauke who lives in Germany and who knows Palestine exclusively through his encounter with Jehad and through the media, will do our best. This chapter is grounded in Jehad's life, country and work. Hauke is the partner to think those experiences in a new direction. With this combination, our intention is to widen (each of) our own imagination(s) and as a potential side-effect provide the reader with impulses for her own imagination. As a result, some of our ideas might be more of an "opening" nature than providing a concluded project. This chapter is written in the hope to provoke feedback from and discussion with our readers.

Palestinians have passports (regardless that it says travel document in a small font underneath the word Passport). They also have a President, a Legislative Council and a Government[1] (from 2007 to July 2014 there were two governments)— regardless that all of these institutions expired by the enacting of the Palestinian Constitution. There are even ministries: Ministry of Education, Ministry of Health etc. More than 135 countries recognise Palestine as a state with two separated areas: Gaza Strip and West Bank. Since the formation of Palestinian Authority (PA) in 1994, 17 governments have been constituted: the shortest was the seventh (1 month, October to November 2003) which was officially entitled "provisional government", and the longest was the 13th (3 years, 2009–2012) which also was officially entitled "provisional government".

The empirical reality on the ground is very different from that formal, "governmental" situation. The unemployment rate in both Gaza and West Bank continues to rise, and so do reports about corruption among the Palestinian institutions. Four and a half million people live in Gaza and West Bank. Around 70 % of Palestinians are under the age of 29, most of them are under 14.[2] In Gaza Strip, the unemployment rate was 45 % in 2014 and the poverty rate was almost 40 % in 2011. At the same time, more than US $24 billion have been "invested" (Wildeman & Tartir, 2014) in Gaza Strip and West Bank since 1993, mainly for governmental institutions as well as non-governmental organisations. These "investments" came mainly from the European Union and the USA. There are many reports about corruption in Palestinian governmental and non-governmental institutions.[3]

There is one more thing that needs to be mentioned, as it colours all aspects of everyday life in Palestine: The Israeli occupation. Since 1967 the Israeli occupation authorities have been controlling Palestinians' life in every detail in Gaza and West Bank. As a result of the Oslo Accord in 1993 and the formation of the PA in 1994, the status of the occupation has changed. There is no Israeli military presence inside

---

[1] After controlling Gaza, Hamas constituted its own government. Thus we had two governments since 2007. Hamas and Fatah (major two political parties) agreed on one government on July 2014.

[2] Find more numbers here: http://www.pcbs.gov.ps/Portals/Rainbow/StatInd/StatisticalMainIndicatorsE.htm.

[3] See for example: http://www.aman-palestine.org; https://www.middleeastmonitor.com/downloads/reports/20131214_CorruptioninthePalestinianAuthority.pdf.

the main Palestinian cities but around them and it is still present in villages and other territories. However, the occupation authorities still control the life of Palestinians indirectly inside the Palestinian cities. The situation remains the same in controlling the borders for travel abroad, travelling being impossible without Israeli authorisation. Having said that, the Israeli occupation is not the main focus of the discussion about mathematics education in Palestine but we highlight its presence in the activities we propose later.

With this brief description, we intend to roughly contextualise the difficulties that education in Palestine faces in the current political situation. Education in general, and mathematics education in particular will be the focus of the next section.

Education, School Mathematics and Agency

In this section, we set out to provide a rough overview of the current situation of (mathematics) education in Palestine. As we have outlined in the beginning, this overview necessarily remains a brutal simplification, due to the inconsistency of Palestine's empirical reality. We start by providing an overview of education in general that afterwards tapers to mathematics education. We then discuss Jehad's current research on the relation of agency and mathematics textbooks as an illustration of the current situation.

Education

There are approximately 2800 schools with more than one million students and 54,000 teachers. There are 14 universities (and some other institutions) which offer first and second degrees (Bachelor and Masters). Some of these universities started recently (after 2000) to provide a Ph.D. in chemistry, physics, mathematics, Arabic, Religion (Hadith) and social sciences. If wished, these numbers can be read as a sign of progress in the development of Palestinian education.

The Palestinian educational system is centralised and controlled by the Palestinian Ministry of Education, which is responsible for producing textbooks and distributing them to governmental schools. Alternative types of schools (private and UNRWA[4]) are obliged to use the same textbooks.[5] There are two stages in the

[4] *United Nations Relief and Works Agency* for Palestine Refugees. There are, however, some private schools that use foreigner system with different textbooks such as International Baccalaureate–IB, IGCSE and SAT.

[5] We feel the urgency to legitimate why the section on education is so heavily concerned with textbooks. Different to other (e.g. European) countries, textbooks remain the biggest force in reforming and shaping (school) education in Palestine. While other countries have developed much more complex systems (see Jablonka and Bergsten in this volume), we speculate that the centrality of textbooks is due to the inimitability of the Palestinian context, as we describe it in the first section of the chapter.

general educational system: basic (and compulsory) stage (1–10 Grades) and secondary stage (11–12 Grades).

The Palestinian Ministry of Education (and other institutions of the government) was formed after the Oslo Accord between The Palestine Liberation Organisation (PLO) and Israel in 1993. Between 1967 and 1994, textbooks "were severely censored by the Israeli occupation authorities until 1994: The word "Palestine" was removed, maps were deleted, and anything Israeli censors deemed nationalist was excised" (Moughrabi, 2001). After 1994, new textbooks for most subjects had been designed that were meant to lay the grounds for a Palestinian self-perception that "should be creative, pragmatic, and truthful without having to engage in historical falsifications" (Abu-Lughod, 1997; cited in Moughrabi, 2001, p. 7). Five principles were suggested to facilitate the development of such self-perception:

> The first of these principles is that the curriculum should be predicated not on giving students facts as if they were eternal truths that must be memorised, but on encouraging them to become critical thinkers. Second, students should be encouraged to make independent judgments and intelligent choices, with careful attention to be paid to individual differences within the classroom. Third, the new curriculum should generate a concept of citizenship that emphasizes individual rights and responsibilities and that establishes a linkage between private interests and the public good so as to encourage responsible and intelligent political participation. Fourth, democratic values such as justice, personal responsibility, tolerance, empathy, pluralism, cooperation, and respect for the opinions of others should be emphasized. Fifth, students should be taught how to read primary texts, to debate, link ideas, read maps, interpret statistics, and use the Internet as well as how to verify facts, sources, and data critically and scientifically (Moughrabi, 2001, p. 7).

Even though new textbooks and curricula reflected the "Palestinian narrative" of a people faced with a settler colonial movement, they avoided dealing with unresolved issues, e.g. waiving to display a map of Palestine (cf. Moughrabi, 2001, p. 7). Nevertheless, Mouhgrabi's analysis reveals how textbooks that actually "make a special effort to promote tolerance, openness, and democratic values" (p. 18) happened to be framed in the US American and Israeli political discourse as "anti-Semitic" and supporting "terrorism".

This, of course, has its consequences on the possibilities that official Palestinian institutions have to act. Such external monitoring and critique made the PA even more vulnerable to the political context, pushing the PA into a state of passivity concerning the self-determined reformation of education: With each reform towards a stronger emphasis on the "Palestinian narrative", the PA has to reckon with jeopardising its financial aids at least potentially. Mathematics education was not beyond such vulnerability as we see in the next section.

Mathematics Education

In Palestine (as in most countries including "Western" countries), there are growing concerns about the poor performance in mathematics among students. Mathematics education in Palestine (and we would generalise that to all subjects) faces serious challenges in schools and in universities. The achievements of Palestinian students in mathematics are modest as reflected in international studies such as the Trends in International Mathematics and Science Study (TIMSS) (see for example Mullis, Martin, Foy, & Arora, 2012). Alshwaikh (2005), found that Palestinian students have severe difficulties in learning geometry based on van Hiele's model (2004/1959). Some local studies attempted to investigate the reasons behind such performance. Al-Ramahi (2006), for instance, assessed teachers' geometric reasoning using van Hiele's model and found that teachers' geometric knowledge is weak. She also found that these shortcomings are presented in Palestinian mathematics textbooks. The analysis suggests that the textbook does not offer a systematic introduction into geometry. Rewadi (2005) found some differences between Palestinian textbooks and the National Council of Teachers of Mathematics (NCTM) standards. For example, Palestinian textbooks usually do not require students to justify their reasoning during solving problems. Thus, from the point of view of what is expected, the performance of Palestinian students as well as the mathematics education offered to them appear as inadequate. However, under the conditions in which students, teachers and mathematics educators have to live, it is possible that all three of these groups will always appear inefficient. Such a continuous labelling of inefficient is unacceptable. Therefore, our aim with this chapter is to imagine possibilities to interrupt this forward projection. First, however, we need to unpack the supposed low performance of students, asking how they come into being as a contingent result of social practices in the first place.

As with most educational systems in the world, mathematics is a compulsory subject for Palestinian students from the first grade (6-year-olds) through to the 12th grade (18-year-olds). Despite the attempts of the Ministry of Education described by Moughrabi (2001, see above), a teacher-centred form of teaching that is focused mainly on rote learning remains a characteristic feature of Palestinian education in general, and school mathematics in particular. Jehad's experience in designing enrichment programmes, training teachers, and reforming curriculum in the Palestinian educational system exposed him to what might be called a fatalistic and ineffective approach to education. When he visits schools in remote places, what he sees is often reminiscent of Paola Valero's description of a Colombian teacher called Mercedes (Valero, 2008a): No windows, "dirty" walls, 28 students in 9 m squared area, a greenboard (in an attempt to change the blackboard approach) that was almost damaged or at least not qualified for writing on it, and a teacher with a "stick". But: Students full of life and energy, eager for exploration. Teachers, however, teach as if students would naturally lack agency and should simply "absorb" the information transmitted to them. We believe that this approach to education is holding back students, especially students from marginalised communities who may be particularly sensitive to the image of the learner as it emerges from textbooks and teacher practice.

These observations also call for a shift of perspective on what it is, that we call mathematics. During the last decades, a variety of international researchers set out to challenge what can traditionally be called the "common" approach towards mathematics: that mathematics is "mind-based" and a pure mental activity (e.g. Fasheh, 1997; Skovsmose, 2011; Valero, 2008b; Watson, 1990). These perspectives focus on the social, cultural and political dimensions of mathematics and mathematics education (see also the work of the Criticalmathematics Educators Group and ethnomathematics: Frankenstein & Powell, 2002; Powell, 2002).

For the case of Palestine, Fasheh (1997), for instance, argues that the mathematics curriculum is isolated from the living reality in the "Occupied Palestinian Territories" (OPT). While Palestinians have always lived under different foreign authorities (Ottoman, British Mandate, Jordan, Egypt and Israel) where different educational systems were applied and consequently changed, these transformations have never in the least affected the status of mathematical knowledge. The reason behind mathematics' "resistance" to change, Fasheh continues, is that the mathematics curriculum has always been held in separation to the reality that Palestinians live in. Mathematics, in this way, remained irrelevant for the particularities of the specific world of Palestinian life—"dead", as Fasheh (1997) put it. He gives a few glimpses on how mathematical knowledge could be made relevant for Palestinians, *if this actually was desired*. For example, he argues, the Israeli occupation imposes a very restrictive policy against the movement of Palestinians within the Palestinian territories. In order for Palestinians to go from Ramallah to Bethlehem without passing Jerusalem, they have to go through a different route called the "valley of fire", which "is almost three times the distance, takes three times the time, and costs three times the amount we used to spend before the 'peace process'!" (p. 26). All of such incidents are not included in the Palestinian mathematics curriculum. If they were, the mathematics curriculum would confer the status of "officially legitimised knowledge" upon daily struggles with the effects of the occupation. One can easily imagine, how quickly this would call suspicions about "anti-Semitic" (Moughrabi, 2001, see above) resentments into action.

While we argue in this article that mathematics educators should emancipate from the restrictive power of such suspicions, we are well aware that bringing the Palestinian situationality into mathematics education brings ideology into play. We are entering an incredibly sensitive terrain. Though limited in number, there are examples for selective ideological interpellation by the PA through school mathematics. Figure 12.1 displays an example of a maths Grade 6th textbook (Part 2, p. 74):

In this case, the textbook is instrumentalised in order to justify unpopular measures implemented by those in power. As such problems are mainly taught by asking children to "translate" the related "keywords" to mathematical procedures (e.g. Hegarty, Mayer, & Monk, 1995) the political aspects—the raise of the price of bread—are posited as a necessary matter of fact. The political dimension is a taken for granted matter that can be translated into numbers, but that is not subject to debate. In this case, it is expected that the quarter and the percentage would attract students' attention, unfortunately neglecting the fact that it is the measure of raising the price of bread, which requires to be challenged and problematised. Such challenge and problematisation require more engaged students in the process of learning mathematics. The mode of politicising is here one that builds on students' passivity and exploits it in order to place its questionable ideological interpellation. In the next section, we present an analysis of Palestinian textbooks that illustrates

أضطرت الدولة لرفع سعر الخبز بسبب ارتفاع سعر القمح عالمياً وقلة الإنتاج
المحلي منه بمقدار الربع ، فما النسبة المئوية للارتفاع؟

Fig. 12.1 [Translation:] The State was forced to raise the price of bread because of the raise of the global price of wheat and the lack of the local production of it by quarter, what is the percentage of that raise?

how such passivity is systematically built into the discourse of school mathematics textbooks.

Passivity and Textbooks: An Illustration of the Current State of Mathematics Education in Palestine

One of the established ways in mathematics education to unpack the sociocultural contingency of school mathematics is to analyse the discourses with which students are confronted when they learn mathematics (see e.g. Dowling, 1998; Lerman, 2000; Morgan, 2006; Straehler-Pohl, Gellert, Fernandez, & Figueiras, 2014), particularly when discussing issues of equity (Herbel-Eisenmann, Choppin, Wagner, & Pimm, 2012). Discourse analysis with a focus on multimodal communication appears as a profound framework for understanding the experienced malaise of Palestinian school mathematics.

In a joint research with Candia Morgan, we (Alshwaikh & Morgan, 2013) looked at the Palestinian mathematics textbooks and analysed geometry lessons in different grades using a framework developed by Tang, Morgan and Sfard (2012). Mathematics tends to be represented as a specialised discourse in these textbooks, reflecting the mainstream view of mathematics, which is that mathematics is absolute, timeless and impersonal (Morgan, 2001). Thus, the mathematical discourse bears a close affinity to the discourse of academic mathematics (O'Halloran, 1996). However, there is an ambivalent message of textbooks concerning engaging students in doing mathematics. On the one hand, the textbooks put an emphasis on proof. Proof itself can be considered an activity, where an agent produces the "truth" of a statement. On the other hand, however, the textbooks do not seriously encourage or enable students to actually engage in the activity of proving. In the way the problems are posed, a "proof" is rather a reified procedure, almost a thing (Sfard, 2008). Students are supposed to mimic an offered solution to the provided example. In this way, the focus on proof supposedly prepares students for the discourse of academic mathematics. However, this is just half of the story, as according to Rotman (1988), an (academic) mathematician's activity can be understood as a dialectic of two apparently antagonistic activities: scribbling and thinking. Scribbling and thinking "are mutually constitutive: each causes the presence of the other; so that mathematicians at the same time think their scribbles and scribble their thoughts" (p. 29). In the Palestinian textbooks the supposedly academic discourse leaves students with mimicking a pure formalism, that is scribbling without actually thinking what is scribbled: a pseudo-activity that actually rather resembles passivity.

These results were supported in other studies (e.g. Alshurafa, 2015) and in a recent chapter in which Jehad expanded the initial work with Candia Morgan on geometry lessons: "An expected consequence of such presentation [scribbling] is that students are not encouraged to approach mathematics (and maybe other subjects) critically" (Alshwaikh, 2015, pp. 132–133). Detailed discussions and examples about how Palestinian mathematics textbooks produce passive learners confined to "scribbling" can be found in Alshwaikh and Morgan (2013), and Alshwaikh (2015).

A common structure within Palestinian school mathematics textbooks is that they present the theoretical part first, followed by an example or two, then activities in classroom, and then presenting the exercises and problems. The processes induced by textbooks that require human agency are almost exclusively affiliated with what Rotman classifies as "scribbling". When mental ("thinking") processes are involved, the textbook takes over these processes in the students' stead. What is expected from the learner is mostly to "copy" or at least imitate the procedures in the examples in order to engage in the activities and in solving the problems at the end of the lesson. The textbook thus constructs the student as active in scribbling activities, but passive in thinking.

Finally, in our attempt to imagine possibilities to revitalise the school mathematics, that Fasheh (1997) has diagnosed as "dead", we follow two main strategies: (1) we intend to challenge mathematics with the particularities of life in Palestine, and (2) we intend to challenge the model of a passive student and citizen. This attempt and these strategies will be demonstrated in the following section.

Imagining the Interruption of Passivity

When we set out to imagine activities that could potentially interrupt student passivity by introducing the Palestinian sociopolitical context into the mathematics classroom, we need to be very aware of three risks:

1. When "real-life"-contexts are introduced into the mathematics classroom, they are more often than not subject to a process that Dowling (1998) calls the "esoteric gaze": When public domain activities are blended with mathematical learning activities *for the sake of learning mathematics*, the public domain itself is subordinated to the domain of mathematics. Instead of learning to actually use mathematics for participation in real life, students are lead to believe in the "myth of participation" (pp. 7). This myth sets out to blind students for the fact that instead of participating in life, they are participating in the distortion of life itself. Our task is then to blend (political) public domain activities with mathematical learning activities *for the sake of engaging with politics*.
2. Time spent in the mathematics classroom on engaging with politics implies less time spent on learning school mathematical knowledge and problem solving skills.

In the Palestinian context, (school) mathematics is a privileged subject as in many other countries. It is a crucial part of the "credit system" (Pais, 2012) that school erects and it is effective in selecting and allocating students, distributing prospective chances in life selectively (see for example Straehler-Pohl & Gellert, 2015). Even though this may sound cynical from an idealistic stance, we need to be aware that when students are given less time for acquiring "credit", this will potentially discriminate them in the accreditation system of schooling.

3. At the same time mathematics is not only a privileged form of knowledge that distributes credits. It is a decontextualised system of meanings, that bears the potentials for making a change to immediate contexts by allowing to think possibilities that appeared as yet "unthinkable" (Bernstein, 2000) from within the context. Blending political public domain activities with mathematical learning activities for the sake of engaging with politics must not replace nor displace disciplinary mathematical learning.

The field of tension that is opened by these three risks illustrates the intrinsic dilemma of politicising Palestinian school mathematics. We will necessarily operate on a thin line between proper emancipation and its exact opposite. However, the delicacy of our venture should not discourage us and set ourselves into a position of passivity. It is obvious that the Palestinian life-world-context requires an "interruption" to the way things go at the moment: It is important to think new possible pedagogical practices that could promote students adopting a role and a position of active agents, even at the risk of counteracting our own intentions.

In the following we firstly describe the methodological frame which we used for imagining activities in awareness of the risks that come along with it, and secondly illustrate the imagined activities and the potentials that we tried to build into them.

Interrogating the Passive Representation of Learners of Mathematics

In order not to waste these potentials of contributing to more active and critical citizens, we wish to suggest an active model of the learner inspired by critical pedagogy:

> The aim of critical pedagogy, according to Freire, is to access the complex ethical dilemmas and power relations inscribed within a given context in order to trigger "moral outrage" (Iyer et al., 2004, p. 356) and increase student participation and social action. The emotion that fuels outrage, unlike that which underpins guilt, can become a source of political agency in the service of the disadvantaged. (de Freitas, 2008, p. 80)

To be clear, the aim is not to transform the sociopolitical context through school mathematics. We are not so naive to believe in school mathematics being the sublime object to finally realise "the high ideals of democracy, social justice and equality" (Pais & Valero, 2012, p. 20). What we believe, however, is that school mathematics can be used to trigger a form of moral outrage, which as de Freitas

Table 12.1 Strategies of representation (Dowling & Burke, 2012, p. 89)

Expression	Orientation to pattern: Consonance	Orientation to pattern: Dissonance
Connotative (tacit)	Invisibility	Tokenism
Denotative (explicit)	Stereotype	Interrogation

suggests, is not oriented towards violence or hate, but towards dialogue, disputation and social action.

Searching for strategies for triggering moral outrage while simultaneously taking the risks described above into account, we take advantage of a scheme developed by Dowling and Burke (2012). We find this scheme particularly productive for our endeavour, as Dowling and Burke explicitly used it for the sake of critiquing the way in which blending political activities and mathematical learning activities undermine the "blender's" very own intentions. Dowling and Burke (2012) use this scheme rather to question the potentials of critical pedagogy in school mathematics, than fostering it, when they conclude: "We can be both mathematics educators and political activists, just not at the same time" (p. 101). Thus, the scheme should be highly sensitive to the risks that we have described above.

Dowling and Burke design a four-field matrix (Table 12.1) that distinguishes strategies of representation in two dimensions: (1) strategies of representation can *express* Otherness by a connotative representation and thus handling difference in a tacit and implicit way, or they can express Otherness by denotative representations by making difference visible and explicit; (2) strategies of representation can *orient* the addressee towards consonance with expected patterns, that is approving her perception of reality, or they can orient the addressee towards dissonance with expected patterns, that is challenging her perception of reality. The combination of these two dimensions produces four different strategies of representation. Invisibility and stereotype strategies can be considered as constituting action that contributes to maintenance of given hierarchies (cf. p. 99). "Tokenism and interrogation, then relate to the maintenance of anti-patriarchal alliances that stand in opposition to patriarchy" (p. 99). Dowling and Burke (2012, p. 100) "expect these strategies [of dissonance] to be most effective where the dissonance becomes explicit, which is to say in interrogation."

Mathematics classroom activities would then be most likely to irritate the common pattern of passivity (see analysis above) and sociopolitical irrelevance (Fasheh, 1997), when they combine dissonance and denotation. In the following, we want to suggest two such activities shown in Figs. 12.3 and 12.4. The activities seek to provoke dissonance with the common pattern by discussing sociopolitical issues, where they are not expected: in mathematics lessons. While the activities are designed as classroom activities, they are not (yet) intended for application in mathematics classrooms, but as a source of interrogation in the training of prospective mathematics teachers. In this way, they are meant to simultaneously sensitise for the urgency of making mathematics relevant to Palestinian life and sensitise for the risks that come along with this.

Fig. 12.2 Original Fermi-problem from Ruwisch and Schaffrath (2009). The heading says "Common meals" and the statement under consideration says "During your time of primary school, you and your family share more than 4000 meals"

For the design of the activities we intentionally estranged "Fermi"-problems that were designed for primary schools (Ruwisch & Schaffrath, 2009).[6] These problems have a common structure (see Fig. 12.2): On top, they ask "Is this possible?", followed by a heading that provokes associations with the context, a statement that could either be true or false, and an image that further contextualises the problem. The statements are formulated in a way that verification and falsification can be reached by estimated quantification, an "educated guess" that either builds on information available on the picture, or information that can be made accessible to the students in another way (e.g. research on the Internet, newspapers, nonfiction). Additionally, each problem is classified in a category (in the original by Ruwisch & Schaffrath, 2009, A: school, B: leisure time, C: me and my body, D: food and drinks, E: nature; we added a new category, F: Politics and Society), each category having its own colour for the frame. We chose this format as it offers different channels for unfolding a political context: The combination of heading + statement + picture allows to highlight enough facets in order not to simplify the context too much,

[6] Needless to say that the mathematical concepts and procedures necessary to solve the problem *mathematically* "lag behind" the sociopolitical awareness needed to *systematically reflect the sociopolitical problem*. We, however, see no problem with giving such problems to, say, a 17-year-old, despite the according mathematical concepts and procedures being on the level of advanced primary school.

when confined in the format. We hope that through the combination of these sources politics can be brought to the agenda. The category we added explicitly signals that the problem is rated as about politics and society. Thus the format facilitates a representation that Dowling and Burke (2012) classify as "denotative". Further, there is no way to engage with the problems, but in an "active mode" as the problems lack a fixed procedure and often also lack crucial information. Finally, the question "Is this possible?" allows to radically break the meaning of the whole problem, interpreting it as "Are you really asking me this?" and thus opening the possibility of refusing to solve the "problem" by mathematical means at all (see Straehler-Pohl in this volume). In this way, the problems are intended to allow for "reflective knowing" of mathematics, that is a form of knowing that addresses its own status as knowledge (Keitel et al., 1993).

Sample Activity 1: "Making Sure People Do Not Starve"

Figure 12.3 shows the first activity about food consumption and food imports into Gaza. The picture on the left side displays a family in Gaza.[7] The drawing on the right side is in black and white and depicts British citizens who board a ship for

Fig. 12.3 "Making sure, people do not starve"

[7] Source: http://gisha.org/image-gallery/2163, Photograph: Karl Schembri/Oxfam GB.

emigration to British North America (Canada) during the Industrial revolution (1750–1850).[8]

This activity provokes different possible discussions in relation to the main question: "Is this possible?" The title and the statement "justify" Israel's act as for the sake of people through the use of the words "not starve" and "protect". So the title picks up the Israeli rhetoric that are used to publicly justify their measures on regulating the import of food-supplies. It is easy to imagine that such representation will trigger "moral outrage" by Palestinian prospective teachers and provoke contestation. The comparison with Britain in 1750 adds another dimension to this provocation. According to the critical Israeli newspaper "Haaretz", the average amount of calories available to Palestinians (per capita per day) after import regulations was 2279 kcal in 2012 (Hass, 2012). According to Roser (2015), the average consumption in Britain (per capita per day) was 3432 kcal in 2009 and was 2237 kcal in 1750. This comparison puts in question what it actually means "not to starve" and is thus likely to provoke debate among discussants. In this way the activity sets out to produce a strong and visible dissonance with the way reality is usually perceived. As the data from Haaretz and Roser are not (yet) available to the prospective teachers, we anticipate some questions such as: How much did British citizens consume in 1750? And how much now? What sort of food do they (British or Palestinians in Gaza) eat? How do the values for consumption relate to values for daily dietary energy requirement for adults? How can we explain the huge gap between the two numbers in Britain today? Given the daily requirements, are the 2279 kcal imported to Gaza really enough? Are they really not enough? Is it really true that Israel did not only ensure a minimum import of food, but restricts on a maximum? How did Israel calculate the amount of food needed for people in Gaza, and why? Do they still? If I eat more than 2279 cal per day, does that mean that another Palestinian is forced to eat less?

In order to debate about these issues on an informed basis, the prospective teachers will need to do some research on statistics, collecting and interpreting numbers. Further questions about adequate mathematical models could arise: How can we calculate how many calories does an adult need per day? How can we include the difference between men, women and children in their needs of calories in the model? However, questions about mathematical models should go beyond the focus on description: Is there a formula that prescribes the restrictions on food imports? Actually there is.[9] This could shift the focus to the "formatting power" (Skovsmose, 2014) of mathematical models, namely to the fact that mathematical models do not only describe a reality, but rather prescribe it (see also Straehler-Pohl in this volume). Is it ethically justifiable to restrict food imports based on mathematical models in this case? Is it ever?

---

[8] Source: http://mrjovanoskisocialstudies.blogspot.de/p/immigration-city-of-montreal.html.

[9] See http://gisha.org/UserFiles/File/HiddenMessages/DefenseMinistryDocumentsRevealedFOIAPetition.pdf.

Sample Activity 2: "Concrete Lines"

The picture[10] in Fig. 12.4 shows the wall that goes through the city of Jerusalem and that divides the Palestinian part of the city from the Israeli part of the city.

Obviously, this problem is much less complex than the problem "Making sure, people don't starve" in terms of mathematical modelling. All information can be derived from the picture directly, the statement can be verified or falsified by simply estimating how many humans are needed to reach the height of the wall (which includes deciding on a model of a human pyramid). It is, then, easy to check whether your family provides enough human material to build a human pyramid of that height. At the same time the orientation pattern of dissonance is pushed to an extreme, so that the question "Are you really asking me to solve this by mathematical means?" almost becomes unavoidable. Once the absurdity of solving this problem mathematically is stated, attention can be directed to other dimensions, e.g. How do family sizes differ on either side of the wall? Is it actually possible to build a human pyramid in front of the wall? Does that differ depending on which side of the wall you are? How come it appears to be plenty of space for building a human pyramid on one side of the wall, while on the other side, the houses are built up to extremely close of the wall?

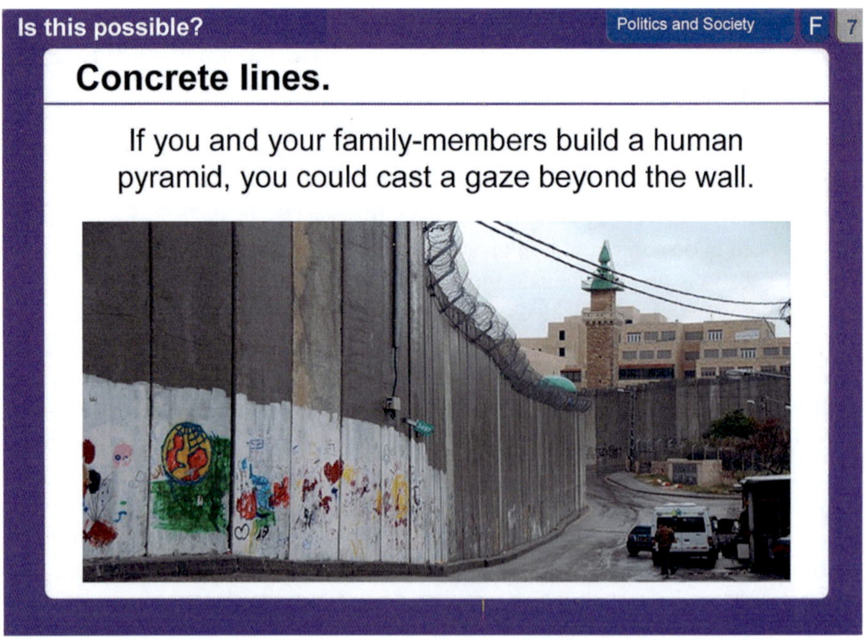

Fig. 12.4 "Concrete lines"

[10] Source: http://www.taz.de/!111950/, Photograph: Johannes Weber.

When Hauke discussed this problem with his students during a teacher training in Berlin, where the political dimension of this particular problem was much less explicit for the students (although some thought about the Berlin Wall in the beginning), all these questions came up in the discussion, which finally lead towards the most important questions of all: Is it *allowed* to raise these questions in a mathematics classroom? Who *decides* whether it is allowed? Without finding consensus, the prospective teachers intensively and controversially discussed this issue, finally identifying politics even in original Fermi-questions they were initially given, on for example water consumption, or on how many common meals are taken with the family. Concerning the original problem in Fig. 12.2, questions posed by the students were, e.g. Where is the father? Why is society organised in a way that families hardly manage to have common meals on a regular basis? Should such common meals be the norm at all? Why are we asking about the absence of the father and not about a second mother? Why are there only females on the picture? And so on. The discussion produced a quite dense deconstruction of the taken for granted assumptions about what a family is supposed to be and do in German society. Beforehand, these original Fermi-problems were perceived by the students as simply being opportunities to learn about mathematics and life simultaneously.

Conclusion

This chapter aims to explore possible alternatives for the relationship between mathematics education and the sociopolitical structure and conditions in the Palestinian society. In brief, we first describe the political context and looked at mathematics education and its relationship to, and as product of, that context. We then present an illustration of Jehad's analysis of school mathematics textbooks focusing mainly on how mathematics is represented and what are the expected roles of the learners of mathematics. Finally, we suggest two activities as possible alternatives that could serve to provoke an imagination of a different form of mathematics education, more sensitive to the Palestinian sociopolitical context.

We try to give a glance of what it could mean to construct the image of mathematics in relation to the contemporary conditions of life in Palestine. This image includes a Palestinian learner of mathematics that is agentive (not passive) not only in doing mathematics and in using it to understand her social and political situation, but further in questioning how mathematics is employed to sustain the unbearable situation they have to live in. We maintain that such critical pedagogy has the potentials to trigger moral outrage in a productive way that reinforces the building of solidarity, not only among Palestinians, but among humans, including Israelis who suffer from the current situation. While our activities address the necessity of overcoming the Israeli military occupation, we want to highlight that we intend in no way to join in simply blaming only Israel for the malaise of the Palestinian situation—as this would actually reinforce the image of passive and oppressed Palestinians who are *subject to* rather than *the subject of* social change. Activities that address the corrupted systems that were built over the last 22 years after the formation of the PA by Palestinians are another step to be taken.

With attempts to imagine a different form of Palestinian mathematics education, we do not argue that presenting the learner of mathematics as active will directly unfold an impact towards the realisation of a better life in Palestine. However, we think that what a critical and activating mathematics education can offer, is making a small contribution to increasing the chance for Palestinian students to be more engaging in the transformation of their own society. In this endeavour, mathematics is an example of learning as any other subject; education is just one example among a wider variety of aspects of life that need to change; but as mathematics has its prestigious power in the Palestinian society, as in many others, making a change here is a part of making a change in the whole picture. Interrupting passivity is surely not yet the desired change itself, but it is a contribution. We wish that the two activities presented here could inspire Palestinian mathematics teachers and Palestinian mathematics teacher trainers to design similar activities, to elaborate the proposed activities and to reflect the relation between mathematics and the sociopolitical. Our hope is for a mathematics education that encourages the learners of mathematics in Palestine to engage more in their immediate sociopolitical problems and have more agency in transforming their society.

Acknowledgments We would like to thank Paola Valero, Christer Bergsten, Alexandre Pais and Jill Adler for their helpful comments.

References

Al-Ramahi, R. J. (2006). *Levels of geometric thinking for teachers and in mathematics school textbooks.* (Masters dissertation), Birzeit University, Ramallah, Palestine.

Alshurafa, H. (2015). *Analysing geometry in the Palestinian mathematics textbooks using a linguistic approach.* (Masters), Birzeit University, Ramallah, Palestine.

Alshwaikh, J. (2005). *Patterns of the Palestinian students' geometric thinking.* (Masters Master's Thesis), Birzeit University, Ramallah, Palestine.

Alshwaikh, J. (2015). Image-writing relations in Arabic mathematical textbooks. In G. R. S. E. A. A. E. Breuer (Ed.), *Studies in writing: Multimodality in writing* (Vol. 30, pp. 117–135). Brill: Leiden.

Alshwaikh, J., & Morgan, C. (2013). Analysing the Palestinian school mathematics textbooks: A multimodal (multisemiotic) perspective. *Proceedings of the British Society for Research into Learning Mathematics, 33*(2), 70–75.

Bernstein, B. (2000). *Pedagogy, symbolic control, and identity: Theory, research, critique.* New York: Rowman and Littlefield.

de Freitas, L. (2008). Critical mathematics education: Recognizing the ethical dimension of problem solving. *International Electronic Journal of Mathematics Education, 3*(2), 79–95.

Dowling, P. (1998). *The sociology of mathematics education: Mathematical myths/pedagogic text.* London: RoutledgeFalmer.

Dowling, P., & Burke, J. (2012). Shall we do politics or learn some maths today? Representing and interrogating social inequality. In H. Forgasz & F. Rivera (Eds.), *Towards equity in mathematics education: Gender, culture, and diversity* (pp. 87–103). Heidelberg: Springer.

Fasheh, M. (1997). Is math in the classroom neutral-or dead? A view from Palestine. *For the Learning of Mathematics, 7*(2), 24–27.
Frankenstein, M., & Powell, A. B. (2002). *Paulo Freire's contribution to an epistemology of ethnomathematics*. http://andromeda.rutgers.edu/~powellab/docs/proceedings/paulofriere_epis.pdf.
Hass, A. (2012, October 17). 2,279 Calories per person: How Israel made sure Gaza didn't starve. *Haaretz*. Retrieved from http://www.haaretz.com/israel-news/2-279-calories-per-person-how-israel-made-sure-gaza-didn-t-starve.premium-1.470419.
Hegarty, M., Mayer, R. E., & Monk, C. A. (1995). Comprehension of arithmetic word problems: A comparison of successful and unsuccessful problem solvers. *Journal of Educational Psychology, 87*(1), 18–32.
Herbel-Eisenmann, B., Choppin, J., Wagner, D., & Pimm, D. (Eds.). (2012). *Equity in discourse for mathematics education: Theories, practices, and policies*. Dordrecht, The Netherlands: Springer.
Keitel, C., Kotzmann, E. & Skovsmose, O. (1993): Beyond the Tunnel-Vision: Analysing the Relationship Between Mathematics, Society and Technology. In: C. Keitel; K. Ruthven (Eds.), Learning from Computers: Mathematics Education and Technology. Berlin: Springer, 243–279.
Lerman, S. (2000). The social turn in mathematics education research. In: Boaler, J. (Ed.) Multiple perspectives on mathematics teaching and learning, pp. 19-44.
Morgan, C. (2001). Mathematics and human activity: Representation in mathematical writing. In C. Morgan & K. Jones (Eds.), *Research in mathematics education volume 3: Papers of the British Society for Research into Learning Mathematics* (pp. 169–182). London: British Society for Research into Learning Mathematics.
Morgan, C. (2006). What does social semiotics have to offer mathematics education research? *Educational Studies in Mathematics, 61*(1), 219–245.
Moughrabi, F. (2001). The politics of Palestinian textbooks. *Journal of Palestine Studies, 31*(1), 5–19.
Mullis, I. V. S., Martin, M. O., Foy, P., & Arora, A. (2012). *TIMSS 2011 international results in mathematics*. Chestnut Hill, MA: TIMSS & PIRLS International Study Centre, Boston College.
O'Halloran, K. L. (1996). The discourses of secondary school mathematics, unpublished dissertation, Murdoch University, Perth, Australien. Rotman, B. (1988). Towards a semiotics of mathematics. Semiotica, 72(1/2), 1–35.
Pais, A., & Valero, P. (2012). Researching research: Mathematics education in the political. *Educational Studies in Mathematics, 80*(1–2), pp. 9–24.
Powell, A. B. (2002). Ethnomathematics and the challenges of racism in mathematics education. In P. Valero & O. Skovsmose (Eds.), *Proceedings of the Third International MES Conference* (pp. 1–15). Copenhagen: Centre for Research in Learning Mathematics.
Rewadi, F. (2005). *A comparison of mathematical reasoning in the Palestinian curriculum with the National Council of Teachers of Mathematics Standards 2000.* (Masters dissertation), Birzeit University, Palestine.
Roser, M. (2015). *Food per person*. Retrieved from http://ourworldindata.org/data/food-agriculture/food-per-person/.
Ruwisch, S., & Schaffrath, S. (2009). *Fragenbox Mathematik: Kann das stimmen?* Dortmund: vpm Lernbuch Verlag.
Sfard, A. (2008). *Thinking as communicating: Human development, the growth of discourses, and mathematizing*. Cambridge, UK: Cambridge University Press.
Skovsmose, O. (2011). *An invitation to critical mathematics education*. Rotterdam: Sense.
Skovsmose, O. (2014). Mathematization as social process. In S. Lerman (Ed.), *Encyclopedia of mathematics education* (pp. 441–445). Heidelberg: Springer.
Straehler-Pohl, H., & Gellert, U. (2015). *Pathologie oder Struktur? Selektive Einsichten zur Theorie und Empirie des Mathematikunterrichts*. Heidelberg: Springer.
Straehler-Pohl, H., Gellert, U., Fernandez, S., & Figueiras, L. (2014). School mathematics registers in a context of low academic expectations. *Educational Studies in Mathematics, 85*(2), 175–199.

Tang, S., Morgan, C., & Sfard, A. (2012). *Investigating the evolution of school mathematics through the lens of examinations: Developing an analytical framework.* Paper presented at the 12th International Congress on Mathematical Education, Topic Study Group 28 on Language and Mathematics, Seoul, Korea.

Valero, P. (2008a). *In between reality and utopia: A socio-political research agenda for mathematics education in situations of conflict and poverty.* Paper presented at The Fifth International Mathematics Education and Society Conference, Lisbon: Centro de Investigação em Educação, Universidade de Lisboa.

Valero, P. (2008a). Discourses of power in mathematics education research: Concepts and possibilities for action. *PNA, 2*(2), 43–60.

Van Hiele, P. M. (2004/1959). The child's thought and geometry. In T. P. Carpenter, J. A. Dossey & J. L. Koehler (Eds.), *Classics in mathematics education research (English Translation of Selected Writings of Dina van Hiele-Geldof and Pierre M. van Hiele)* (pp. 60–67). Reston, VA: National Council of Teachers of Mathematics.

Watson, H. (1990). Investigating the social foundations of mathematics: Natural number in culturally diverse forms of life. *Social Studies of Science, 20*(2), 283–312.

Wildeman, J., & Tartir, A. (2014). Unwilling to change, determined to fail: Donor aid in occupied Palestine in the aftermath of the Arab uprisings. *Mediterranean Politics, 19*(3), 431–449.

Chapter 13
Assessment in Mathematics Education: A Gatekeeping Dispositive

Lisa Björklund Boistrup

Abstract This chapter aims at shedding light on a governing assessment dispositive in mathematics education, which effects that some students are provided with affordances to learn and engage in mathematics, while others are not. Through such a dispositive, the system of school is governing the act of gatekeeping and selection of students, which is contradictory to what is stated in official documents. While drawing on findings from previous classroom studies and action research, a tentative assessment dispositive is presented. It consists of different assessment discourses for students to experience, or not, affordances for learning mathematics; and regulatory decisions which affect assessment practices on classroom level. The purpose of presenting such a dispositive is twofold: (a) to contribute to an understanding of how such an assessment dispositive may look like; and (b) to provide a starting point for further research and discussions among teachers, students, and decision makers.

Introduction

This chapter aims at shedding light on a governing assessment dispositive in mathematics education, which effects that some students experience affordances for the learning and engagement in mathematics, while others do not. I take the mathematics classroom, age group 7–15 years, as a starting point to analyse how students' and teachers' assessment practices on a daily basis are governed within the institution of school. A Foucauldian framework is adopted to describe effects of an assessment dispositive in terms of governing. Simultaneously I cast a normative pedagogical stance on this description outgoing from the assumption that it is of interest to study if and how students in mathematics education experience affordances, not only for learning mathematics, but also for engaging in mathematics education as active

L. Björklund Boistrup (✉)
Department of Mathematics and Science Education, University of Stockholm, Stockholm, Sweden
e-mail: lisa.bjorklund@mnd.su.se

H. Straehler-Pohl et al. (eds.), *The Disorder of Mathematics Education*,
DOI 10.1007/978-3-319-34006-7_13

agents. This interest does not derive from a neoliberalist performative perspective (discussed by for example Ball, Maguire, Braun, Perryman, & Hoskins, 2000), such as worries due to "bad" results on surveys of PISA (e.g. Skolverket, 2013) or a naïve understanding of "a school for all" (Valero, 2013). The interest rather derives from an assumption that learning mathematics, for students, may be beneficial in different ways. One example is getting access to knowledge which may be adopted in various situations in life (see for example FitzSimons, 2014, for a discussion on mathematics in relation to working life contexts); as qualifications, which open opportunities for future education and work (discussed by Gellert in this volume); or just for the possibility to experience mathematics as "fun". The "usefulness" of mathematics may be problematised in various ways (e.g. Straehler-Pohl in this volume) but such a discussion is beyond the scope of this chapter. There is a tension between governing in the sense of Foucault versus the normative pedagogical stance described above. I return to this tension in the discussion.

Assessment is in this text understood as a broad notion incorporating feedback in classroom interactions as well as feedback, which is communicated through testing of various kinds. From a pedagogic or didactic perspective, it might be argued that feedback is more about teaching than about assessment. However, in this text, assessment is seen as something that is present in teaching in any moment. A teacher makes some kind of assessment already when meeting a group of students for the first time. The teacher may, then, assess students in comparison to what s/he perceives as what this age group normally is able to accomplish in mathematics, or what could be expected from students in a certain neighbourhood. This chapter argues that discourses of assessment, to a certain extent, determine the expectations of student accomplishment. In these discourses, assessments are displayed to students through various kinds of feedback. A point made in this chapter is that whether or not students are experiencing affordances to learn and engage in mathematics, heavily depends on experienced assessment discourses, as well as regulatory decisions and administrative measures, which govern assessment practices in mathematics classrooms.

Perspectives on Classroom Assessment in Mathematics

As a background, I here develop a perspective of assessment as part of a critical perspective in mathematics education research. I also briefly provide an orientation of the context of the study: Sweden.

Classroom Assessment and Politics in Mathematics Education

In the common discussion in research literature on assessment, assessment is often treated as something inherently "good" (see Cizek, 2010, as an example on such a discussion on formative assessment), while not acknowledging how assessment in fact is a practice of the exclusion of students within the institution of school (e.g. Ball et al., 2012; Emanuelsson, 2002). Frequently, research on assessment is focused on "efficient" techniques, (see Wiliam & Thompson, 2007, where five assessment strategies are presented, as an example). In this chapter, I challenge such assumptions, and I provide a description of assessment in mathematics education as part of an overall governing dispositive. Also, I claim that teaching without any assessment is empirically not possible, since feedback is inherent in teaching, and assessment, in turn, is inherent in feedback. Moreover, as this chapter will illuminate, the kind of assessment that different students encounter may not only be more or less inviting to the learning and engagement in mathematics, but moreover allocating (see Gellert in this volume) students to different positions in relation to their immersion in that subject. The chapter sets out to explore how the contemporarily governing assessment dispositive confines the possibilities for developing assessment practices that could invite more students to learn mathematics.

An interest in investigating assessment practices from a social and critical perspective is not new. One example is Bernstein (1973), where assessment is one of three central features maintaining current social orders. Curriculum then determines what counts as valid knowledge, and pedagogy determines what counts as a valid transmission of knowledge. Assessment determines what counts as a "valid realisation of that knowledge on the part of the taught" (Bernstein, 1973, p. 85). Another example is how Foucault (2003) wrote about the role of formal assessments in education arguing that, in assessment, surveillance is combined with normalisation. Through assessment, there is qualification and classification taking place simultaneously, as well as the exercise of power and education of a specific knowing (see Torrance & Pryor 1998).

Also in the field of mathematics education there has for a while been an interest in assessment from a critical and social perspective. One example is Morgan's (2000) critique of mainstream traditions of mathematics assessment research. In her conclusion, Morgan emphasised the necessity for research that adopts a social perspective, arguing that a main concern of research from a social perspective is to understand how assessment works in mathematics classrooms and more broadly in education systems. As a consequence, it is essential to analyse potential effects of classroom assessment in a way that relates the local teacher-student perspective to the governing context of society.

One area critiqued in the general literature on classroom assessment with political implications is equity issues (Broadfoot, 1996; Gipps, 2001). These issues can be identified on a system level, where it can be argued that assessment serves in the selection, certification and control of groups of students (Broadfoot, 1996; Jurdak, 2014). These processes are in a few studies identified in classroom work. In Björklund Boistrup (2015) it is shown how different students in the same classroom may experience different assessment practices during the same lesson, which can be compared to Watson (2000) where it is shown that the same students would be assessed differently by different teachers. In Mercier et al. (2000) too, there are findings indicating that the feedback students receive from the teacher in the mathematics classroom varies. In their findings they give account for how teachers' assessment of students' actions are affected by each student's social position.

Different grounds for inequities in relation to assessment in mathematics are possible to investigate, e.g. socio-economical factors, gender, language and culture. Cooper and Dunne (2000) investigated in their influential study how children from different social backgrounds were given non-equal opportunity to perform well on national tests in terms of what constituted a "realistic" test item and for whom. Similar inequalities were examined for boys and girls. In McGrady and Reynolds (2013), analyses of assessments of students with different ethnical background on data from a longitudinal study confirm that the effects of mismatch often depend on the racial/ethnic statuses of both the teacher and the student.

Ball et al. (2012) focus on assessment in general when they adopt Foucault's work in order to explore the "pressures" to "deliver", which bear upon English secondary schools in relation to required performances on national tests. They examine the standards agenda in terms of a set of practices specified by policy: "As a policy standards 'works' through a simple but effective and public technology of performance—made up of league tables, national averages, comparative and progress indicators" (Ball et al., 2012, p. 514). This is similar to what Forsberg and Wallin (2006) describe for Sweden when they illuminate how Swedish teachers and students are increasingly being controlled (or under surveillance using a term from Foucault).

Another aspect of effects of assessments is different expectations from different groups of students. Straehler-Pohl, Gellert, Fernandez, and Figueiras (2014) address this in a study where students have been streamed into three different ability groups at the beginning of secondary school. During the last years, Swedish mathematics education has more and more become characterised by pedagogical segregation, where teachers adopt different teaching and assessment practices according to their perceptions of their groups' social and linguistic composition, leading to lower achievement for children from low socio-economic backgrounds or particular immigrant groups (Hansson, 2010).

The Swedish Context

The societal governing context to which teacher-student interactions are related in this chapter is located in Sweden. The Swedish steering system of teaching includes no external examination throughout compulsory school and upper secondary school such as, for example, the GCSE in the UK. Since 1940, the marks (for example, E as a mark signifying passed and A as the highest) a student receives in the end of a semester are determined by the teacher (Pettersson & Björklund Boistrup, 2010). Currently this kind of marking firstly takes place in the 6th year of schooling and the marks are mainly used as selection for further studies (Skolverket, 2011). During the first 10 school years, Swedish students are not streamed into achievement levels. This means that students structurally are not divided into different study levels or directions until after compulsory school, when they enter upper secondary school. However, since 1992 it is possible for individuals and organisations to start schools, and parents may choose other schools than the closest municipality driven school for their children. Still the same national curriculum applies for all students up to the age of 16. It is stated in the Swedish Compulsory School Ordinance (SFS, 2010, p. 800) that all students should encounter equal opportunities for learning in school.

There is a variety of steering documents, although these documents do not, in fact steer teaching in full (Björklund Boistrup, 2010; Skolverket, 2003). Instead, the textbooks, which are not obliged to follow steering documents, are shown to dominate the teaching practices in mathematics classrooms (Skolverket, 2014). There are no mandatory official textbooks and the decision of what teaching material to purchase is made on a local level. The overall national criteria are stipulated by the Swedish Parliament and Government (Skolverket, 2011).

Theoretical Considerations and Methods

As an analytical framework I have drawn on Foucault's terms *dispositive* and *discourses*. For the operationalisation of this I adopted four previously construed assessment discourses and reanalysed data from previous research projects.

Analytical Concepts

In order to allow for investigating assessments in mathematics education as part of an overall assessment technology that is related to the sociopolitical sphere beyond a particular classroom, I built my analysis around the Foucauldian term *dispositive* (Bussolini, 2010; Foucault, 2003). This term further allows to perceive assessment as a technology of governing. While we can use the metaphor of an apparatus to imagine the functioning of a dispositive, it is simultaneously to be crucially distinguished from it, as it is "more distributed" (Bussolini, 2010, p. 86) and less centralised in the source of its effects.

Raffsnøe, Gudmand-Høyer, and Thaning (2014) have scrutinised the work of Foucault, also taking into account the meaning of *dispotif* in French everyday language: "the term *le dispositif* often describes an arrangement set up for a specific purpose, also designed to have immediate effect" (p. 6). This is in line with Agamben (2009, p. 2) who writes:

> The hypothesis that I wish to propose is that the word *dispositif*, or "apparatus" in English, is a decisive technical term in the strategy of Foucault thought. He uses it quite often, especially from the mid 1970s, when he begins to concern himself with what he calls "governmentality."

Governmentality (Foucault, 2008b) is constituted by the institutions, procedures, and tactics that allow for the execution of power. Foucault includes in the term a tendency of steering—governing—that is common in society. This governing is by

Jørgensen and Klee (2014) described as "conducting the conduct" of, in the case of this chapter, teachers and students. "Conducting the conduct" is the "aim" of a dispositive. In this way, teachers and students are not only governed (conducted), but also governed to perform the conduct on themselves (e.g. Fejes, 2008).

The term *discourse,* according to Foucault (1993, 2003), establishes a relation between language (taken in a broad sense in this text), knowledge and power. The institution of school can be considered as an institution establishing discourses. The entanglement of language, knowledge and power that a discourse creates within mathematics classrooms has its origins beyond any particular mathematics classroom's communication. For the people who are part of a discursive practice, like teachers and students, the "rules" of the discourses affect how it is possible to act and what is possible, or not possible, to communicate (Foucault, 1993, 2003, 2008a).

Discourses can be considered as elements of *dispositives* that are, among other entities, effective in establishing a dispositive and constituting its power:

> What I am trying to pick out with this term [the dispositive; le dispositif] is … a thoroughly heterogeneous ensemble, consisting of discourses, institutions, architectural planning, regulatory decisions, laws, administrative measures, scientific statements, philosophical, moral and philanthropic proportions—in short, the said as much as the unsaid. Such are the elements of the dispositive. The dispositive itself is the network that can be established between these elements. (Foucault, 1977, p. 299, translation to English from Raffsnøe et al., 2014, p. 7)

I draw on Foucault in the sense of this quote when describing an assessment *dispositive* within the *institution* of school through the construal of *discourses*. I have also analysed instances in data from mathematical classrooms, where parts of *regulatory decisions*, school mathematical *statements*, and *administrative measures* are present. Similar to Raffsnøe et al. (2014), I interpret Foucault to analytically include not only words as means of communication but also other means, such as gestures, and artefacts (Foucault, 1993, 2008b).

In his own analyses of dispositives and discourses, Foucault focused on much broader discursive formations that operate on a much higher level of abstraction than for example classroom communication, which can be localised quite concretely in time and space.[1] In order to be able to adequately address the particularities that result from such concrete localisation, I adopted social semiotics with a multimodal approach as a more fine-grained analytical framework. In social semiotics, assessment of knowing and learning is an instance of communication, a matter of acts taking place between teacher and student, or student and student. From this perspective, when students demonstrate that they have learned something by means of a variety of communicative resources, such as speech, pictures, and symbols; this is an essential part of the established communication (and as a consequence the discourse). Kress (2009, p. 21) writes:

[1] Inspiration for adopting a Foucauldian perspective on such "small elements" as a mathematics classroom, can, among others, be found in the work of Walkerdine (1988) who construed a "testing discourse" where the teacher posed questions to which she already knew the answer.

> [Multimodality and social semiotic theory] together enable an account of communication, of meaning, of learning and, with that, of assessment, in which these issues can be treated as distinct and yet remain connected, in theory and in practice."

Van Leeuwen (2005) describes multimodal communicative acts as a way to outline a social semiotic approach to the "how" of communication. In all communication, meaning is produced through different semiotic resources that are co-present and interrelated in multimodal ensembles (Kress, 2009). Communicative practices affect the meanings produced through the semiotic resources, as do the specific situations. With respect to classroom assessment in mathematics, the meanings possible to construe are affected by the semiotic resources that constitute the acts, existing assessment systems and procedures, and the assessment practices in the various classrooms. Kress (2009) emphasises the importance of understanding multimodal communication in order to fully understand a phenomenon like assessment.

Data and Analysis

The data for this text derive from two research projects. One is a case study where five mathematics classrooms with a total of 25 lessons were visited with an interest in assessment and feedback between teachers and students (Björklund Boistrup, 2010). The transcripts from the classroom films were made multimodal (software used: Videograph). Written data from the same classrooms were additionally collected. A preliminary analysis was performed earlier, but in terms of "institutional traces" (Björklund Boistrup, 2010). The second set of data derives from a project where I and a colleague together with mathematics teachers performed action research projects with an overarching interest in assessment and feedback (Björklund Boistrup and Samuelsson in preparation). Discussions with on a total 16 participating teachers about political aspects of classroom assessment in mathematics were summarised in the discussions of four research reports written in Swedish (e.g. Björklund Boistrup et al., 2013). These summaries are analysed in this text while outlining a tentative assessment dispositive in mathematics education.

I have reanalysed the data from the two previous research projects, searching for instances where the mathematics classroom's institutional context was present in assessment acts and feedback. I have focused the analysis on governing elements interpreted from classroom communications and summaries from discussions with teachers. Such elements may be textbooks or assessment material from municipalities. In the operationalisation of the theoretical concepts described above I have adopted four previously construed assessment discourses (Björklund Boistrup, 2010, 2015):

1. Do it quick and do it right.
2. Anything goes.
3. Openness with mathematics.
4. Reasoning takes time.

They are presented in short below. What finally is presented as a tentative assessment dispositive is constituted by the four discourses and their relations to other governing elements in the sense of how Foucault conceptualised the term dispositive.

Four Assessment Discourses

In the following, I briefly describe the four discourses from Björklund Boistrup (2010, 2015), which I adopted as analytical tools. Feedback is here interpreted as conveying the teacher's assessments to the student, through words (for example "Well done") and/or body movements (for example a nod with a smile). As previous research suggests that these four discourses differ crucially concerning the extent to which they provide affordances, or not, for students' learning in mathematics education, this extent will be consulted as a horizon to judge assessment practices in relation to their social effects. This justified injection of normativity will be discussed after the presentation of the discourses.

The first discourse, (1) "Do it quick and do it right" has similarities to a traditional discourse of assessment described in the literature where the main "rule" is that the work should be done quickly and what is counted is whether an answer is right or not (e.g. Broadfoot & Pollard, 2000). In this discourse, the teacher's feedback focuses on procedures with limited mathematical content, for example on whether an answer is mathematically correct or not, instead of why and how the answer may be counted as mathematically relevant. Another typical feedback focus concerns how many items from the textbook the student has accomplished. Students are not really invited to engage in any aspect of mathematics through the feedback. An example is the feedback "17 correct answers out of 25" on a test. Here it is important to keep in mind that items on a test may well be mathematically rich and also inviting to the students. What is analysed here is mainly the subsequent feedback. In this discourse, teacher's feedback is usually short and describes whether the student's work is correct or not or whether the student is doing the "right" thing. The affordances for students to be invited to learn mathematics through the feedback are limited in this discourse.

The second discourse, (2) "Anything goes", is a discourse where students' performances, which can be regarded as mathematically inappropriate, are left unchallenged. There is not much articulated feedback apart from general approval. There is a presence of open questions, but challenges do not really occur. There are no critical discussions about students' solutions, and answers considered to be mathematically incorrect can be left unchallenged. The students are invited by the teacher to use whatever communicative resources they want, without any considerations by the teacher or the students on what resources that have most affordances for their learning at that specific occasion. A teacher usually values the students' performance often through general praise. Consequently, the teacher takes the role as the main agent, as "the one that is evaluating". Sometimes the teacher takes a more passive role in the discourse. S/he then does not interfere with students' reasoning even though something wrong is demonstrated. The affordances for students' learning in this discourse are low.

The third discourse, (3) "Openness with mathematics", has more of an open focus on mathematical processes. Also, the teacher displays an openness to students' feedback on the mathematics teaching. Occasionally, goals for the learning are present. Quite often the questions posed are open. The teacher and student often show interest in mathematical processes, and there is also an awareness of students' alternative interpretations of tasks. The focus is mostly on mathematical processes and sometimes on the student's own reflection of her/his own learning. "Wrong" answers are used as starting points for discussions, but it is always clear what can be considered mathematically correct. Different semiotic resources are acknowledged and at times the teacher promotes, whilst at other times restricts, the use of semiotic resources dependent upon the meaning making and learning process demonstrated by the student(s). This discourse holds affordances for students' active agency and learning of mathematics.

Finally, the fourth discourse, (4) "Reasoning takes time", takes the characteristics of "Openness with mathematics" one step further with a slower pace and an emphasis on mathematics processes such as reasoning, problem-solving and defining/describing. There are often instances of recognition of the students' demonstrated knowing, which are sometimes in relation to stated goals, and the questions posed are mostly open ones. At times feedback as interest and engagement are communicated by the teacher to the student and vice versa. The students are often challenged towards new learning with the focus mainly on mathematical processes. In this discourse, silences in teacher-student interactions are common, and the possibility to be silent serves the depth of the communication and the mathematical focus. Various kinds of feedback from teacher to student are often communicated, sometimes through open questions. Both the teacher and student can be active for longer periods of time. In this discourse the affordances for students to take active agency are high. The possibility to be quiet and think for a while promotes this potential agency. Similarly, the affordances for students' learning of mathematics are high and include a wide range of mathematical processes.

Considerations Regarding the Four Assessment Discourses

The four assessment discourses presented above are not displaying a complete picture of possible assessment discourses in mathematics classrooms. All four discourses are clearly part of an overall school mathematical discourse where expected outcomes are easily aligned with what is claimed as important mathematical knowing in international frameworks of today, such as PISA (Skolverket, 2013) as well as in the Swedish national syllabus (Skolverket, 2011). Other assessment discourses are possible to imagine, such as discourses where the focus in the assessments is on critical considerations about the use of mathematics in society (Gellert & Jablonka, 2009; Skovsmose, 2005; Straehler-Pohl in this volume). However, such explicitly critical discourses have not been possible to construe in the original data, although students' and teachers' possibilities to take active agency in mathematics education have been elaborated on in action research with mathematics teachers (Björklund Boistrup and Samuelsson in preparation).

An Assessment Dispositive on the Level of the Mathematics Classroom

The potential assessment dispositive in mathematics education presented in this chapter derives from data from teacher-student communications in mathematics classrooms and summaries from discussions with mathematics teachers. In this section, firstly, findings from analysis of classroom data are presented with a focus on the relation between the four assessment discourses and governing elements such as textbooks. Secondly, an analysis of summaries from the action research projects is presented in terms of governing elements which the teachers claimed to be resources or obstacles in their assessment practices. Finally, these findings are brought together into a tentative governing assessment dispositive in mathematics education.

Discourses and Other Governing Elements in the Mathematics Classroom

Governing elements are present in the assessment practices from which the four assessment discourses in mathematics classrooms are construed. At the same time, the discourses can be viewed as governing elements on their own. In the analysis leading to the findings described below, I used the four discourses as an analytical frame within which governing elements could be identified. Excerpts from classroom data are shown and discussed for each discourse.

Elements Governing Towards "Do It Quick and Do It Right"

Figure 13.1 displays an excerpt of a test on numbers and shapes, which the teacher Cecilia (T) has had her students take. The excerpt displays the student Cilla's (S) answers on the test and Cecilia's (T) comments.

Cecilia (T) has marked Cilla's (S) solutions to several of the items with an "R" as being correct, which provided feedback on whether an answer is considered right or wrong. Additionally, Cecilia (T) has written guiding feed forward: "Use the squares [on the paper] to make it easier to make a number axis" referring to Cilla (S) not using the squares. This comment is focusing on procedures rather than on mathematical processes.

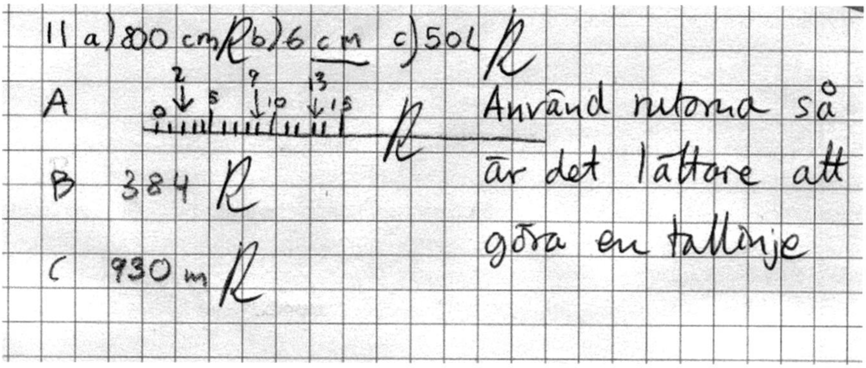

Fig. 13.1 Excerpt (from written material). Part of Cilla's (S) paper. The teacher comment translated to English is "Use the squares [on the paper] to make it easier to make a number axis"

This focus on procedures corresponds with the use of the *textbook* in several situations, particularly when the teacher followed the textbook closely, which is quite common in Sweden (Skolverket, 2003, 2014). This culture of following the textbook includes a notion that "all" items must be accomplished, which leaves little room for the teacher's own planning and also for the students' possibilities to affect the mathematics teaching. In such situations, it was expected that all students should work on the same tasks, at least in the first pages of each text. The students were expected to solve these tasks rather fast and students in some classrooms competed to be the first one to accomplish these tasks. Hence, the textbook can be considered an artefact that affects the interaction towards "Do it quick and do it right".

Further, *extra teaching material* was used with the textbook, including *diagnostic tests* and more substantial *tests* in the end of a period. The situation above with Cecilia (T) and Cilla (S) derives from the taking of such a test. A number of items are then expected to be solved, sometimes within a certain time limit. The guidelines expect the teacher to summarise the results as the number of correct answers. These tests then get a similar role in the interaction affecting affordances both for the teaching, and students' engagement and learning.

In the data I also identified governing elements that derived indirectly from defaults by the Swedish government. One example in the data was a *template for student/teacher/parent meetings*. These meetings must, according to the Compulsory School Ordinance, be held with each child and her/his caretakers twice a year. The data in one such template was developed on the school level, and all teachers were expected to follow the document. In the template, the questions for the students concerned whether they were "good at maths" and provided no room for a more elaborated discussion about processes of learning mathematics. Student answers were supposed to be held short in words. These meetings take place under conditions of high *time constraints* as there are many school subjects to cover, and the periods for these meetings are usually heavy in terms of workload.

Elements Governing Towards "Anything Goes"

In a classroom situation in the video data, Britta (T) discusses in this situation a diagnostic test with Belinda (S). What seems to create problems for Belinda (S) is to solve a task on fractions through drawing and dividing circles. The task was about jars containing coloured marbles, and to finding out in which jar the fraction of white marbles was the highest. After several minutes, Britta (T) brings out manipulatives. These manipulatives consist of "poles" on which coloured blocks are stacked (Fig. 13.2).

In analysing the role of *manipulatives* here, one aspect is that Belinda (S) seems to have advanced rather far through the textbook's "levels". Nonetheless, she has problems with several tasks in the diagnostic test and, in a few cases, Britta (T) and Belinda (S) end up using manipulatives. As an example Belinda (S) finds out the number of thirds that equals 4/12 when comparing 4 "twelfths-pieces" on one pole with one "third-piece" on another pole. There is no discussion about whether Belinda (S) should more and more solve the tasks without manipulatives. Instead, she goes on working at the next "level", following the textbook system. The assessment that Britta (T) communicates here is considered to be that Belinda (S) has demonstrated sufficient knowing during the discussion. This situation is exemplary for interactions where the manipulative piloted the student to a correct answer during assessment acts. That is, the manipulative makes the essential mathematical reasoning redundant and also takes over parts of the assessment acts. The task is solved faster, but without the student engaging in mathematics with any depth. In the case of Britta (T), the emphasis on manipulatives is a *decision* made on the institutional level of the school.

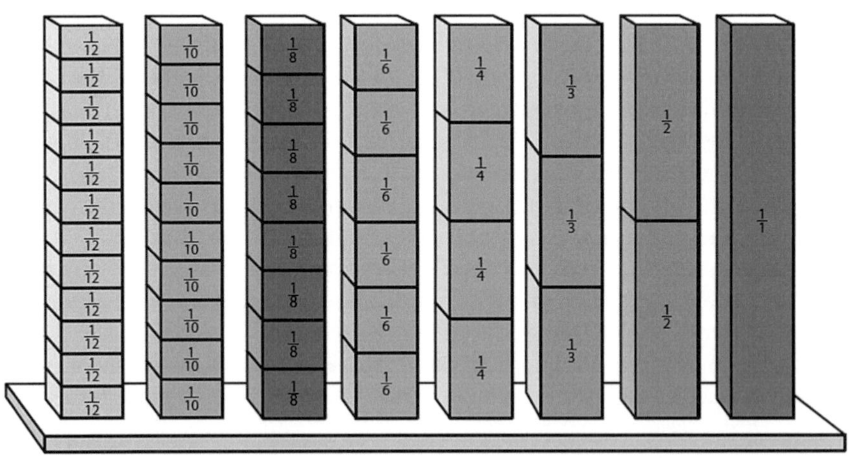

Fig. 13.2 Excerpt with manipulatives used by Britta (T) and Belinda (S) (Björklund Boistrup et al., 2013, p. 137)

A similar tendency to pass over students' problems with mathematical reasoning can be found in the *textbook* on several occasions. The items of the diagnostic tests in textbooks in these cases do not serve to reveal student reasoning that might be important to capture. For example, one was constructed in a way that students may arrive at the key's correct answer while actually performing an incorrect reasoning. Here the textbook does not afford a discussion between the teacher and the student about the understanding of the particular concept, but rather covers incompatibilities under the veil of supposed correctness.

Organisational *rules* stipulated by the school were also considered as governing elements affecting assessment practices. One example is the organisation of learning in *time slots*. There are sequences in the data where students were solving a problem in a way where a feed forward from the teacher probably would have helped them in the mathematical process. The teacher, however, did not linger with the students, for example because the lesson was about to finish. Time and pace are parts of the multimodal "ensemble" that constitutes the classroom interaction. The lack of time, governed by the institution of school, here provided constraints in this ensemble for what kind of assessment practice the teacher and the student could engage in.

Elements Governing Towards "Openness with Mathematics"

Figure 13.3 is from a document concerning *parent/teacher/student meetings*. On the first two pages of the document there are two pages where the student is asked questions. Then there are pages for the teacher to fill out before the meeting. The last document is filled out during the actual meeting, and this is what mainly is analysed here. Part of the document is spaces for comments on both short-term and long-term goals. The same structure is used for all such meetings in every class at Anna's (T) school. The comments on long-term goals are found in Fig. 13.3.

Content	*School's contribution*	*Student's contribution*	*The contribution from home*
Stärka dina matte-kunskaper	Ge uppgifter som passar Ali.	Jobba bra Koncentrera sig	Hjälpa Ali med läxan och påminna honom
Strengthen your maths knowledge	*Provide assignments suitable for Ali*	*Work well. Concentrate*	*Help Ali with homework and remind him*

Fig. 13.3 Transcript from written material. Comments on long-term goals

The document offers possibilities for clarifying the future work in mathematics to students and parents, as well as to the teacher. In the writings in Fig. 13.3, the mathematical content is vaguely written, but it is clear that the discussion is focusing on the learning of a mathematical content of some kind. The joint planning is construed as clear to the participants of the meeting and the planning concerns the student's as well as teacher's (school's) course of action. The student has possibilities to take active agency, for example through the questions posed to the student in the overall document. This meeting was followed by an interaction between Anna (T) and Ali (S) at a later occasion, where Ali (S) acted to change a small part of the teaching: He raised his hand during classroom work, and when Anna (T) approached, he asked her for more difficult tasks (related to Anna's (T) contributions in Fig. 13.3). These kinds of documents are regarded as governing elements since they have a direct effect on what takes place during the student/teacher/parent meetings at this school. Often they included *rules* explicitly presented in the form of local and/or national goals.

The *textbook*, in this case, was influential as a governing element towards discourse 3. In the video material, there were examples of students working on textbook tasks, where subsequent teacher feedback focused on the student's mathematics learning and engagement. Often the teacher-student communication was short, but focused on mathematical processes instead of procedures.

Another governing element was *self-assessment materials* coming from textbook supplements in the written material of the data. The students were invited to describe their learning in terms of mathematics but, in these forms, students were also asked questions about what they thought of the teaching and if they wanted changes in the future. Hence, there were possibilities for students communicating feedback to the teacher about the teaching in mathematics. Whether this was possible at all, was dependent on the overall assessment practice in the classroom. In the case of Anna (T) and Ali (S) it is likely that such a document would actually govern towards discourse 3.

Elements Governing Towards "Reasoning Takes Time"

In the following situation, Erika (T) and Enzo (S) had an assessment discussion about Enzo's (S) learning in the teaching unit now ending. They have worked in the class on a theme in which mathematics has been a substantial part. The theme was about baking, and the goals for mathematics that were presented at the beginning of the theme are related to measurement and fractions. These goals are regarded as local goals since they are articulated at this particular school (however, they are related to national goals as well). At the beginning of the lesson, the students were given an *assessment matrix* indicating different levels of knowing in terms of the local goals. The students were asked to look at the matrix but not to mark it until the teacher arrived, since they were going to fill it out together. The first part of the matrix is shown in excerpt 4 (Table 13.1).

Table 13.1 Transcribed assessment matrix from written data

Areas in mission baking *Områden inom uppdrag bakning*	On the way to the goals *På väg mot målen*	Reaches the goals *Når målen*	Reaches the goals well *Når målen väl*
Volume *Volym*	Knows what litre and decilitre are *Vet vad liter och deciliter är*	Also knows how many dl go into a litre *Vet också hur många dl som får plats i en liter*	Knows how many cl go into a l [litre] and dl *Vet hur många cl som får plats i en l resp. dl*

Original Swedish transcript in italics
Assessment—Mission baking
Bedömning—Uppdrag bakning
Name:
Namn:

The basis for the assessment discussion between Erika (T) and Enzo (S) consisted of several materials: (1) Enzo's assessment matrix (Table 13.1), (2) a diagnostic test taken earlier, (3) a summary that Enzo (S) wrote as homework for the theme, and (4) Erika's (T) notes on Enzo's (S) demonstrated knowing. At the beginning of the sequence, Enzo (S) was prompted to take active agency when he read the first goal in the matrix and marks it. During the communication both Enzo (S) and Erika (T) were occasionally silent. They were producing the assessment together and both were agents in the discussion. Enzo (S) and Erika (S) recognised that Enzo (S) had demonstrated the knowing described in the goal. A similar communication was involved with the second cell. When they looked at the third cell about volume, Enzo (S) said that he knew the first part, that one hundred centilitres go into one litre. In the following communication, it became apparent that Enzo (S) was not sure about how many centilitres go into a decilitre. Erika (T) finished this part of the sequence by taking the marking pen and marking the first part of the cell and left the last part unmarked. Going through such a matrix takes time, and it is clear that, when the student participates, s/he is indicating her/his own demonstrated knowing in the matrix. The matrices may constitute governing elements where multimodal aspects such as the possibility to follow a structure through columns and rows, as well as possible knowing made explicit in words, figures and symbols hold affordances in line with "Reasoning takes time".

There were some instances in the data where the *textbook* was a governing element towards the discourse "Reasoning takes time". One example was when students were working in pairs for several lessons with the same textbook problem. The assessments acts in the communication with the teacher that were connected to the work on this problem often reflected this discourse. Some problems invited to mathematical reasoning and problem-solving with different ways of solving the problem. These processes were emphasised by the teacher in several assessment acts.

Affording and Restricting Elements for Mathematics Teachers' Assessment Practices

The documentations from four action research projects have been summarised in the following table (see Table 13.2). The four action research projects were performed within the system of school but with a specific aim to investigate and challenge boundaries within this system. A question that the teachers (and researchers) answered was to identify affordances and constraints from their institutional context when engaging in developmental work focusing on assessment practices in their mathematics classrooms. In the four research projects, we (teachers and researchers) identified the assessment discourses 3 and 4 as something to strive towards, but with a specific focus on students' possibilities to affect the teaching, and with an interest in critically scrutinising frames such as the national syllabus in mathematics.

Table 13.2 Summarised governing elements described by teachers in action research

Level of decision	Constraints for a development of the assessment practice	Affordances for an assessment practice with affordances for students' learning of and active agency in mathematics
National	The system of grading with an increasing amount of high stake assessments which takes time	National syllabus and curriculum making criteria for assessment clear
Municipality	Salary Status	Mathematics education study group Mathematics education developers Time for engaging in assessment project
School	Hard to find time for change Work load Limited response from colleagues. Reluctance towards change (colleagues) Dominating tradition of discourse 1 Tradition that only the textbook counts Competing demands on teachers, for example on extensive documentation Hard to meet all students' needs School culture where engagement is questioned by colleagues Non-supporting head who does not provide time to discuss assessment practices	Team of mathematics interested teachers experiencing guidance Meetings at school where mathematics education is discussed Interested, positive, supporting school head Interested and supportive colleagues Extra money for buying teaching materials Maths ed responsible at school Discussions among colleagues on research literature

Going back to Foucault's (1977) own description of a dispositive, it is possible to interpret different kinds of elements from Table 13.2. Additional to the four assessment discourses, I construed other kinds of discourses concerning teachers' ways of communication about a change of assessment practices which would provide affordances to students' learning and active agency in mathematics. One such teacher discourse is labelled "Reluctance towards a development of assessment practices". This discourse is here interpreted as encompassing the following statements in Table 13.2: "Reluctance towards change", "Limited response from colleagues" and "School culture where engagement is questioned by colleagues". Another discourse construed from the content of the table is "Engagement in development of assessment practices". This discourse encompasses "Interested and supportive colleagues" and "Discussions among colleagues on research literature".

I have interpreted also other elements in an assessment dispositive from the content in Table 13.2. These are not discourses but what Foucault (1977) labels regulatory decisions and administrative measures. Some constraining elements are interpreted as part of an assessment dispositive such as competing demands on teachers, non-supportive heads, and the system of grading. Other elements are interpreted as supporting elements in an assessment dispositive: organised study groups for teachers, external facilitators such as mathematics education developers, time for professional development, and supportive school heads.

A Summarised Picture of an Assessment Dispositive

In Table 13.3, I summarise the elements of the assessment dispositive that I have construed from both classroom data, and data from teacher discussions in action research projects. As Foucault (1977) describes a dispositive it is "a thoroughly heterogeneous ensemble".

The governing assessment dispositive in Table 13.3 is not meant to be a complete structure. Rather it is a tentative collection of discourses and other elements constituting a likewise tentative assessment dispositive in mathematics education. What is clear though, is that in the same dispositive there are both constraints and affordances for students learning and active agency in mathematics. In this way, the assessment dispositive also is a way of picturing a gatekeeping dispositive where students are given un-equal possibilities to learn and engage in mathematics.

Concluding Discussion

In this chapter, a potential governing assessment dispositive in mathematics education was presented. In the introduction I addressed the tension between a description of the governing of school through the lens of Foucault, and the normative

Table 13.3 A summary of elements constituting a tentative assessment dispositive

	A governing assessment dispositive			
	Governing towards an assessment practice where students are not invited to learn mathematics		Governing towards an assessment practice where students are invited to learn mathematics	
Assessment discourses	"Do it quick and do it right"	"Anything goes"	"Openness to mathematics"	"Reasoning takes time"
Regulatory decisions, administrative measures, and the like construed from classrooms (examples)	*Textbooks* where accomplishing tasks becomes most important *Diagnostic tests* focusing on correct/incorrect answers *Templates* for student/teacher/parent meetings with a limited focus *Decisions* made on other decision levels than the classroom	*Diagnostic tests* in *textbooks* not revealing students knowing *Manipulatives* piloting students in assessment situations *Decisions* made on other decision levels than the classroom *Rules* such as *time slots* affecting possible assessment practices	*Textbooks* when the focus is on the mathematical processes in tasks *Self-assessment materials* from textbook supplements where students also may comment on the teaching *Templates* for student/teacher/parent meetings with a focus on student agency and processes in the learning of mathematics	Some problems in *textbooks* which provide for feedback on problem-solving and the like The use of *matrices* where teacher and students could reason about the students learning in relation to *goals*
Teacher discourses	"Reluctance towards a development of assessment practices"		"Engagement in development of assessment practices"	
Regulatory decisions, administrative measures, and the like mentioned by teachers	Competing demands on teachers Non-supportive heads A new system of grading		Organised study-groups for teachers External facilitators Time for professional development Supportive heads	

pedagogical stance I also took, that it is of interest to study if and how students experience affordances to learn and engage in mathematics. I concur with Valero (2013, adopting Foucault) that the problems with the reproduction of social inequalities, produced within the system of school, are not best solved within mathematics education. However, I still think it is of interest to describe how this production of inequalities works, and I argue that assessment is a key aspect here. When discussing inequalities I have, albeit not following a strict Foucauldian view, chosen to accept the rules of "the game" of schooling in the meaning that students that succeed in school mathematics get a broad set of future opportunities. The proposed assessment dispositive, construed adopting concepts from Foucault, is a description of a significant part of the "how" of such an inequality production.

The dispositive consists of four assessment discourses together with other governing elements. The assessment discourses are construed from Swedish data, but have been presented internationally with reactions which show that they, at least to a large extent, resonate with other western countries (e.g. Björklund Boistrup, 2015). Still, they are meant to be conceived as tentative and temporary, open for alterations if adopted for analysis. This goes also for the overall dispositive, at least in its details which may differ between contexts. Having said that, I argue that similar dispositives are likely to be construable in many school systems. Of particular interest with regards to a production of inequalities is that the dispositive is not coherent when it comes to how students are invited, or not, to learn and engage in mathematics.

The notion of both the inclusion and exclusion being present at the same time in the dispositive is a way to capture a complex picture. As Raffsnøe et al. (2014) points out, the dispositive represents "dualisms replaced by the perception of a 'both-and' approach that permits a demonstration of how elements of binary oppositions appear in their interrelatedness as part of the same correlation" (p. 3f.).

The "order" claimed in steering documents (e.g. Skolverket, 2011), where all students should have the possibility to learn mathematics for future gain is hereby challenged. Even though the system has equality claims, statistics speaks clearly that this equality does not exist empirically (e.g. Skolverket, 2013). In this sense, the equality claimed by the system is rather a disorder. Moreover, since different students experience different assessment practices, the assessment dispositive is an essential part of a gatekeeping practice, where social structures in society are perpetuated.

As an opening to this dark picture, I draw on Foucault (2003) and point to the claim that where there is surveillance, such as through a governing assessment dispositive, there also is resistance. Jørgensen and Klee (2014) write something similar when they argue against Agamben's (2009) critical and dark reading of Foucault. I concur with Jørgensen and Klee that even though we need to be reminded how bad things may be, we "should also be careful not just to accept the conclusion with the inevitable effect that our analyses will turn into a desubjectifying gaze on people" (p. 20). An elaborated account of a tentative assessment dispositive in school mathematics, as in this chapter, offers researchers, teachers, students and decision makers means to grasp essential aspects of assessment practices in mathematics classrooms. I argue that there is positive power in an increased awareness of what assessment practices hold in terms of how students are experiencing affordances, or not, for learning and engaging in mathematics in school. The dispositive given account for in this text may serve as a tool in a work of resistance from within school where the current social order could be, at least locally, challenged.

Acknowledgments I would like to thank Nina Bohlmann and Kenneth Mølbjerg Jørgensen for helpful comments during the peer-review process. I also want to express my gratitude to Hauke

Straehler-Pohl for helpful suggestions during the revision of the chapter. The work on the chapter was funded by Linköping University (with the municipalities of Linköping and Norrköping), and Stockholm University.

References

Agamben, G. (2009). *What is an apparatus? And other essays*. Stanford: Stanford University Press.

Ball, S. (2000). Performativities and fabrications in the education economy: Towards the performative society? *The Australian Educational Researcher, 27*(2), 1–23.

Ball, S., Maguire, M., Braun, A., Perryman, J., & Hoskins, K. (2012). Assessment technologies in schools: "Deliverology" and the "play of dominations". *Research Papers in Education., 27*(5), 513–533.

Bernstein, B. (1973). *Class codes and control* (Vol. 2). London, UK: Routledge.

Björklund Boistrup, L. (2010). *Assessment discourses in mathematics education: A multimodal social semiotic study*. PhD thesis. Stockholm: Stockholm University.

Björklund Boistrup, L. (2015). Governing through implicit and explicit assessment acts: Multimodality in mathematics classrooms. In M. Hamilton, R. Heydon, K. Hibbert, & R. Stooke (Eds.), *Negotiating spaces for literacy learning: Multimodality and Governmentality* (pp. 131–148). London: Bloomsbury Publishing.

Björklund Boistrup, L. & Samuelsson, J. (In Preparation). *Power relations in participatory action research in mathematics education.*

Björklund Boistrup, L., Samuelsson, J., Bengtsson, K., Bertilsson, U., Grundström, M., & Järvstråt, M. (2013). *Betydelsen av tystnad. Aktionsforskning om bedömning i matematik*. Linköping: Linköpings kommun.

Broadfoot, P. M. (1996). *Education, assessment and society*. Buckingham, UK: Open University Press.

Broadfoot, P. M., & Pollard, A. (2000). The changing discourse of assessment policy. In A. Filer (Ed.), *Assessment: Social practice and social product* (pp. 11–26). London, UK: RoutledgeFalmer.

Bussolini, J. (2010). What is a dispositive? *Foucault Studies, 10*, 85–107.

Cizek, G. J. (2010). An introduction to formative assessment: History, characteristics, and challenges. In H. L. Andrade & G. J. Cizek (Eds.), *Handbook of formative assessment* (pp. 3–17). Abingdon, UK: Routledge.

Cooper, D., & Dunne, M. (2000). *Assessing children's mathematical knowledge. Social class, sex and problem solving*. Buckingham, Philadelphia: Open University Press.

Emanuclsson, I. (2002). I bchov av särskilt stöd—och ändå godkänd? [In need of special support—but still get passed?]. In H. Andersson (Ed.), *Att bedöma eller döma: Tio artiklar om bedömning och betygsättning* (To assess or judge. Ten articles on assessment and grading, pp. 27–38). Skolverket: Stockholm, Sweden.

Fejes, A. (2008). To be one's own confessor: Educational guidance and governmentality. *British Journal of Sociology of Education, 29*(6), 653–664.

FitzSimons, G. (2014). Commentary on vocational mathematics education: Where mathematics education confronts the realities. *Educational Studies in Mathematics, 86*(2), 291–305.

Forsberg, E., & Wallin, E. (Eds.). (2006). *Skolans kontrollregim—Ett kontraproduktivt system för styrning?* (School's regimen of control. A counter-productive system of directing?). Stockholm, Sweden: HLS förlag.

Foucault, M. (1977). Le jeu de Michel Foucault. In M. Foucault (Ed.), *Dits et écrits* (Vol. III, pp. 298–329). Paris: Gallimard.

Foucault, M. (1993). *Diskursen ordning* (The order of the discourse). Stockholm, Sweden: Brutus Östlings Bokförlag Symposium.

Foucault, M. (2003). *Övervakning och straff* (Discipline and Punish). Lund, Sweden: Arkiv förlag.

Foucault, M. (2008a). Diskursen ska inte uppfattas som… [The discourse should not be interpreted as…]. In T. Götselius & U. Olsson (Eds.), *Diskursernas kamp* (The battle between discourses, pp. 181–182). Stockholm, Sweden: Brutus Östlings Bokförlag Symposium.

Foucault, M. (2008b). Regementalitet [Governmentality]. In T. Götselius & U. Olsson (Eds.), *Diskursernas kamp* (The battle between discourses, pp. 183–204). Stockholm, Sweden: Brutus Östlings Bokförlag Symposium.

Gellert, U. (This volume). Revisiting mathematics for all: a commentary to pais's critique.

Gellert, U., & Jablonka, E. (2009). The demathematising effect of technology. Calling for critical competence. In P. Ernest, B. Greer, & B. Sriraman (Eds.), *Critical issues in mathematics education* (pp. 19–24). Charlotte, NC: Information Age Publishing.

Gipps, C. (2001). Sociocultural aspects of assessment. In G. Svingby & S. Svingby (Eds.), *Bedömning av kunskap och kompetens: Konferensrapport* (Assessment of knowledge and competence: Proceedings, pp. 15–68). Stockholm, Sweden: PRIM-gruppen.

Hansson, Å. (2010). Instructional responsibility in mathematics education: Modelling classroom teaching using Swedish data. *Educational Studies in Mathematics, 75*(2), 171–189.

Jørgensen, K. M., & Klee, N. (2014). Artisan storytelling and management "dispositifs". In K. M. Jørgensen & C. Largacha-Martinez (Eds.), *Critical narrative inquiry—storytelling, sustainability and power* (pp. 15–33). Hauppauge, NY: Nova.

Jurdak, M. (2014). Socio-economic and cultural mediators of mathematics achievement and between-school equity in mathematics education at the global level. *ZDM: The International Journal on Mathematics Education, 46*(7), 1025–1037.

Kress, G. (2009). Assessment in the perspective of a social semiotic theory of multimodal teaching and learning. In C. M. Wyatt-Smith & J. Cummings (Eds.), *Educational assessment in the 21st century* (pp. 19–41). London, UK: Springer.

McGrady, P. B., & Reynolds, J. R. (2013). Racial mismatch in the classroom: Beyond black-white differences. *Sociology of Education, 86*(1), 3–17.

Mercier, A., Sensevy, G., & Schubauer-Leoni, M.-L. (2000). How social interactions within a class depend on the teacher's assessment of the students' various mathematical capabilities. *ZDM: The International Journal on Mathematics Education, 32*(5), 126–130.

Morgan, C. (2000). Better assessment in mathematics education? A social perspective. In J. Boaler (Ed.), *Multiple perspectives on mathematics teaching and learning* (pp. 225–242). Westport, CT: Alex.

Pettersson, A., & Björklund Boistrup, L. (2010). National assessment in Swedish compulsory schools. In B. Sriraman, C. Bergsten, S. Goodchild, G. Palsdottir, B. Dahl Søndergaard, & L. Haapsalo (Eds.), *The sourcebook on Nordic research in mathematics education* (pp. 373–385). Charlotte, NC: Information Age Publishing.

Raffsnøe, S., Gudmand-Høyer, M., & Thaning, M. S. (2014). Foucault's dispositive: The perspicacity of dispositive analytics in organizational research. *Organization, 2014*, 1–27.

SFS. (2010). *Svensk författningssamling 2010:800. Skollag.* Utbildningsdepartementet.

Skolverket. (2003). *Lusten att lära—med fokus på matematik: Nationella kvalitetsgranskningar 2001-2002.* [The desire to learn—with a focus on mathematics: National quality audits 2001-2002]. Stockholm, Sweden: Skolverket.

Skolverket. (2011). *Läroplan för grundskolan, förskoleklassen och fritidshemmet.* Stockholm: Fritzes.

Skolverket. (2013). *PISA 2012—15-åringars kunskaper i matematik, läsförståelse och naturvetenskap.* Stockholm: Skolverket.

Skolverket. (2014). *Grundskolan i internationella kunskapsmätningar: Kunskap, skolmiljö och attityder till lärande Rapport 407.* Stockholm: Skolverket.

Skovsmose, O. (2005). *Travelling through education: Uncertainty, mathematics, responsibility.* Rotterdam. The Netherlands: Sense Publisher.

Straehler-Pohl, H., Gellert, U., Fernandez, S., & Figueiras, L. (2014). School mathematics registers in a context of low academic expectations. *Educational Studies in Mathematics, 85*(2), 175–199.

Torrance, H., & Pryor, J. (1998). *Investigating formative assessment. Teaching, learning and assessment in the classroom.* Buckingham, UK: Open University Press.

Valero, P. (2013). *Mathematics for all and the promise of a bright future.* Paper presented at the 8th Congress of European Research in Mathematics Education (CERME 8), Antalya, Turkey.

Van Leeuwen, T. (2005). *Introducing social semiotics.* London, UK: Routledge.

Walkerdine, V. (1988). *The Mastery of reason. Cognitive development and the production of rationality.* London: Routledge.

Watson, A. (2000). Mathematics teachers acting as informal assessors: Practices, problems and recommendations. *Educational Studies in Mathematics, 41*(1), 69–91.

Wiliam, D., & Thompson, M. (2007). Integrating assessment with instruction: What will it take to make it work? In C. Dwyer (Ed.), *In the future of assessment: Shaping teaching and learning* (pp. 53–82). Mahwah, N.J., USA: Lawrence Erlbaum Associates.

Chapter 14
Expressions of the Commodity Form: Alienation and Mathematics Education

David Swanson

Abstract This chapter refocuses Marxist debates on alienation and mathematics education around the unifying factor of the commodity. The various complex paths from the organisation of the economy to the mathematics classroom are traced in outline. This tracing touches on the historical birth of modern schooling in the UK, the roles that schooling plays in capitalist society, recent debates on the commodification of education itself, and dominant understandings of the nature of knowledge. If the form of economy does influence the classroom in these ways, it should be possible to find some expressions of the relationship within the world of school mathematics. An experimental methodological approach to finding and illustrating such expressions is adopted here, inspired by the work of the cultural critic Walter Benjamin. Extracted fragments of interviews with school students are presented, initially without commentary, in a montage format. As this approach to evidence is unusual, perhaps seeming closer to art than science, some attention is paid to explaining its usage. This explanation touches on the origins of the data and Benjamin's particular approach to fragments, montage and cultural expressions of the commodity form. Following the data, there is a brief analysis of the fragments, individually and collectively, in relation to the themes of the earlier sections, and methodological questions are returned to. Finally, implicit in the chapter's critique of mathematics education is a call for change, and a potential pathway to this is suggested in the conclusion.

Introduction

This chapter is a contribution to recent debates on alienation and education (see Jones, 2011; Lave & McDermott, 2002; Williams, 2011). Its aim is to refocus Marxist understandings of alienation in relation to education, and mathematics education in particular, around a unifying explanatory factor—the dominance of the

D. Swanson (✉)
Manchester Institute of Education, The University of Manchester, Manchester, UK
e-mail: david.swanson@manchester.ac.uk

H. Straehler-Pohl et al. (eds.), *The Disorder of Mathematics Education*,
DOI 10.1007/978-3-319-34006-7_14

commodity form of production within capitalism. It does this by first describing the wide range of influences that commodity production can have on education, which extend from capitalism's need for a compliant workforce through to the atomisation of knowledge. For reasons of space, and to avoid repetition of points made more eloquently by those cited above, these various paths from the commodity form to the classroom are traced primarily in outline.

Although the influence of the form of the economy on the classroom is not taken to be one of crude mechanical causality, if such influence exists, it should nonetheless be possible to find some expression of it within the world of school mathematics. An experimental methodological approach to finding and illustrating such expressions is adopted here, inspired by the work of the cultural critic Walter Benjamin. Interviews with school students on their experiences of, and attitudes toward, mathematics education were mined for fragments, which expressed aspects of the theoretical perspective outlined, and they are presented, initially without commentary, in the form of a montage. As this methodological approach to evidence is somewhat unusual, perhaps seeming closer to art than science, some attention is paid to explaining its usage before getting to the presentation of data. This explanation touches on the origins of the data and Benjamin's particular approach to fragments, montage and cultural expressions of the commodity form.

Following the presentation of the montage of fragments, a brief analysis relates the fragments, individually and collectively, to the theme of the commodity. The methodological question of how much this process can claim to provide in the way of evidence is also further discussed. Finally, implicit in the chapter's critique of mathematics education is a call for change, and a potential pathway to this is suggested in the conclusion.

Commodities, Alienation and Education

At the heart of Marx's theory of alienation lies the commodity. In the 1844 manuscripts Marx (1992) famously discusses some of the implications of satisfying human needs through a system of generalised commodity production. Instead of a process of engaging in activity /activities activity to satisfy needs, as in, say, humanity's pre-historic hunter-gatherer past, the current form of social organisation of production mediates the link between activity and needs. Workers produce, but the product they produce is not theirs to use. The majority sell their time and effort to the minority who have the means to organise the production of commodities, in essence turning themselves, or their ability to work, into a commodity. In exchange workers get money, with which they can buy other commodities. Marx argues that this alienation from the product of labour also means alienation from the process of labour (through its disconnection from immediate needs, and the lack of control of the process), from what it is to be human (i.e. conscious social producers), and therefore from other humans (whether within production or between either side of the production consumption divide). Some of these aspects of alienation could also be argued to apply in earlier forms of class society where the product of labour is taken from the producer, for example, when a peasant works part of the year on the estate of their feudal lord (see Sayers, 2011, for a wider discussion). However, as Marx (1982, p. 164ff.) argues in Capital, what differentiates capitalist society is that the social relations involved in production become hidden:

> The mysterious character of the commodity-form consists therefore simply in the fact that the commodity reflects the social characteristics of men's own labour as objective characteristics of the products of labour themselves.
>
> It is nothing but the definite social relation between men themselves which assumes here, for them, the fantastic form of a relation between things.
>
> [T]hey do not appear as direct social relations between persons in their work, but rather as material relations between persons and social relations between things.

In societies which are dominated by commodity production, and where human beings are themselves objectified in the commodity labour power, this *commodity fetishism* is argued by (some) Marxists to have a profound effect on all aspects of human culture, including education. For example, Lukacs (1971, p. 83) argues: "the problem of commodities must not be considered in isolation or even regarded as the central problem in economics, but as the central, structural problem of capitalist society in all its aspects." This radical perspective on the role of commodities is adopted here. This is not to deny the relevance or importance of other perspectives on education, but rather to aid the drawing out of the possible influences of commodity production on mathematics education, and to present an additional contribution to the critical debates which this volume exemplifies.

In order to discuss education from the perspective of commodity production, one additional relevant factor is necessary and one which itself arises from the alienation of genuine social relations and the competitive nature of capital, that is, the state. Marx & Engels (1974, p. 83) call the state the "illusory community", in that it gives the impression of standing neutrally over society representing the common good, whereas, in practice, it represents the interests of capital against other classes, and mediates conflicts between capitals when this becomes necessary. It is also seen to perform, or attempt to perform, tasks that are in the general interests of capital, the things that companies cannot, or do not, want to do individually, such as ensuring the general availability of the commodity labour power at sufficient levels.

Before capitalism, in what was to become the UK, institutionalised education was very much a minority pursuit. In 1072, for example, even the king of England still had to sign documents with a cross (Poplawski, Morrissey, Kitson, Frawley, & Brannigan, 2008, p. 15). The earliest schools were religious vocational schools attached to cathedrals and monasteries, at a time when the church was a central part of the ruling class (Williams, 1961, p. 148), but from the thirteenth century independent schools such as Winchester and Eton were formed, and these became the schools of the rising capitalist class (Williams, 1961, p. 151). The poor went largely uneducated. For example, even with an increase in schooling through the industrial revolution, only around half of children attended school in 1816, and this generally only on 1 day a week, focussed on moral education and for a brief period (the average duration of schooling was 1 year even by 1835 (Williams, 1961, p. 157). There was rising pressure from industry to avoid the burden of training the minimally literate and numerate supply of labour power required to remain competitive, but this met resistance. For example, one failed attempt to increase the spread of schooling in 1807 met this response in the UK parliament:

> However specious in theory the project might be of giving education to the labouring classes of the poor, it would, in effect, be found to be prejudicial to their morals and happiness; it would teach them to despise their lot in life, instead of making them good servants in agriculture and other laborious employments to which their rank in society had destined them; instead of teaching them the virtue of subordination, it would render them factious and refractory, as is evident in the manufacturing counties; it would enable them to read seditious pamphlets, vicious books and publications against Christianity; it would render them insolent to their superiors; and, in a few years, the result would be that the legislature would find it necessary to direct the strong arm of power towards them. (Giddy, 1807, p. 798. See also Graff, 1991, p. 22; Lincoln, 1859, in relation to parallel arguments in the U.S.)

This was not far from the truth as pressure for education also grew from below, from a radicalised working class developing its own educational practice, particularly following the rise of the Chartist movement. One small illustrative example is the Lord Street Working Men's Reading Room in Carlisle, where,

> [F]ifty men, anxious to read about the European revolutions of 1848, clubbed together to buy newspapers. A year later, with 300 members and 500 books, it had far outgrown its premises, a borrowed schoolroom. A new Elizabethan-style building was constructed in 1851, with congratulatory messages from Charles Dickens and Thomas Carlyle. Governed by a committee of workingmen, it charged a subscription of only 1 day a week, and even that was waived for the unemployed. (Rose, 2002, p. 65)

The needs of industry and pressure from below (Simon, 1974) fed into the development of the mass compulsory education seen today, emerging through various parliamentary acts from the 1870s onward. The needs of factory production, in this sense, played a strong role in shaping mass schooling from its beginning, and schools arguably reflect a similar influence today. The needs of companies and the profitable selling of commodities are still seen by the state as the primary motivation for education, as can generally be seen in speeches on the subject by politicians. To give a flavour, here is recent UK Conservative Prime Minister David Cameron (2011, online) explaining the main reason for aiming to lower truancy levels:

> We want to create an education system based on real excellence, with a complete intolerance of failure. Yes, we're ambitious. But today, we've got to be. We've got to be ambitious if we want to compete in the world. When China is going through an educational renaissance, when India is churning out science graduates…any complacency now would be fatal for our prosperity.

The Usefulness of Education for Capital

To say that education has particular uses for capitalism is not to claim that education therefore exists only because of those uses (even if they did consciously play a role in the formation of schooling). However, these uses can be argued to help explain the stable existence of schooling within capitalist society. These functions can be seen to extend well beyond the suitable development of the commodity labour power through the narrow learning of skills. The other main social functions of schooling listed by Reimer (1971, p. 23), during a previous wave of radical thought in education, remain valid. These are: custodial care, social-role selection and indoctrination. Each of these can be related directly to commodity production.

First is the importance of custodial care, that is, the effective babysitting of children so that their parents are in a position to go to work and produce commodities.[1] It is estimated that if schools in the UK were closed for 1 week, the economic cost due to absenteeism from work would amount to £1 billion (Sadique, Adams, & Edmunds, 2008), and industrial action by teachers is often seen to have a more generalised impact on the economy than is seen in other sectors due to its childcare implications (e.g. see Mason, 2011).

The second function, role selection, relates directly to the adequate distribution and differentiation of the commodity labour power. A societal division of labour requires a mechanism to distribute people across different levels of jobs. Schools can be viewed as an efficient machine for taking people in at whatever class position they are at in society—and throwing them out at roughly the same position—but with the illusion that the end result is based on individual merit, making people believe they themselves are the cause of their own failure (or success) rather than the system (see particularly, Bourdieu & Passeron, 1990). Exams and streaming are all essential to this task, with their consequent impact on individual self-belief and teacher expectations (see for example the classic experiments relating to rats, and children, by Rosenthal & Jacobson, 1968).

This "learning your place" can also be viewed as vital for capitalism in relation to the third function of schooling; indoctrination. In Marxist terms, the commodity labour power is unusual. In purchasing any other commodity the price is fixed and money and commodity are exchanged. If someone's ability to work is purchased, however, the amount and intensity of work remains open to ongoing renegotiation by either party to the exchange (since human beings are not really objects even when they are treated as such).[2] Learning your place in school can play a part in instilling a subordination, which can undermine the confidence to engage in individual and collective attempts at renegotiating when later in the workplace. The culture of sitting still, being quiet, learning to follow rather than create, and learning to compete with others rather than co-operate, arguably all help adjust children to their future role as a good worker/commodity. Although attempts at more explicit indoctrination, in the sense of consciously convincing individuals of the dominant ideology (Marx & Engels, 1974, p. 64), do take place within schools (e.g. Mansell, 2013), the shaping of expectations through lived experience is taken here as more fundamental.

[1] See Lundin and Christensen (in this volume) for an analysis of the mechanisms of defence that adults develop in order to neglect and productively mis-recognise this intrinsically economical function of schooling.

[2] See for example Hamper (1991) for some entertaining examples of this constant renegotiating.

Education as Commodity

Recent years have seen a relentless drive towards privatisation across all education sectors of the UK.[3] Even where education is not formally a business it is increasingly run as if it were. But in education, what is the product? What is the commodity? As Jones (2011) and Williams (2011) point out, students themselves are not engaged in productive labour which generates surplus value and the grades and certificates they achieve are not commodities produced by them. However, certificates can be seen instead as the commodities which are being sold *to* them. (In practice, students are sold the process of education and potential success in an exam, rather than the certificate directly, but this is similar to other commodities such as say film, where customers do not get their money back if they fail to pay attention when watching the film). The view that "certificates belong to the student, not the teacher or school, and are not sold or exchanged" (Jones, 2011, p. 369) also would not exclude them as commodities from this perspective. This could be said about many things that would not be denied the status of commodity, haircuts being one random example. Although, perhaps more than haircuts, certificates do potentially influence the future exchange rate of the commodity labour power for those holding them.

According to Jones (2011, p. 369), teachers in the state sector are not productive in the Marxist sense. However, it is certainly arguable that they are *indirectly* productive, in that state education increases the potential productivity of the system as a whole (Harman, 2009, p. 135) through reducing the socially necessary labour time required for production within the capitalist firms serviced by the particular state. Teachers within private education, on the other hand, are directly involved in the production of commodities (Rubin, 1973, p. 265). The impact of the neoliberal era, with the development of transitional semi-public, semi-private forms in schools, colleges and universities alongside the pseudo-marketisation of publicly funded education, has been to increase the transfer of the realities of commodity production into the state sector. It is in this sense that it is meaningful to talk about the commodification of qualifications, such as in Warmington (2007). The logic of this process has led to exams increasingly shaping how education is "delivered", an increase in managerialism and control in schools, fixed and atomised curricula, set lesson plans for teachers, and set ways to teach disciplined by inspection regimes.[4] In the words of the radical left slogan, "the school is a factory." Even if teachers want to produce a different product, for example, well-rounded, critical thinking, self-confident, social human beings with a depth of understanding of their subject, there are increasing mechanisms to ensure they focus on exam results instead (Paton, 2013).

---

[3] For recent developments in Higher Education see McGettigan (2013), and for the impact on schools, and beyond, see Ball (2007).

[4] See Jablonka and Bergsten (in this volume) for an analysis of how a similar trend can be identified in mathematics education research.

Knowledge as Object

From a Lukacsian-Marxist perspective, this commodification seen in education, the reification of a social process of learning into its outcome object, the qualification, is an example of a generalised feature of capitalism affecting wide areas of life (for example, Badiou, 2012, on aspects of the commodification of love via dating websites). This generalised process of reification is seen to extend even to how we think, and how we think we think. There are two key elements to this. First, as Labriola (1966, p. 155) puts it "ideas do not fall from heaven and nothing comes to us in a dream": Similar to the indoctrination function of schooling resting primarily on the lived experience of schooling, more generally too, ideas are fundamentally shaped by practice (Marx & Engels, 1974, p. 47). Generalisations made from experience are then intertwined with the dominant ideology (the generalisations from practice made from the perspective of those who dominate society, Marx & Engels, 1974, p. 65) to form what Gramsci (1971, p. 419) calls "common sense". Although the process is complex and non-mechanical, it is argued to be far-reaching in terms of its impact on ideas within society. For example, the dominance of the commodity form of production and the resultant objectification, alienation and atomisation are argued to underpin modern conceptions of the individual, and theories which take the individual as their starting point (Meszaros, 1970, p. 254). They are also seen to reinforce the rationalism and reductionism of the scientific revolution, which accompanied the rise of capitalism (Lukacs, 1971, p. 230). And finally, they are understood to reduce our understanding of knowledge, an active relationship with the world, to that of an object, something that can be taken out of one person's head and slotted into another's (for example, Freire, 2005, p. 71).

Mathematics has played a central role in many of these developments:

> It is anything but a mere chance that at the very beginning of modern philosophy the ideal of knowledge took the form of universal mathematics; it was an attempt to establish a rational system of relations which comprehends the totality of the formal possibilities, proportions and relations of a rationalised existence with the aid of which every phenomenon—independently of its real and material distinctiveness—could be subjected to an exact calculus. (Lukacs, 1971, p. 129)

Mathematics education is therefore particularly shaped by this objectification of knowledge and the accumulated effects of the processes of alienation described above, with many forces acting in a direction to encourage transmissionist pedagogy and limit the potential for more meaningful approaches to teaching. For example, the division of labour into mental and manual, and the further separation of knowledge into individual subjects, discourage the use of meaningful problems and the relating of mathematics to everyday knowledge or physical and social experience. The needs of capital for a stratified and differentiated labour force, the reification and commodification of knowledge in certificates and a view of knowledge as object shapes atomised curricula which divide mathematics into narrow process skills, and loses the systemic connections which are central to understanding. The individualisation of the commodity labour power and the competitive demands of exams discourage emphasis on the social dialogue which is, at least from a Marxist approach to practice, necessary in concept development. And the practicalities of custodial care, the time pressures of the "production line", the training of submission and the idea of knowledge as transferable object encourage passive drill and memorisation, rather than active and reflective thought.

These are just some examples of how the influence of commodity production could be seen to work its way through to the classroom and pedagogy. If the description of the multiple paths from the commodity form to the negative aspects of education is correct, it should be possible to find evidence within schooling of the various aspects of the analysis presented here. Data which arguably represent such expressions of the influence of the commodity form will be presented below, in the form of fragments of interviews with students about their schooling, teaching and mathematics. First, however, some ideas of Walter Benjamin related to fragments, montage and cultural expressions of the commodity form are explored, in order to help justify how those interview fragments are used and presented here.

Walter Benjamin, Fragments, Montage and the Commodity

In presenting data to illustrate the influence of commodity production on mathematics education, this chapter adopts an approach inspired by Walter Benjamin's (1999) Arcades Project, a sprawling (and unfinished) book on Paris in the nineteenth century. His approach in that work is at heart a montage of carefully selected fragmentary quotations. It is explored here to help explain the unusual form of data presentation, which follows. Where possible, Benjamin's own words are used due to the many possible interpretations of his intentions.

Benjamin's writing has a recurring relationship with fragments at many different levels. To give a flavour of this structural style, a sample of the contents page of the Arcades Project (Benjamin, 1999, p. 29) includes:

> (A) Arcades, Magasins de Nouveautes, Sales Clerks; (D) Boredom, Eternal Return; (E) Haussmannization, Barricade Fighting; (M) The Flaneur; (N) On the Theory of Knowledge, Theory of Progress; and (T) Modes of Lighting.

This categorisation is reminiscent of the random disconnectedness of Borges' (1999, p. 231) famous (and assumed to be fictional) Chinese encyclopedia, the "Celestial Empire of benevolent Knowledge", in which animals are divided into categories such as mermaids, stray dogs, those that tremble as if they were mad and those that at a distance resemble flies (see also the introductory chapter by Straehler-Pohl, Pais and Bohlmann, in this volume). Zoom a little further in to Benjamin's writing and the fragmentation continues. In his work One Way Street (1979), for example, seemingly disconnected sections that vary in length from a few lines to a few pages sit next to each other. Zoom in still further to the sentences on the page, and like the self-similarity of a fractal, the same pattern emerges:

> His sentences do not seem to be generated in the usual way; they do not entail. Each sentence is written as if it were the first, or the last… [Quoting Benjamin:] "A writer must stop and restart with every new sentence". (Sontag, 1979)

It is while writing his "Origins of German Tragic Drama", in 1924, that Benjamin (1994, p. 256) becomes conscious of his emerging method and reports in a letter to a friend,

> What surprises me most of all this time is that what I have written consists, as it were, almost entirely of quotations. It is the craziest mosaic technique you can imagine… .

In "the Origins" itself, a year later, he expands on this metaphor of the mosaic, and his methodology,

> Just as mosaics preserve their majesty despite their fragmentation into capricious particles, so philosophical contemplation is not lacking in momentum. Both are made up of the distinct and the disparate; and nothing could bear more powerful testimony to the transcendent force of the sacred image and the truth itself. The value of fragments of thought is all the greater the less direct their relationship to the underlying idea, and the brilliance of the representation depends as much on this value as the brilliance of the mosaic does on the quality of the glass paste. The relationship between the minute precision of the work and the proportions of the sculptural or intellectual whole demonstrates that truth-content is only to be grasped through immersion in the most minute details of the subject-matter. (Benjamin, 1998, p. 28)

So, for Benjamin (1999, p. 461), his fragments are not just random scattered thoughts, they are part of an "intellectual whole". Later, in the "Arcades Project", he expresses something similar:

> In what way is it possible to conjoin a heightened graphicness [Anschaulichkeit] to the realization of Marxist method? The first stage in this undertaking will be to carry over the principle of montage into history. That is, to assemble large-scale constructions out of the smallest and most precisely cut components. Indeed, to discover in the analysis of the small individual moments the crystal of the total event.

For the Arcades Project, this crystal of the total event is "the fetish character of commodities" (Benjamin, 1994, p. 482). In 1924, although still at the early stages of a development towards Marxism, Lukacs's sophisticated philosophical analysis with its emphasis on totality resonated greatly with Benjamin (1994, p. 248). By 1935, the key to unlocking that totality, shared by both Lukacs and Marx, also became central to Benjamin's (1999, p. 460) attempts to describe and explain wider culture. As he puts it:

> Marx lays bare the causal connection between economy and culture. For us, what matters is the thread of expression. It is not the economic origins of culture that will be presented, but the expression of the economy in its culture. At issue, in other words, is the attempt to grasp an economic process as perceptible Ur-phenomenon, from out of which proceed all manifestations of life in the arcades (and, accordingly, in the nineteenth century).

Such an approach to culture is often criticised as reductionist and mechanical, but Benjamin (1999, p. 392) argues this is based on a misunderstanding,

> At first it appears as if Marx only wanted to establish a causal relation between the superstructure and the base. But even the observation that ideologies of the superstructure mirror

> relations falsely and distortedly points beyond this. The question is this: if the base determines the superstructure, in what might be termed the material of thought and experience, and if this determination is not a simple mirroring, how—irrespective of the question of how it arises—should it then be characterized? As its expression. The superstructure is the expression of the base. The economic conditions, under which society exists, are expressed in the superstructure; just as an overfull stomach, although it causally conditions the sleeper's dream content, does not find therein its reflection but its expression.

In summary, Benjamin's Arcades Project (at least in one key aspect) can be seen as an attempt to unfold the impact of commodity production on the nineteenth century, recreating the complex totality shaped by the commodity, and drawing attention to that shaping, through a montage of carefully chosen fragments. Attempting to apply such a method to educational research, as is done here, raises many questions. One relates to an over-influence of the author in selecting and presenting the quotations, an influence which perhaps seems closer to art than science. While accepting that science has no monopoly on truth, and that artistic approaches may have much to offer research,[5] it should be noted that Benjamin himself did not appear to see the Arcades Project as art, but rather as philosophy of history (Benjamin, 1994, p. 333). Instead, what is argued for here is a move beyond a naive, narrow view of science which pretends to exclude human agency in its processes, and beyond a view that objective truth in social science is dependent on neutrality. In discussing cultural criticism, Benjamin (1979, p. 45) argues:

> Opinions are to the vast apparatus of social existence what oil is to machine: one does not go up to a turbine and pour machine oil over it; one applies a little to hidden spindles and joints that one has to know.

This interrelation between "knowing" (or theoretical understanding), "opinions" and authorial shaping is possible within educational research too, and is arguably common. The montage methodology deliberately makes the role of the author more explicit. If one takes the perspective that the test of any qualitative research lies in the explanatory value of the theoretical perspective in relation to the data (for example, Bhaskar, 1998, p. 194; Silverman, 2005, p. 136), then this explicitness can only assist in making that judgment.

A second, interrelated question lies in the necessity of an active role for the reader in relation to montage. In one sense this could be seen to counteract the first potential problem of the author imposing an interpretation, but it can also be argued that two layers of subjectivity take the work even further from objectivity. Awareness of such contradictions does not, of course, resolve them. However, the stance taken here is to embrace these contradictions and to view the method as posing exactly such questions for the reader, adding another layer to the active role it proposes.

Like Benjamin's (1994, p. 505) attempts to capture history through "the most inconspicuous corners of existence—the detritus of history", what follows aims to capture the influence of the commodity form on mathematics education through the detritus of the research process, a montage of fragments taken from interviews.

---

[5] See Appelbaum (in this volume) for a welcome plea for "educational space as studio" (p. XX).

The source of the data used here is the Teleprism project, a large scale ESRC funded mixed-method project (award ref: RES-061-25-0538 www.teleprism.com) investigating the relationship between pedagogy and the beliefs and attitudes of students in relation to mathematics.[6] The data initially arose from an attempt to challenge the potential rigidity of early categorisations of qualitative data in a mixed-method project. A process of looking for data which lay outside or went beyond the categorisation, unearthed fragments which appeared to speak of alienation and this initiated a process of interrelation between theory and the wider dataset. The interview fragments are presented in a spirit of solidarity with those involved; despite the humorous nature of some of the quotes, no criticism or mockery of the students, their teachers, (or their researchers), is intended.

The montage of fragments is presented without commentary and should be read with the themes of Sect. 14.2 in mind, that is, in relation to: alienation from product, process, self and others; the development of the individualised commodity labour power; education as commodity; school as factory; and knowledge as object. Following the quotations, some connections with alienation and the commodity are drawn out, and questions of methodology are returned to. In what follows, R=a researcher, S=a student, and student age and gender are indicated in brackets. The active role required of the reader in relation to the fragments, in reading them as excerpts of interviews, as fragments related to the theoretical themes, and as elements of a montage is a heavy burden. The unfolding analysis of the effect of alienation is likely to induce feelings not too distant from the content of the analysis, and is intended to do so. The effectiveness of the montage affect is to a certain extent contingent upon the reader's willingness to let herself be affected by a mildly intimidating unease.

The Commodity Form and Alienation: The Montage

School

1.	S (13F)	I've really enjoyed school like most people, like, they probably don't like school or they dislike it but I've never hated school because if we don't have it there's really nothing else for us to do

Self

2.	R	Tell us a little about yourself
	S (15f)	… I work well in a team
3.	R	Ok. So…do you think in general you're a good student?
	S (12f)	Yeah, generally I always hand my homework in on time, present myself smartly and always try my hardest with my work

[6] See Kollosche (in this volume) for an analysis of students' perceptions of and relations to mathematics that builds on the work of Foucault.

Career

4.	R	So what do you want to do with your life after school, have you thought about that?
	S (11m)	Well I have always wanted to be a pioneer, but there is nowhere to pioneer any more is there? … I don't know what else I can do apart from that, because I don't want to be here for the rest of my life sat a desk, just sorting out files like asking someone to fax this to the next country; I don't want to do that… That is not my type of thing—sitting at a desk with coffee doing that all day and then coming home… Yes, because if you want to be a pioneer … you set your mind to it and OK I have set my mind to it, I want to be a pioneer, you do your exams and then it is can I be a pioneer? "No." How do you be a pioneer? Do you have to do this sort of exam? Then you have to do that and that, and it is just like, I am only going to countries, just looking around
5.	S (14f)	I have been set on being a lawyer for about three quarters of a year now
6.	R	Does his [the student's father] work involve Maths?
	S (13m)	No I don't think so
	R	What does he do?
	S	Suitcases
7.	R	Do you want to go to university?
	S (12m)	Yeah
	R	And what would you do there?
	S	I don't even know what you do at university
8.	S (11f)	I'm definitely going to college but I don't know about university because it's really expensive

Exams

9.	R	Why don't you think you're going to get As?
	S (13m)	It's just like mission impossible for me
	R	Why?
	S	"Cause I'm not bright, I'm not intelligent"
	R	Why do you say these things?
	S	"Cause I'm not"
	R	When did you start thinking of yourself like that?
	S	Like if you get something wrong you think I knew that but I weren't thinking I weren't concentrating
10.	R	Hmm how do you feel about GCSEs now?
	S (13f)	I'm scared I'm so scared
	R	Yes but you are quite good really, you are progressing very well aren't you?
	S	In that way I don't believe in myself, GCSEs, I am a lot more confident in maths but GCSEs I just don't believe in myself
	R	You have another 2 years you know before you, you know
	S	I know that's the scary thing

11.	S (13f)	I don't think I'm going to get a job to be honest
	S (13m)	Why?
	S (f)	Because I'm getting low grades in everything
	S (m)	You don't think you're going to get a job? You're still going to get a job just work in McDonalds
12.	R	And do you think Mathematics will be useful in your future lives?
	S (15m)	Yeah, it's big, GCSE, isn't it

Relationships

13.	S (13f)	Some teachers, the way they, like, speak to kids, the way they approach them and the way they speak, and, you know, shout the words. I've never believed in any of that, like you shout at me I'll shout back you, speak to me calmly I'll give you a calm response, do you know what I mean?
14.	S (11f)	I've been told off a couple of times, but that was only because I've not understood it and I speak to people if I've not understood it, but now I know not to speak to people and ask the teacher
15.	S (11f)	I think I have had my ideal lesson, it wasn't intentional, it was when our teacher, who has really bad asthma, so one day she came in and said she wanted everyone to be quiet. She explained it, because it was something she had done from the lesson and she gave us sheets on what to do; and everyone just quietly finished their work; it doesn't have to be anything fancy I can work with noise, but everyone was just quiet, so that was nice

Primary to secondary

16.	R	...Primary school, did you have any good memories or bad memories?
	S (15f)	Yeah, I think I have good memories, but I can't remember
17.	R	So what about school, what can you remember from primary school, can you remember much?
	S (11m)	When we were in year 4 from reception you would get four breaks, but when you go into juniors—above year 4 you get three breaks, morning, dinner and end
18.	R	How did it change with the transition from that school then to this school, the secondary school?
	S (11f)	It changed because the lessons are shorter, you have to hurry up to do more work and produce what you can
	R	Can you give me an example?
	S	Well in English we used to write the date, the title and the learning objective in a maximum of 2 min, whereas I used to write the date, then underline it, then write the title, then underline it. So now I just write it and do it, and she moves on to the next slide, so it is just getting into that way of what to do

Other subjects

19.	S (11f)	My worst subject...I like art but I don't like the lesson. I mean I like doing my own art, I don't like people telling me what to do in art

20.	S (13m)	I decided to pick drama, science and French…I took drama for a lesson for me to get you know, you know a free lesson, "cause I like Drama…"
	R	So it's easy for you?
	S	Yeah, cause with too much on… cause I thought a lot of work to do on paper, I needed a break so…
21.	S (13f)	In Citizenship they were saying, erm, some things you want to change and I said homework and I got the highest score out of the whole class saying you should ban homework and I did really good reasons and I got a 6A

Mathematics

22.	R	Okay can you describe to me the lesson a little bit?
	S (13m)	She puts like an objective on the board and then like she reads it out and then she goes through examples what we've got to do and then she explains what we've got to do and then she asks us to try and work out the answer and at the end she like shouts one of us up and then we've got to write the answer up on the board and then if we get it wrong she like explains how we got it wrong and stuff
23.	S (14f)	My old teacher used to explain why something works, but my new teacher doesn't do that, I quite like to see how things work. Because now when we are told something we just have to accept that it works
24.	S (11f)	…On "Fun Friday's" which is today, erm, Miss gets us like sheets where we've got to work out what we've been doing in class like practice on times tables or something else
25.	R	So can you think of examples where you use this algebra in real life?
	S (12m)	…Like…I don't know actually, I think like…you can do it for sharing but I don't know how, if you add three pizzas…no actually I can't think of anything
26.	R	In an ideal world as we say, how would you like the lesson to be?
	S (15m)	A lot more interactive
	R	Can you tell me an example?
	S	Like…in the classrooms they've got that board, like…that…I don't even know what it's called, you know what I mean?
	R	The interactive whiteboard, yes
	S	Teachers just sit there and they write on that and then you've got to copy, but it'd be better if like other people could come up and attempt stuff on it if you know what I mean
27.	S (11m)	The other thing I find hard is when you have to get like say minus seven add minus four I don't get it it's like you have to swap them around and like add it up and take it away and it's just a bit…
	R	So do you know why you're doing swapping them around and adding them up and taking them away?
	S	I kind of know it's like the method but it's just…
	R	So you don't mean all the…
	S	No
	R	Like the real?
	S	Yeah

28.	S (14f)	Yeah there's like different methods and things so that could be quite difficult to get used to
	R	So what is the difference between this teacher and before?
	S	Um, well for factorising, use that as an example, I know the teacher I had in year 7 used a different method than the one I use now
	R	Can you tell me what exactly was different?
	S	Well when you have like the two brackets next to each other
	R	Yeah
	S	You like supposed to factorise it out and my teacher now uses eyebrows and smiley face and it's like it looks like that but the one in year 7 used a crab claw and it looked like a crab claw
29.	R	Which is your favourite or best topic in mathematics?
	S (12m)	Probably times
	R	And the most difficult one, the least?
	S	Divide
30.	R	Are there any topics you like more than others?
	S (12m)	… I don't mind doing brackets

Some Thoughts on the Fragments, the Montage and the Methodology

The Fragments

Individually, many of the quotations can be interpreted as expressing some of the educational system's functional roles for capital: custodial care—"There's really nothing else for us to do"; role selection—"just work in McDonalds"; learning your place and blaming yourself (indoctrination)—"I'm not intelligent"; and the lived experience of subordination—"you have to hurry up to do more work and produce what you can." What makes the function of indoctrination so effective?—"I'm scared." These quotes evoke the sense of school as factory, which has arguably increased as a result of the quasi-commodification of state education. There is even an example of awareness (at age 11) of the actual commodification of education—"university [is] really expensive."

The theme of reification also runs through the quotations including experiences of failure being taken as objective characteristics of the self, or, the importance of mathematics being reduced to the importance of passing the exam. As does the connected theme of atomisation, from the formation of the competitive individual ("I work well in a team" is a phrase from the genre of job applications) to atomisation in learning—"now I know not to speak to people." This reification, and atomisation is also seen when it comes to mathematical knowledge, with mathematics separated from the real world—"I can't think of anything", from reasoning—"we just have to accept that it works" and from other mathematics (in a world where "times" and "divide" are unconnected, and "brackets" becomes a topic in itself).

Although this seems unrelentingly negative, there are also signs within the quotes of the genuine processes and relationships which are being alienated, such as when a teacher's asthma suspends normal classroom relations. There are even signs of resentment and resistance, for example, "Speak to me calmly I'll give you a calm response", or "I don't like people telling me what to do." Or even the student who says "I don't mind doing brackets." The grudging "I don't mind" rather than "I like" showing a healthy psychological distance from the absurdity of school mathematics. Some generalise their unease to their future life—"I don't want to be here for the rest of my life sat a desk, just sorting out files like asking someone to fax this to the next country." Some show the signs of an active resistance, similar to the renegotiating of the expenditure of labour power seen in the workplace, including one who has formulated conscious demands (an end to homework). These small instances offering hope are returned to and built upon in the conclusion.

These are a small selection of the possible connections with the themes of alienation and the commodity in the data. Of course, additional, very different, or even opposite interpretations of the presented fragments are possible. Other stories could be told from the interview data and other quotations chosen to tell it. Any evidential claims for the data lie first in the fact that *this* story *could* be told, but also in the explanatory power of the theory in making sense of some of what is said.

The Montage

The process of montage involved a deliberate de-contextualisation of particular fragments, and their re-contextualisation under the theme of the influence of commodity production. The aim being to capture the relationships between those influences through their juxtaposition, and, through this process, to recreate the complex totality of influences. One key element of that is in juxtaposing aspects of alienation in relation to the self, or the individual commodity labour power, with those in relation to the object of school mathematics itself. For both of these, Marx's description of alienation from product, process, self and others is seen. In the production of the self, there is very little say for the individual in the nature of that self (the path to being a commodity is laid down), or in the details of what and how that self becomes (witness the frustrations of "the pioneer" or one student's certainty of attending university without knowing what will happen there). Similarly, many of the quotes relating to mathematics suggest the subject is something done to students rather than by them and for them (e.g. "we just have to accept that it works").

Of course, given the active role of the reader, it cannot be stated what any individual's experience of the montage will have been (some may even have given up in despair). The existence of subsequent analysis and preceding theory represents one strategy to overcome that potential problem, through providing some support for the reader. At the same time, the analysis and theory can also be seen as reinforcing the preferred interpretations of the fragments by the author. This tension in the method between excess influence of both author *and* reader can be seen as its strength however. The tension it aims to provoke is itself an invitation to challenge the passivity, lack of control, atomisation and reification described in its content. This marks the method as not just experimental, but as subversive, aiming to challenge the effects of alienation within educational research.

Conclusions and Implications

The aim of this chapter is to refocus Marxist perspectives on alienation in education on a unifying explanatory factor, the commodity. Various paths are traced from commodities to the classroom: the needs of capital for the availability of adequately developed and sufficiently differentiated labour power; the need for that labour power to accept its differentiation as fair, to view itself as competitive, individual and subordinate; the need for custodial care of the children of workers; the increasing commodification of education itself; and the impact of cultural understandings of the individual, knowledge and learning which emerge in a society dominated by commodity production. An unusual form of evidence for these influences of commodity production on schooling is then presented through a montage of their claimed expression in students' discussions of the experience of mathematics education.

Taken together the theory and data are admittedly an attempt to convince the reader that the world is indeed shaped this way, or at least, that this may be an important part of the story of how the world is. However, just as with any individual fragment of the montage, other important parts of the story (or even different stories) about education, other than the negative influence of commodity production, may be told. This volume contains many.

This chapter itself also contains a different story, and one which may be helpful if the need for change implicit in the critique that is the central argument is to be realised. Although it is argued here that processes such as education can be commodified, and, that generalised commodity production can and does shape every aspect of human life, including within the classroom, there is a general contradiction fundamental to all commodities: that between object and the processes and human relationships they mask. The contradictions within the commodity labour power (the potential to continually renegotiate the exchange, treating a thinking, feeling human being as an object, the collective and social reality of individuals) are perhaps the most important, as these can generate resistance. In the quotes above, alongside the expression of alienation there are also expressions of the real processes and relationships, and of such contradictions, including at times resentment and resistance. Their existence allows space within education for the idea and practice of "teaching as a subversive activity" (Postman & Weingartner, 1971). Introducing elements into education which oppose the dominant practice can lead to students and teachers *feeling* less alienated (something worthwhile in itself), and at the same time this can point towards more general, alternative forms of education. To sustain that subversive activity requires organised networks of teachers and researchers, wherever they can be formed, whether focussed narrowly on pedagogical questions or extending towards wider social questions and linked to social movements.

It is in alliance with wider social movements that education has the potential to be reshaped. The evidence for this is in the spread of radical educational ideas and practice that followed the waves of social struggle in the 1960s and 1970s (including some of the literature referenced in this chapter). It is also seen in the attempted educational reforms that followed the Russian revolution in 1917, the most serious attempt to end the domination of the commodity form seen so far. The brief period before the revolution unwound saw experiments in reuniting education with meaningful social and practical activity, homework was prohibited, and for 10 years, exams were banned across education (Karp, 2012). Alternatives to alienated education are possible.

Acknowledgments This work was supported by the Economic and Social Research Council [grant number ES/J500094/1 and RES-061-25-0538]. The author would also like to acknowledge the contributions of Maria Pampaka, Afroditi Kalambouka and Sophina Qasim to the TELEPRISM project.

References

Badiou, A. (2012). *In praise of love*. London: Serpent's Tail.

Ball, S. J. (2007). *Education plc: Understanding private sector participation in public sector education*. London: Routledge.

Benjamin, W. (1979). *One way street and other writings*. London: NLB.

Benjamin, W. (1994). *The correspondence of Walter Benjamin 1910–1940* G. Scholem & T. Adorno (Eds.). Chicago: University of Chicago.

Benjamin, W. (1998). *The origins of German tragic drama*. London: Verso.

Benjamin, W. (1999). *The Arcades project*. London: Harvard University Press.

Bhaskar, R. (1998). *The possibility of naturalism: A philosophical critique of the contemporary human sciences*. London: Routledge.

Borges, J. L. (1999). *Selected non-fictions*. NewYork: Penguin.

Bourdieu, P., & Passeron, J.-C. (1990). *Reproduction in education, society and culture*. London: Sage.

Cameron, D. (2011). Speech on free school education. http://www.westminster-briefing.com/news-detail/newsarticle/cameron-free-school-education-speech-in-full/.

Freire, P. (2005). *Pedagogy of the oppressed*. London: Continuum.

Giddy, D. (1807). *The parliamentary debates: Parochial schools bill, 2nd reading, 13 June* (Vol. IX, p. 798). London: Hansard.

Graff, H. J. (1991). *The literacy myth: Cultural integration and social structure in the nineteenth century*. Brunswick: NJ Transaction Publishers.

Gramsci, A. (1971). *Selections from the prison notebooks* (Q. Hoare & G. N. Smith, Eds., Trans.). New York: International Publishers.

Hamper, B. (1991). *Rivethead*. New York: Warner.

Harman, C. (2009). *Zombie capitalism: Global crisis and the relevance of Marx*. London: Bookmarks.

Jones, P. E. (2011). The living and the dead in education: Commentary on Julian Williams. *Mind, Culture, and Activity, 18*(4), 365–373.

Karp, A. (2012). Soviet mathematics education between 1918 and 1931: A time of radical reforms. *ZDM, 44*(4), 551–561.

Labriola, A. (1966). *Essays on the materialistic conception of history*. New York: Monthly Review Press.

Lave, J., & McDermott, R. (2002). Estranged labor learning. *Outlines. Critical Practice Studies, 4*(1), 19–48.

Lincoln, A. (1859). Address before the Wisconsin State Agricultural Society. *Milwaukee, Wisconsin, 30*, 1859.

Lukacs, G. (1971). History and class consciousness. London: Merlin.

Mansell, W. (2013, June 21). Michael Gove redrafts new history curriculum after outcry. *The Guardian*. Retrieved from http://www.theguardian.com/education/2013/jun/21/michael-gove-history-curriculum.

Marx, K. (1982). *Capital*. London: Penguin.

Marx, K. (1992). Economic and philosophical manuscripts (1844). In *Early Writings* (pp. 279–400). London: Penguin.

Marx, K., & Engels, F. (1974). *The German ideology*. London: Lawrence & Wishart.

Mason, R. (2011, November 18). Striking public sector will cost the UK economy millions. *The Telegraph*. Retrieved from http://www.telegraph.co.uk/news/politics/8900262/Striking-public-sector-will-cost-the-UK-economy-millions.html.

McGettigan, A. (2013). *The great university gamble*. London: Pluto.

Meszaros, I. (1970). *Marx's theory of alienation*. London: Merlin.

Paton, G. (2013). Teacher pay "to be linked to pupils" results and behaviour. Daily Telegraph. Retrieved April 16, 2013, from http://www.telegraph.co.uk/education/educationnews/9998752/Teacher-pay-to-be-linked-to-pupils-results-and-behaviour.html.

Poplawski, P., Morrissey, L., Kitson, P. J., Frawley, M. H., & Brannigan, J. (2008). *English literature in context*. Cambridge: Cambridge University Press.

Postman, N., & Weingartner, C. (1971). *Teaching as a subversive activity*. Harmondsworth: Penguin.

Reimer, E. (1971). *School is dead*. Harmondsworth: Penguin.

Rose, J. (2002). *The intellectual life of the British working classes*. London, UK: Yale University Press.

Rosenthal, R., & Jacobson, L. F. (1968). Teacher expectations for the disadvantaged. *Scientific American, 218*(4), 19–23.

Rubin, I. I. (1973). *Essays on Marx's theory of value*. New York: Black Rose Books.

Sadique, M. Z., Adams, E. J., & Edmunds, W. J. (2008). Estimating the costs of school closure for mitigating an influenza pandemic. *BMC Public Health, 8*(1), 135.

Sayers, S. (2011). Alienation as a critical concept. *International Critical Thought, 1*(3), 287–304.

Silverman, D. (2005). *Doing qualitative research: A practical handbook* (2nd ed.). London: Sage.

Simon, B. (1974). *The two nations and the educational structure, 1780-1870* (Vol. 1). London: Lawrence & Wishart.

Sontag, S. (1979). Introduction. In W. Benjamin (Ed.), *One way street* (pp. 7–28). London: NLB.

Warmington, P. (2007). Popular press, visible value: How debates on exams and student debt have unmasked the commodity relations of the "learning age". In A. Green, G. Rikowski, & H. Raduntz (Eds.), *Renewing dialogues in Marxism and education: Openings* (pp. 215–228). New York: Palgrave Macmillan.

Williams, R. (1961). *The long revolution*. Harmondsworth: Penguin.

Williams, J. (2011). Toward a political economic theory of education: Use and exchange values of enhanced labor power. *Mind, Culture, and Activity, 18*(3), 276–292.

Chapter 15
The Effects of School Geometry in the Shaping of a Desired Child

Melissa Andrade-Molina and Paola Valero

Abstract In this chapter we explore how school geometry becomes a technology for the government of the self, and how the pedagogical devices of school geometry conduct students' ways of thinking and acting. We contend that students, in their working with pedagogical devices, engage in a training process in which they learn to regulate their own conduct so that they perceive space through the trained eyes of reason provided by Euclidean, school geometry. Our contribution is an analysis of the power effects of school geometry in terms of the fabrication of children's subjectivities towards the shaping of the desired child of society.

Introduction

The Organisation for Economic Co-operation and Development, OECD, states that mathematics is a tool to solve everyday life problems (OECD, 2014). The ability of mathematising, measured by OECD's Programme for International Student Assessment (PISA), is considered to be central to solve those everyday life problems because it involves connecting the "everyday-world" and the mathematical world. Schools seem to address the need of developing this ability by introducing everyday life, contextualised problems into the mathematics classroom. Particularly the learning of geometry, geometric modelling and spatial thinking are said to be important competencies for solving those types of problems (Mevarech & Kramarski, 2014; National Council of Teachers of Mathematics [NCTM] 2000).

It is said that school geometry offers ways of interpreting, describing, and reflecting on the world that is reachable through our senses (Clements & Battista, 1992; NCTM, 2000; National Research Council [NRC] 2006). This claim grants spatial

M. Andrade-Molina (✉)
Faculty of Engineering and Science, Aalborg University, Aalborg, Denmark
e-mail: melissa@plan.aau.dk

P. Valero
Faculty of Engineering and Science, Aalborg University, Aalborg, Denmark

Faculty of Science, Stockholm University, Stockholm, Sweden

H. Straehler-Pohl et al. (eds.), *The Disorder of Mathematics Education*,
DOI 10.1007/978-3-319-34006-7_15

thinking and visual-spatial skills the status of a key element to improve students' performances in school settings, for example, while dealing with digital technologies (OECD, 2012). According to the American National Research Council, spatial thinking is a "pervasive and powerful way of thinking that operates across the sciences, social sciences, and even the humanities. [It] is the start of successful thinking and problem solving" (NRC, 2006, p. 131). Therefore, several studies on school geometry have attempted to provide ways for students to enhance and develop their spatial thinking, varying from the introduction of activities involving the manipulation of building blocks to the overall improvement of the school geometry curriculum (Clements, 2008; Hauptman, 2010; Prieto & Velasco, 2010).

It is also recognised that visual-spatial skills improve student's performances while learning geometry or mathematics. These skills are fundamental while reading, interpreting and recognising information contained in diverse images, whether in two or three dimensions (OECD, 2012). In the field of science education, it is believed that visual-spatial skills and spatial thinking go beyond mathematics and have been implicated in several scientific advancements and theories such as the discovery of the DNA structure (Newcombe, 2010), Galileo's laws of motion, Faraday's electromagnetic field theory and Einstein's theory of relativity (Kozhevnikov, Motes, & Hegarty, 2007). In these studies, spatial thinking is referred to as a way of thinking that is as important as verbal and mathematical thinking (Newcombe, 2010; Wai, Lubinski, & Benbow, 2009; Verdine, Golinkoff, Hirsh-Pasek, & Newcombe, 2014).

In the statements sketched above, spatial thinking and visual-spatial skills are positioned as key elements in the shaping of the successful minds of the future, and, therefore, they should be developed or enhanced in schools. It seems that whoever possesses these abilities will have a successful performance in many areas of life in general, and particularly in areas of interest for education, such as international comparative assessments like PISA, the use of digital technologies for learning, and the reading of images that appear in diverse school subjects like physics or biology.

These statements articulate discourses about not only what students should learn in school geometry but also about whom they should become with and through school geometry. Here we bring further our previous work (Andrade-Molina & Valero, 2015), as well as add new insights to recent investigations that use Foucault's tools to think power and governmentality in mathematics education (Diaz, 2014; Kollosche, 2014; Valero & Knijnik, 2015). In this chapter, we focus on the question of how school geometry becomes a technology for the government of the self, and how the pedagogical devices of school geometry conduct students' ways of thinking and acting. We contend that students are not forced to be or to see in a particular way. Rather, in their working with pedagogical devices, they engage in a training process in which they learn to regulate their own conduct so that they perceive space through the trained eyes of reason provided by Euclidean, school geometry. Our contribution is an analysis of the power effects of school geometry in terms of the fabrication of children's subjectivities. More precisely, we shed light on how school geometry is understood as a technology of the self that conducts the conduct of children towards the shaping of the desired child of society (Popkewitz, 2008).

Governing Through Mathematics Education

There are several understandings of the notions of government and governing. Cotoi (2011) describes two meanings of these notions in the work of Foucault. The first refers to an area of human existence and expertise produced by ways of thinking and acting aimed at transforming human behaviour. The second refers to what Rose (1996) describes as the attempt, by political elites, to ensure the wellbeing through the ordering of the affairs of a territory and its population. The former meaning is used in a comprehensive sense to trace the link between forms of power and processes of subjectification (Lemke, 2002). Following Foucault (1993, p. 204),

> Governing people is not a way to force people to do what the governor wants; it is always a versatile equilibrium, with complementarity and conflicts between techniques which assure coercion and processes through which the self is constructed or modified by himself.

Governing understood as the "conduct of conduct" (Foucault, 2008) works through technologies that are systematised, regulated and reflected modes of power that include forms of self-regulation (Foucault, 1997). These technologies, inscribed in a particular form of rationality, enable subjects to change and to develop their thoughts and conduct their ways of being. In this regard, the practices of the self are not invented by the subject. They do not emerge from thin air, but "they are models that [the subject] finds in his culture and are proposed, suggested, imposed upon him by his culture, his society, and his social group" (Foucault, 1984, p. 291). However, the subject also embraces them and acts with them productively. It is in this sense that, from a Foucaultian perspective, subjectivity is formed in the constant tension between subjection and subjectification.

Education has been pointed out as a very important space of Modern government where subjectivity is fabricated with the use of technologies of pedagogy, the curriculum and educational sciences. Within that, the school mathematics curriculum is a way of conducting subjects' conducts since mathematics education practices insert in children norms of reason in both productive and constraining ways (Valero & García, 2014). Mathematics education is not only a process of knowledge objectification but also a process of subjectification. For example, Kollosche (2014, p. 1070) argues that school mathematics is a technology for the government of others through logic and calculation practices, in which the desired student is "able and willing to think and speak logically and act bureaucratically". Diaz (2014) investigates the way in which the emphasis on the adequate teaching/learning of the equal sign in current reforms in mathematics in the USA link with broader meanings in society about equality. The pedagogies of mathematics in the curriculum operate classifications and differentiations of those children who learned the right equality and those who fail to do so. The apparent neutrality and goodness of learning "the equal sign" renders children objects of the calculations of power.

Our interest here is to approach school geometry as a technology of the self that governs students' perception of the spatial and visual according to certain norms, thus allowing them to become the desired child of the curriculum. Since diverse pedagogical devices articulate the technologies of the self, we deploy an analysis of a series of official curricular documents from the Chilean Ministry of Education (MINEDUC). These materials are school textbooks designed to accompany students' learning processes (Del Valle, Muñoz, Santis, 2014; Muñoz, Jiménez, & Rupin, 2013), curricular programmes and guidelines for teachers (Ministry of Education of Chile [MINEDUC] 2004, 2011; Bórquez & Setz, 2012; Ortiz, Reyes, Valenzuela, & Chandía, 2012; Zañartu, Darrigrandi, & Ramos, 2012) and a map of learning progress in geometry (MINEDUC, 2010). Heightened attention is given to this map because it is the official document that displays the desired performances of students while solving school geometry problems, and because it is expected that teachers use this map to trace individual progress in geometry. The examples of students' answers displayed in this map of progress are considered, by MINEDUC, the successful or ideal performance students should achieve in the learning of school geometry. Because this map of progress expresses the successful practices of desired students, it is possible to find in it evidence of the expected power effects of school geometry in terms of the training of the self of children.

Statements about the training of the self were detected in the map of learning progress. Then they were connected to statements about expected behaviour of children in geometry expressed in student's textbooks and teacher's guidelines. To detect expressions of the training of the self implies to find discursive recurrences in the documents regarding the desired expectation of students' performances while learning geometry. For example, students should be able to recognise, to demonstrate, to measure and to locate by themselves (MINEDUC, 2010). These recurrences in the school geometry discourse are the ground for an interpretation of school geometry as a technology of the self and its effects on students' subjectivities.

School Geometry and the Trained Child

The Chilean curriculum states that schools should provide students with the basic knowledge of mathematics to facilitate an understanding of the "real world" (MINEDUC, 2010). The Chilean curriculum for school geometry aims at teaching students to deal with everyday life problems to engage their knowledge and abilities developed in the classroom (MINEDUC, 2011).

> Students should develop logical thinking, deduction skills, accuracy, problem posing and solving, and modelling ability [as/since] mathematics enriches the understanding of reality, facilitates the selection of strategies to solve problems, and contributes to an autonomous and own thinking. (MINEDUC, 2010, p. 3, own translation)

It is believed, as aforementioned, that by developing spatial skills, in two and three dimensions, students would be able to link "everyday life experiences" and school mathematics. For instance, according to MINEDUC (2010), a key component in the development of spatial thinking is measuring, precisely because it enables students to link school geometry to the environment and to other school subjects. In this sense, school geometry and spatial thinking are taken as a tool to link the surroundings and school mathematics, but, also, as a tool to make sense of other school subjects such as physics and geography (Battista, 2007). The teaching of school geometry involves practices in which students should engage themselves in a training process in order to achieve what is desired. The desired child should be able to perform successfully in everyday life problems by using the tools, the abilities, and the skills acquired in the classroom.

In 2010, MINEDUC released a map of learning progress with seven levels that students have to complete along school geometry during the 12 years of compulsory education. This map expresses the successful performances that the desired student should deliver at every level of this map. From recognising the fundamental elements of geometrical figures—side, vertex, perimeter, area and so on—to solving problems by applying geometrical axioms and theorems in a three-dimensional rectangular coordinate system. Since there are many topics in school geometry, in terms of content, the training of the self is going to be illustrated following the route traced in the map, along the levels and tasks the desired child should complete while learning how to navigate in space and how to locate shapes and places.

The Training of the Self

On the first level, students should learn concepts and notions such as vertices and sides of geometrical figures, as well as parallelism and perpendicularity. Students should recognise the basic elements of prisms and geometrical figures. They should also be able to relate these basic elements using the notions of parallelism and perpendicularity. Students should use these elements to describe and represent diverse shapes of their "physical environment". For example, in one task of this level, students should surmise the resulting shapes produced by cutting a cube along the diagonals of one of its faces. They have "to anticipate the resulting shapes correctly; identify them […] and depict them" (MINEDUC, 2010, p. 7, own translation). While performing this task, students should learn that it is more accurate to draw figures and prisms when using a squared paper, as seen in the expected performance of students in Fig. 15.1.

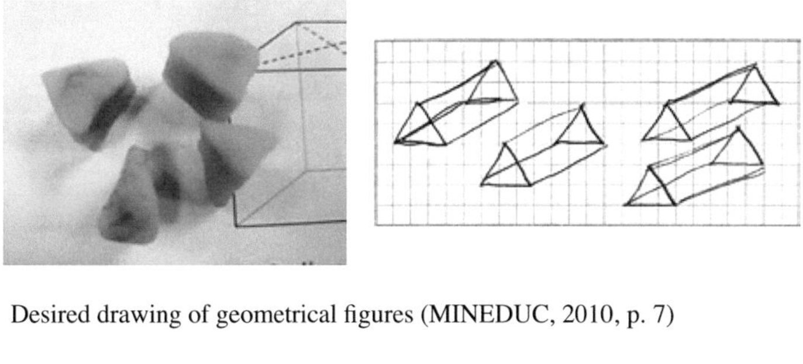

Fig. 15.1 Desired drawing of geometrical figures (MINEDUC, 2010, p. 7)

By acquiring the basic elements of geometrical figures and prisms and realising the usefulness of the squared paper, students should learn how to measure, depict, and operate with these tools. By grasping how to operate with the figures and the squared paper, they can then solve another type of problems. As students advance through the levels of the map of learning progress, they should gain new tools. They should continue to apply these processes to the next school experience or to the next level. For example, identifying vertices and sides of a figure becomes helpful to identify polygon's angles. Then, students will be able to use these tools to operate with the congruence criteria on the third level of the map of learning progress. Each time students encounter a problem, they should use the previously acquired tools and they should find other uses for them, such as sketching a flattened box on the squared paper as a step to build it in three dimensions (Fig. 15.2):

There are many possible ways to solve this problem, but the expectation at the second level of the map of learning progress—the desired solution—requires the use of the squared paper. Students should be able to draw the resulting box on a grid surface, in which each square of the grid has a 1 cm side. Once students draw the figure (Fig. 15.3), the following task is to calculate the dimensions of the resulting box "by counting the squares along each side" (MINEDUC, 2010, p. 9, own translation). In this second level, students should be able to estimate lengths, areas and volumes of geometrical shapes by counting squares. This tool will become applicable in subsequent levels, for example at fourth level, when students learn how the variation of the perimeter of a figure modifies its volume and area.

Once students learn how to handle the 1 cm-squared paper, they can use this tool for solving another type of tasks. Students could apply their knowledge of positioning vertices and sides of geometrical figures to a rectangular coordinate system. For example, at the fifth level they should learn how to depict and transform basic elements of Euclidean geometry into a Cartesian coordinate system (Ortiz et al., 2012). One of the tasks of this level, in which students should use all previous tools—counting squares, locate vertices, trace sides and so forth—is to rotate vertices (Fig. 15.4). This new skill, locate vertices by knowing the rotation centre, will become helpful in the sixth level, when they learn vectors in two and three dimensions.

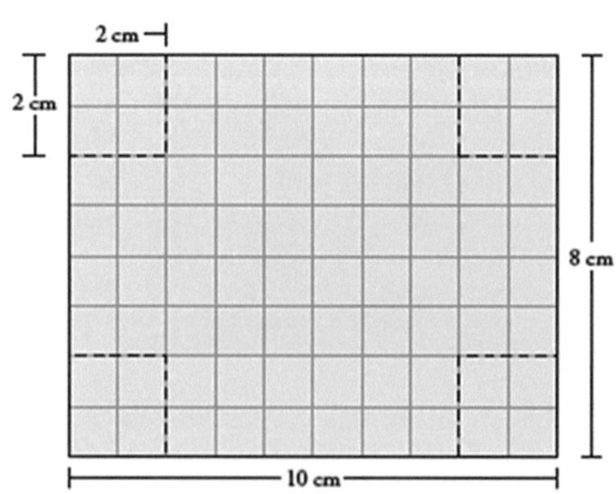

Fig. 15.2 School task using the squared-paper (MINEDUC, 2010, p. 22, own translation)

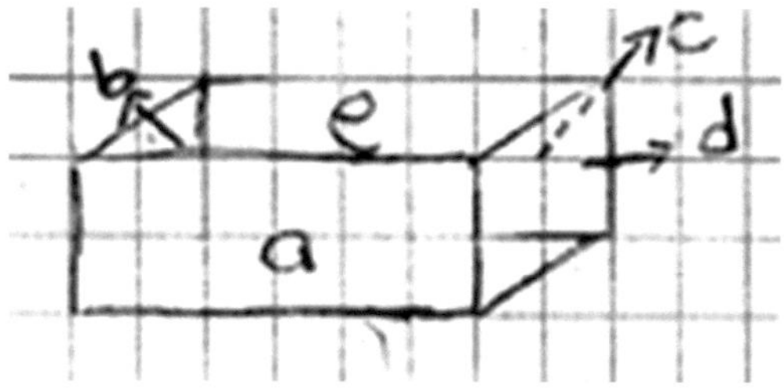

Fig. 15.3 Squaring geometrical shapes (MINEDUC, 2010, p. 9)

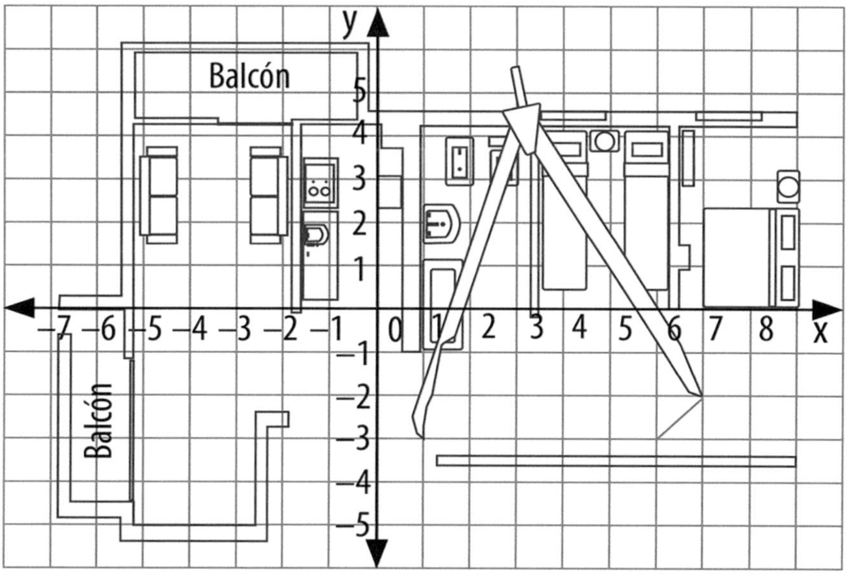

Fig. 15.4 Cartesianised geometrical figures (Del Valle et al., 2014, p. 205, own translation)

At fifth level, students also learn how to transform—rotate, translate, and reflect—geometrical figures, and how to use the congruence and similarity criteria in a Cartesian coordinate system. They have to use the tool acquired before to face these tasks. For example, while determining which transformations a triangle had and which were the rotation centres of each transformation (Fig. 15.5).

At the sixth level, students should learn how to locate specific points, mentioned in everyday life situations, in a Cartesian coordinate system. By doing this, students should realise that the Cartesian coordinate system could be related to the cardinal points—North, West, East, and South (Fig. 15.6). For example, students should be able to estimate the distance between two given points according to their coordinates (MINEDUC, 2010).

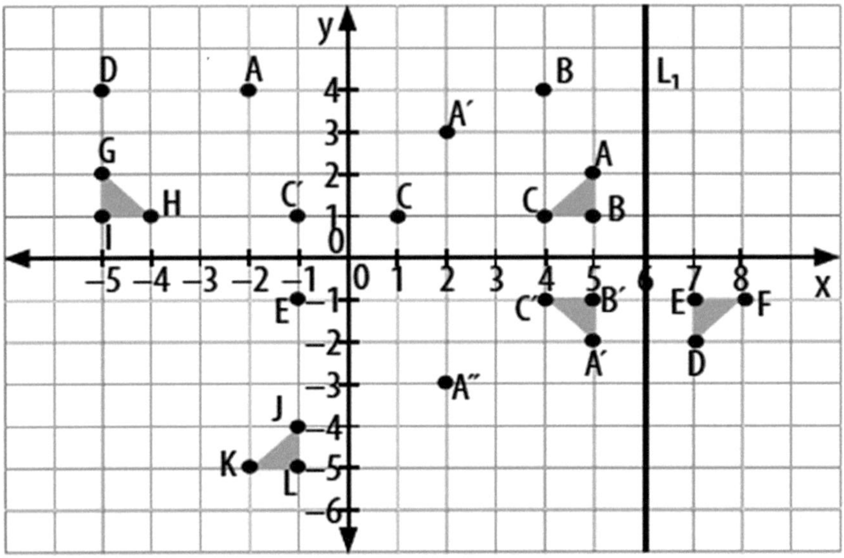

Fig. 15.5 Transformations in a Cartesian coordinate system (Del Valle et al., 2014)

Fig. 15.6 Locate places in a Cardinal coordinate system (Del Valle et al., 2014, pp. 176–177, own translation)

Moreover, students are also prompted to use all these tools while being challenged with another type of problems. In these challenges students "are placed in real three-dimensional situations [which provides] new tools to make spatial and flat depictions, such as the vector model" (MINEDUC, 2004, p. 68, own translation). Students should learn how to navigate in space with Cartesian tools and Euclidean planes, figures and shapes. Students should be able to relate movements in the everyday life with vectors (Fig. 15.7). At the seventh level they should use models as Cartesian and parametric equations to solve these tasks.

Through this training of the self, students should be able to understand their situation, their environment, and experience to master this process. When students solve a problem "they also learn how to act while facing new challenging experiences" (Zañartu et al., 2012, p. 26). While facing a challenge, according to MINEDUC, students should decide for themselves, by following a four-step plan for action and decision-making (Bórquez & Setz, 2012; Zañartu et al., 2012). Firstly, they have to understand the problem. Secondly, they have to create a plan. Thirdly, they have to execute this plan. Finally, they have to reflect on the resulting outcome.

As illustrated above, the desired student should be able to apply the previously acquired tools and the four-step plan strategy each time he/she is being faced with new experiences where he/she has to learn new abilities and strategies. In this sense, the tools students apply to these new experiences at one point stop being "external", but start flowing from an internal, individual source of thinking. These were not familiar tools to the students. They were planted or were acquired—learned—through the technologies of the self in the pedagogical devices for learning geometry. Through this process of training of the self, the desired student should learn how to use school geometry tools to navigate in space.

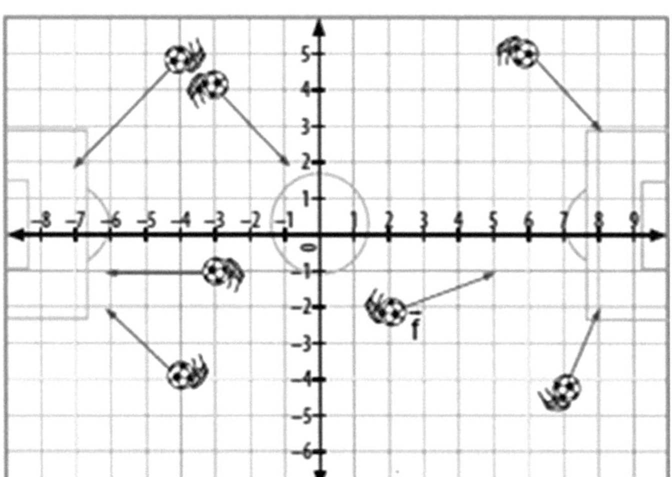

Fig. 15.7 Cartesianised football match (Del Valle et al., 2014, p. 183, own translation)

Students are also prompted to apply the tools acquired in the training process while facing other tasks. For example, the items in the PISA test are another scenario for performance that offers challenging experiences to students. The following problem presents modern architecture to students; it presents a building with "unusual" shapes (Fig. 15.8).

The task students face prompts the use of spatial visualisation and spatial skills. The presentation of the task in words and the illustrations apparently appeal to students resorting to their capacity of visualisation of how the middle of the building would look like. However, when checking the criteria for point assignment to the students' answer to this task, it becomes clear that the prompted behaviour is not valuable. If students rely only in visual information and depict the resulting shape according to the instruction, they will receive no credit at all if, additionally, they do not include information about the rotation centre, the direction of the rotation and the angle:

> *Full credit*: A correct drawing, meaning correct rotation point and anti-clockwise rotation. Accept angles from 40° to 50°. *Partial credit*: One of the rotation angle, the rotation point, or the rotation direction incorrect. *No credit*: Other responses and missing. (OECD, 2009, p. 184, emphasis added)

It becomes evident here that actual use of visualisation and spatial skills is not really part of the tools that the desired child should display when solving everyday life problems. Our point here is not whether there is a mis/match between reality in and outside school, and how such a mis/match is handled in mathematics education.

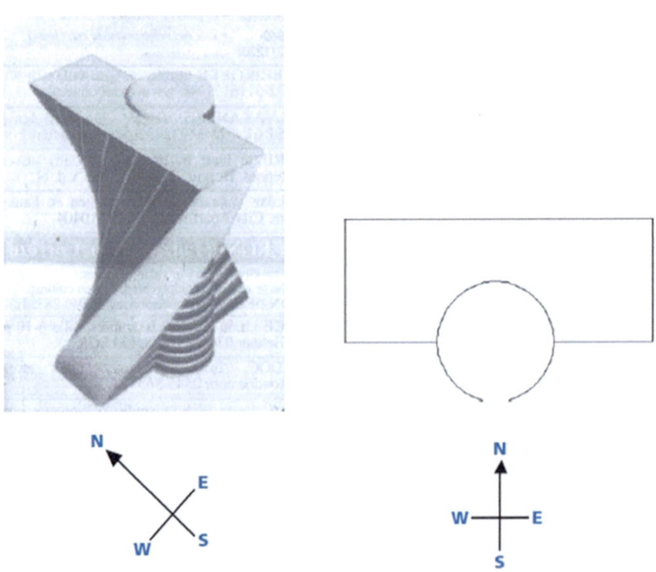

Fig. 15.8 PISA task with unusual shaped building (OECD, 2009, p. 146)

Our point here is that the training of the self in school geometry, although apparently appealing to the usefulness of geometrical thinking in the real world, in fact inserts in children a way of perceiving and seeing that is articulated under the logic of Euclidean metrics, axioms and theorems, and Cartesian coordinate systems.

This has two effects. First, the eye of the body becomes a trained eye through the norms of Euclidean reason (Andrade-Molina & Valero, 2015). Second, for this eye to "see", a new type of space is needed. These technologies simultaneously fabricate a subject and objectify space in particular ways. The space of school geometry emerges. This space has been "chopped" into particular routines, and it has restricted ways of seeing and being:

> The concept of space to be reconstructed in the students' understanding is that of a rational, referential space with fixed points in two or three dimensions. It is assumed that the conceptual development of the child will lead to an internal and abstract representation, which will contribute to making a decontextualized child, freed from the practical capacities of acting with objects in space, particularly of those spaces where everyday life occurs. (Valero, García, Camelo, Mancera, & Romero, 2012, p. 7)

Objectification and subjectification, the two basic mechanisms of mathematical learning according to Radford (2008), are brought together in the making of child who trains himself and his eye to perceiving space and geometrical knowledge as decontextualised, universal and timeless.

The Training of the Untrained Eye

If a feature of the desired child is a trained eye, then untrained eyes would be the navigation of space that the child has at the beginning of the process of training of the self through school geometry. Untrained not because these eyes cannot see, but because they are not trained in terms of school. Although it is difficult to imagine and know exactly how untrained eyes would navigate in space, these eyes would rely on each subject's individual relations with space before school. The type of interaction through the body and the senses of a subjectification outside the school also shape untrained eyes. As Crary (1992, p. 5) states,

> Vision and its effects are always inseparable from possibilities of an observing subject who is both the historical product and the site of certain practices, techniques, and procedures of subjectification.

Outside the training of the self in schools, the subject becomes the observer of an optically perceived world. Optical perception has been considered as a key aspect of the interaction between the body and world. It is believed that humans have a visual dominance, which means that humans tend to rely more on visual information than other forms of sensory information (Gal & Linchevski, 2010; Sinnett, Spence, & Soto-Faraco, 2007).

Optical perception and its relation to geometry have been addressed in research. Some studies have been exploring, among others, the connection between the optical

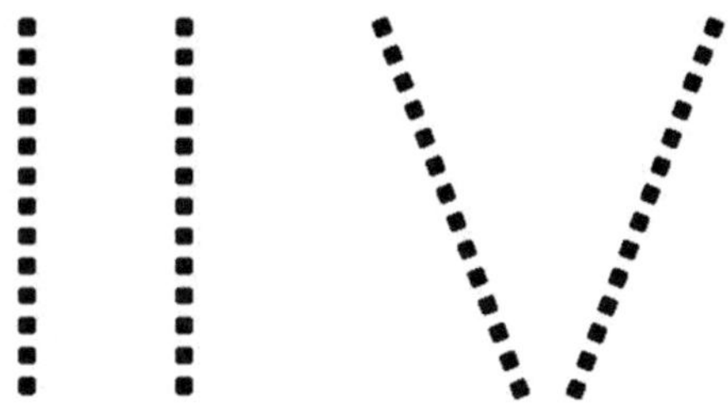

Fig. 15.9 Parallel lines and the experiment result

perception of space and the space described by Euclidean geometry (Blumenfeld, 1913; Boi, 2004; Burgin, 1987). For example, a series of experiments conducted in 1913 demonstrated that phenomenological visual judgments do not satisfy the properties of Euclidean geometry (Hardy, Rand, & Rittler, 1951). It was concluded that physical configurations and Euclidean geometry do not coincide. One of the experiments was to arrange two rows of point sources of lights as straight and parallel to each other as possible. The lights were placed on either side of a plane. While, in Euclidean geometry, parallel lines are equidistant along any mutual perpendicular, the resulting lines in the experiment diverged. The lines were not parallel at all (Fig. 15.9).

If the existing research shows the inconsistency between Euclidean geometry and optical perception (see Suppes, 1977), it is possible to problematise the reduction that may operate in children when the untrained eye starts interacting with the notions and procedures that organise school space.

Space Through Trained Eyes

Lefebvre (1991) challenged the monopoly of mathematics over the concept of space. On the grounds of a critique to the influence of the metaphysical philosophy that had made space and time absolute categories to organise the physical world, he proposes to bring back space to the realm of the social. He argues that space is as a product of concrete practices and the attempt to represent them. Space is experienced in three forms: space as perceived, as conceived, and as lived. The *perceived space* exists as a physical form, a space that is generated and used. The *conceived space* is instrumental; it is a space of knowledge (savoir) and logic. Space becomes a mental construct, an imagined space. The *lived space* is produced and modified over time and through its use, it is a space of knowing (connaissance); a space that is real-and-also imaginary.

However, space becomes the realm of abstraction when it comes to the knowledge that traditionally has dealt with it—geometry. Schools cut the links with the body—senses; the perceived space is only reachable by reason and logic. By doing this, school space turns into an instrumental space, a mathematical space of savoir.

The stated aims of school geometry attempt to connect all three forms of space. However, the pedagogical devices that operate the training of the self lead to a reduction of this link. Thus, conceived space cannot connect with the lived space. In other words, the untrained eye, produced by *connaissance*, is unlinked to the trained eye, produced by *savoir*. As illustrated above, the desired child should navigate in a space in terms of XYZ. Nevertheless, understanding spatiality in terms of a coordinate system restricts the concept of dimensionality of an object. It is argued that this understanding of space leads to students' misconceptions about the perceived space (Skordoulis, Vitsas, Dafermos, & Koleza, 2009).

> In geometry, students are presented with the properties of shapes and theorems for proof [...] all the information needed is given in the problem, and the students are asked to apply the theorems in what has to be proven. [...] The skills needed to solve these types of problems are limited, and teaching these skills usually consists of demonstrating the appropriate technique followed by a series of similar problems for practice". (Mevarech & Kramarski, 2014, p. 24)

The lack of interaction between both eyes in school raises some concerns about failure in geometry. It is believed that the learning of geometry has been difficult for students due to the emphasis that school has been given to deductive processes, which neglect the underlying spatial abilities (Del Grande, 1990). NRC (2006) also addresses this issue by stating that Euclidean geometry interferes with the understanding of other notions, for example, activities involving "specifying locations". This type of activities moves forward to formal Euclidean geometry, in which, for example, a point is a dimensionless location, not a point in space with a small but definite area.

Euclidean axioms and postulates have gained such importance within school geometry in Chile that students during the last grade of compulsory education (17–18 years old) are less prone to use spatial abilities while solving problems. Andrade and Montecino (2011) show the struggles students experienced while accepting, for example, that the interior angles of a triangle do not always add 180°. This type of difficulties emerges while solving problems involving spatial abilities. It restricts students to move from a flat surface to a curved surface. The example illustrates how students have been trained to navigate in an instrumental space dominated by the tools acquired in the process of training of the self through school geometry.

Building on Valero (2011) it is possible to formulate that the forms of subjectivity that school geometry promotes in children contrast sharply with children's experiences of space in their activities out of school. Hence, school geometry promotes the fabrication of a certain type of subject, a desired trained child who is able to see through reason and logic, with trained eyes.

The Horror! A School Nightmare

What if students were not able to detach from the "eyes of the body"? The desired child portrayed has some features. It is known that many students will come close to become such desired child by having appropriated the forms of thinking and being specified through the practices of school geometry. However, does every student become the desired child? What would happen to those who do not succeed? It is said that students encounter difficulties when both eyes meet:

> [Difficulties] that occur as a result of their spontaneous processes of visual perception in cases in which they contradict the geometric concepts/knowledge aimed at by the teacher and the tasks. Students fail to accomplish a dimensional deconstruction of the figures in order to infer mathematical properties in axiomatic geometry. (Gal & Linchevski, 2010, p. 180)

This invites to understand how perception interacts with reason and logic while shaping students' own existence in the training process. Andrade-Molina (2015) shed light on the implications of neglecting perception in school geometry. Students should train themselves to operate within certain discourses. They learn that while solving problems involving the measurement of the height of buildings or objects located perpendicularly to the ground, they can use the Pythagorean theorem or trigonometry. They have a 90° angle to operate with (Fig. 15.10).

As students advance through the levels of the map of learning progress, they are confronted with the challenge displayed in Fig. 15.11. Students should be able to use their tools to solve this problem. They have learned the Pythagorean theorem as a method to calculate the height of an object or the distance between a given point and that object—an object that is perpendicular to the ground. When using this behaviour to overcome this challenge, they should gather all information regarding the situation to understand the problem, and then they have to create a plan, as MINEDUC expects.

Because they trained themselves to use Pythagorean theorem, the plan would be to find a right angle (Fig. 15.12): The height (h) with the ground, the same technique as their previous experiences. Therefore, they have a height (h); they have the right angle and the distance that they should estimate. They are not asked to give an exact measurement, but they are asked to play with the tools they have.

Probably they will use other theorems, even trigonometry to face this challenge and "win the game". But neglecting the curvature of the Earth is a horrifying nightmare,

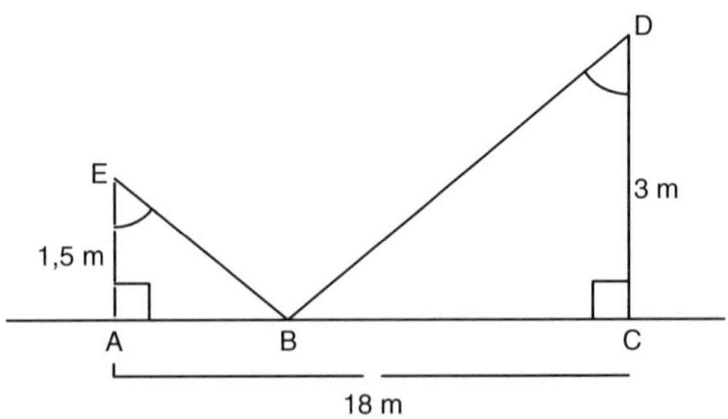

Fig. 15.10 Calculating heights and distances (Muñoz et al., 2013, p. 104)

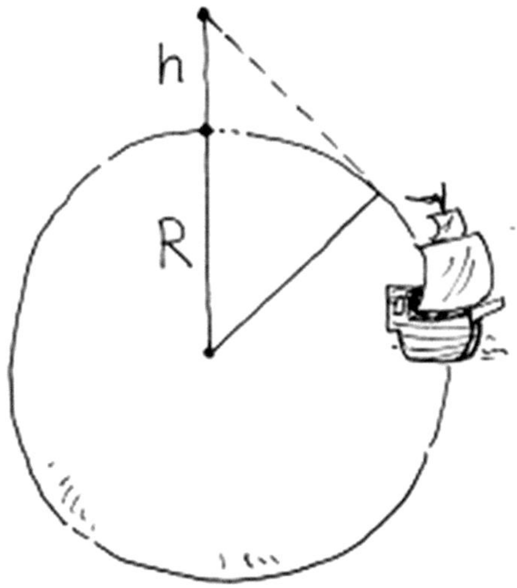

Fig. 15.11 School geometry task (MINEDUC, 2004a, p. 95, own translation)

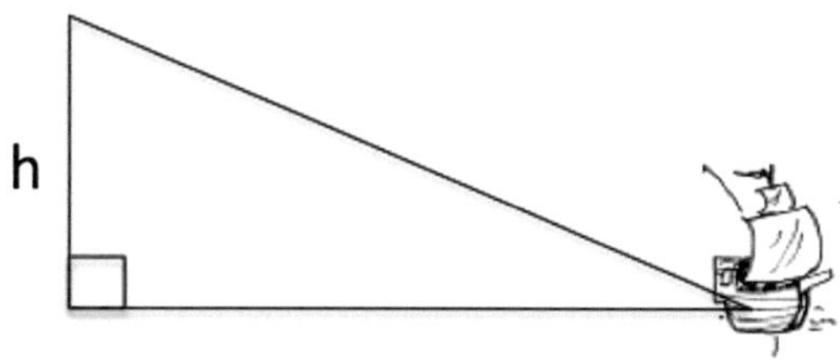

Fig. 15.12 Resulting triangle

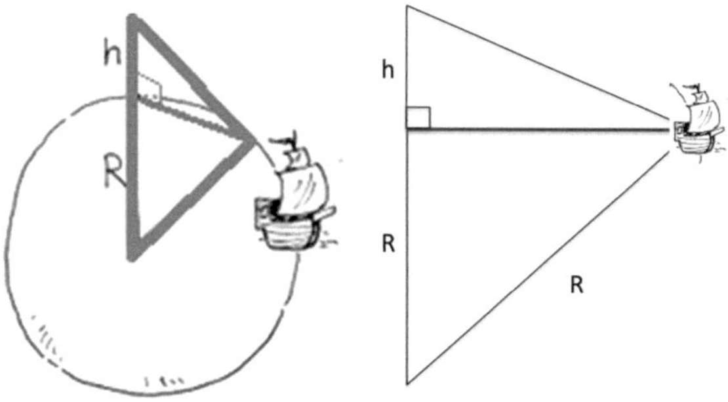

Fig. 15.13 A horrifying nightmare

at least in this case. The students who are able to make the type of configuration in Fig. 15.12 are disregarding the curvature. Why is this horrifying? Because it leads to contradictions of geometrical theorems, for example, that the sum of the length of the legs of a triangle does not add more than the length of its hypothenuse (Fig. 15.13).

> [In this particular problem] it does not seem so irrelevant to neglect the curvature of the earth, through our eyes it is impossible to say that a 90° angle is formed. Visually it does not make sense, but in school we have to accept it. Otherwise, we cannot use the geometry learned in school. We accept it because, locally, it seems like a right angle [...] both leg and hypotenuse have the same measurement, the Earth radius [...] Perception and visual judgements cannot be separated from reason and logic. (Andrade-Molina, 2015, p. 7, own translation)

Students have to agree that, for this type of problems, the earth has to be flat. They should be subjected to this rationality. In this sense, the horrors that may result in the extension of the logic of school space are not to be explained as a fault created by teachers' "mis-implementation" of the curriculum, neither by students' cognitive deficiencies. Rather, they are to be explained in terms of the very same power effects of school geometry to shape students' ways of dealing with space.

Popkewitz (2008) introduces the term "abjection" to state that while certain discourses express the desired—the included—at the same time, they are expressing the undesired—the excluded. The analysis deployed in this chapter reveals the features of the desired child. It also reveals the ones of the "feared child". The feared child can tell the contradiction of the leg and hypotenuse of the triangle having the same length (Fig. 15.13). It is the one who is capable to link both eyes and see school space not only through reason and logic.

Shaping the Child Through School Geometry

> Forms of knowledge are effected and effect power as they bring together knowing and being as two sides of a coin. Forms of knowledge do not only bear the rules of how one knows and what it is to be known, but also impose ways of being on the knowers [...] If knowing and being are inseparable, the question emerges of what the forms of knowing and being are that the mathematics curriculum effects in children, and whether those forms of subjectivities are desirable. (Valero et al., 2012, p. 3)

Chilean school geometry has effects of power in students' subjectivities, not only in terms of oppression and subjection, but also in fabricating productive forms of being in the world through practices of the self. Students should be able to perceive themselves as "agents" (Foucault, 2009) who are responsible for their own learning. They should care for themselves. Students should feel curiosity; students should engage and ask themselves "what if?" while learning geometry. However, it is not only an invitation to think, but it is also an invitation to act, to create plans, strategies, and execute them.

> By introducing at first vectors in the plane—which are easier to imagine and to depict—and then, moving forward to depict and operate with vectors in space, might invite students to ask themselves about other possibilities in higher dimension. (MINEDUC, 2004a, p. 68, own translation)

Reasoning with governmentality, techniques of government aim at controlling human behaviour, but at a distance—a kind of wireless management. This is achieved by using techniques that regulate the habits and desires of students, "arranging things so that people, following only their own self interest, will do as they ought" (Scott, 1995, p. 202). Therefore, the interplay between power and mathematics education is on how the school mathematics curriculum generates cultural and historical subjects (Valero & García, 2014). Understanding school geometry as modes by which power operates to shape subjects, allows unpacking how school geometry inserts students into a form of rationality. In other words, how school geometry becomes a technology of the self that fabricates a "desired child" who is able to see with trained eyes—sightless eyes (Andrade-Molina & Valero, 2015)—generating systems of reason in which forms of life and subjectivity are made possible, organised and constrained.

Henceforth, technologies of the self enable students to change and to develop their thoughts and conduct their ways of being. These technologies enable students to modify, structure and constitute themselves as subjects.

> Permit individuals to effect by their own means or with the help of others a certain number of operations on their own bodies and souls, thoughts, conduct, and way of being, so as to transform themselves in order to attain a certain state of happiness, purity, wisdom, perfection, or immortality. (Foucault, 1988, p. 18)

Since discipline makes individuals (Foucault, 1979), disciplinary technologies are means of producing compliant, meeting the requirements subjects, through the exercise of management techniques, which govern every aspect of life. All in all, "who is subjected to a field of visibility, and who knows it, assumes responsibility for the constraints of power [...] he becomes the principle of his own subjection" (Foucault, 1979, pp. 202–203). Thus, in order to reach a state of happiness, purity, wisdom, perfection or immortality, students must train and modify themselves. Not only they will acquire the skills of the "desired child", but they will also acquire certain attitudes to navigate in space. School geometry, as a technology of the self, shapes and also subjugates students through "relations of power" (Foucault, 1982). Students accept the space deployed by school geometry as well as the tools to operate in it, Within formal school settings, students should accept to neglect their senses; they should accept to model everyday life situations by using geometrical deductions; they should accept to see space through reason and logic.

The analysis deployed in this chapter traces these techniques. The techniques used in the production of the "new subject" from school geometry discourse. It also traces the power effects on students' subjectivities, as a process of cutting the links between the eyes of the instrumental space and the eyes of the lived space. Although this critique emerges from the expressions of a desired particular way of being, however, there is no certainty on how these strategies of power self-govern students. The map of progress analysed does not offer information on how many students performed accordingly to the expectations of MINEDUC, and how many did not. In that sense, it is more the critique of a dream, of a non-existent desired child. It is possible to tell the story about how school geometry becomes a technology of the self because it effects children's subjectivities. Nevertheless, it is not possible to state, by this discourse analysis, how power is effecting the shaping of the self. Only the desired child will be the one shaped by the strategies of power, but not all students become the desired child.

Acknowledgments This research is funded by the National Commission for Scientific and Technological Research, CONICYT PAI/INDUSTRIA 79090016, in Chile and by Aalborg University, in Denmark. This research also makes part of the NordForsk Center of Excellence "JustEd". We would like to thank Elizabeth de Freitas, the reviewers and editors for their comments to previous drafts of this chapter.

References

Andrade, M., & Montecino, A. (2011). La problemática de la tridimensionalidad y su representación en el plano. In A. Ruiz & E. Mancera (Eds.), *Proceedings of XIII Conferencia Interamericana de Educación Matemática (CIAEM-IACME)*. Brasil: Recife.

Andrade-Molina, M. (2015). Geometría escolar: Una batalla entre razón versus lógica y razonamiento. *Proceedings of XIV Conferencia Interamericana de Educación Matemática (CIAEM-IACME).* http://xiv.ciaem-iacme.org/index.php/xiv_ciaem/xiv_ciaem/paper/viewFile/1193/477.

Andrade-Molina, M., & Valero, P. (2015). The sightless eyes of reason: Scientific objectivism and school geometry. *Proceedings of 9th Congress if European Research in Mathematics Education.*

Battista, M. T. (2007). The development of geometric and spatial thinking. In F. Lester (Ed.), *Second handbook of research on mathematics teaching and learning.* (Vol. 2, pp. 843–908).

Blumenfeld, W. (1913). Untersuchungen über die scheinbare Größe im Sehraume. *Zeitschrift für Psychologie und Physiologie der Sinnesorgane, 65*, 241–404.

Boi, L. (2004). Questions regarding Husserlian geometry and phenomenology. a study of the concept of manifold and spatial perception. *Husserl Studies, 20*(3), 207–267.

Bórquez, E., & Setz, J. (2012). *Matemática cuarto año de Enseñanza Media. Guía didáctica del docente.* Santiago: Santillana.

Burgin, V. (1987). Geometry and abjection. AA Files, 35–41.

Clements, M. A. (2008). Spatial abilities, mathematics, culture, and the Papua New Guinea experience. In P. Clarkson & N. Presmeg (Eds.), *Critical issues in mathematics education: Major contributions of Alan Bishop* (pp. 97–106). New York: Springer.

Clements, D. H., & Battista, M. T. (1992). Geometry and spatial reasoning. In D. A. Grouws (Ed.), *Handbook of research on mathematics teaching and learning* (pp. 420–464). New York: Macmillan.

Cotoi, C. (2011). Neoliberalism: A Foucauldian perspective. *International Review of Social Research, 1*(2), 109–124.

Crary, J. (1992). *Techniques of the observer: On vision and modernity in the nineteenth century.* Cambridge, MA: MIT Press.

Del Grande, J. (1990). Spatial sense. *The Arithmetic Teacher, 37*(6), 14–20.

Del Valle, J., Muñoz, G., & Santis, M. (2014). *Matemática primer año de Enseñanza Media. Texto del Estudiante.* Santiago: Ediciones SM Chile.

Diaz, J. D. (2014). Governing equality. *European Education, 45*(3), 35–50. doi:10.2753/EUE1056-4934450303.

Foucault, M. (1979). *Discipline and punish* (A. Sheridan, Trans.) New York: Pantheon.

Foucault, M. (1982). The subject and power. *Critical Inquiry, 8*(4), 777–795.

Foucault, M. (1984). The ethics of the concern of the self as a practice of freedom. In P. Rabinow (Ed.), *Ethics, subjectivity and truth, essential works of Michel Foucault, 1954-1984* (pp. 281–302). London: Penguin.

Foucault, M. (1988). Technologies of the self. In L. H. Martin, H. Gutman, & P. H. Hutton (Eds.), *Technologies of the self* (pp. 16–49). Amherst: University of Massachusetts Press.
Foucault, M. (1993). About the beginning of the hermeneutics of the self: Two lectures at Dartmouth. *Political Theory, 21*(2), 198–227.
Foucault, M. (1997). Technologies of the self. In M. Foucault & P. Rabinow (Eds.), *Ethics: Subjectivity and truth* (pp. 223–251). New York: The New Press.
Foucault, M. (2008). *The birth of biopolitics: Lectures at the Collège de France, 1978-79*. (G. Burchell, Trans.) New York: Palgrave Macmillan.
Foucault, M. (2009). *Security, territory, population: Lectures at the College de France 1977-1978* (G. Burchell, Ed., M. Senellart, F. Ewald, A. Fontana, Trans.). New York: Macmillan.
Gal, H., & Linchevski, L. (2010). To see or not to see: Analyzing difficulties in geometry from the perspective of visual perception. *Educational Study of Mathematics, 74*, 163–183.
Hardy, L. H., Rand, G., & Rittler, M. (1951). Investigation of visual space: The Blumenfeld alleys. *A.M.A. Archives of Ophthalmology, 45*(1), 53–63.
Hauptman, H. (2010). Enhancement of spatial thinking with Virtual Spaces 1.0. *Computers and Education, 54*(1), 123–135.
Kollosche, D. (2014). Mathematics and power: An alliance in the foundations of mathematics and its teaching. *ZDM, 46*(7), 1061–1072.
Kozhevnikov, M., Motes, M. A., & Hegarty, M. (2007). Spatial visualization in physics problem solving. *Cognitive Science, 31*(4), 549–579.
Lefebvre, H. (1991). *The production of space* (Donald Nicholson-Smith Trans.) Oxford, UK, Cambridge, USA: Blackwell.
Lemke, T. (2002). Foucault, governmentality, and critique. *Rethinking Marxism: A Journal of Economics, Culture and Society, 14*(3), 49–64. doi:10.1080/089356902101242288.
Mevarech, Z., & Kramarski, B. (2014). *Critical maths for innovative societies. The role of metacognitive pedagogies*. Paris: OECD Publishing. doi:10.1787/9789264223561-en.
Ministerio de Educación de Chile. (2004). *Matemática. Programa de Estudio. Tercer Año Medio*. Santiago: MINEDUC.
Ministerio de Educación de Chile. (2004a). *Matemática. Programa de Estudio. Cuarto Año Medio*. Santiago: MINEDUC.
Ministerio de Educación de Chile. (2010). *Mapas de Progreso del Aprendizaje. Geometría*. Santiago: MINEDUC.
Ministerio de Educación de Chile. (2011). *Matemática. Programa de Estudio. Segundo Año Medio*. Santiago: MINEDUC.
Muñoz, G., Jiménez, L., & Rupin, P. (2013). In F. Muñoz (Ed.), Texto Educación Matemática 2, Educación Media. Texto del estudiante. Edición especial para el Ministerio de Educación. Santiago: Ediciones SM Chile.
National Council of Teachers of Mathematics. (2000). *Principles and standards for school mathematics*. Reston, VA: National Council of Teachers of Mathematics.
National Research Council. (2006). *Learning to think spatially: GIS as a support system in the K-12 curriculum*. Washington, DC: The National Academies Press.
Newcombe, N. S. (2010). Picture this: Increasing math and science learning by improving spatial thinking. *American Educator, 34*(2), 29–35.
Organisation for Economic Co-operation and Development. (2009). *Take the test. Sample questions from OECD's PISA assessments*, OECD Publishing.
Organisation for Economic Co-operation and Development. (2012). *Connected minds: Technology and today's learners, educational research and innovation*. OECD Publishing. http://dx.doi.org/10.1787/9789264111011-en.
Organisation for Economic Co-operation and Development. (2014). *PISA 2012 results: What students know and can do—student performance in mathematics, reading and science* (revised ed., Vol. I, February 2014). Pisa: OECD Publishing. http://dx.doi.org/10.1787/9789264201118-en.
Ortiz, A., Reyes, C., Valenzuela, M., & Chandía, E. (2012). *Matemática primer año de Enseñanza Media. Guía didáctica del docente*. Santiago: McGraw-Hill Interamericana.

Popkewitz, T. S. (2008). *Cosmopolitanism and the age of school reform: Science, education, and making society by making the child*. New York: Routledge.
Prieto, G., & Velasco, A. D. (2010). Does spatial visualization ability improve after studying technical drawing? *Quality and Quantity, 44*(5), 1015–1024.
Radford, L. (2008). The ethics of being and knowing: Towards a cultural theory of learning. In L. Radford, G. Schubring, & F. Seeger (Eds.), *Semiotics in mathematics education: Epistemology, history, classroom, and culture* (pp. 215–234). Rotterdam: Sense.
Rose, N. (1996). Governing "advanced" liberal democracies. In A. Barry, T. Osborne, & N. Rose (Eds.), *Foucault and political reason* (pp. 37–64). Chicago, IL: University of Chicago Press.
Scott, D. (1995). Colonial governmentality. *Social text., 43*, 191–220.
Sinnett, S., Spence, C., & Soto-Faraco, S. (2007). Visual dominance and attention: The Colavita effect revisited. *Perception and Psychophysics, 69*(5), 673–686.
Skordoulis, C., Vitsas, T., Dafermos, V., & Koleza, E. (2009). The system of coordinates as an obstacle in understanding the concept of dimension. *International Journal of Science and Mathematics Education, 7*(2), 253–272. doi:10.1007/s10763-008-9130-2.
Suppes, P. (1977). Is visual space Euclidean? *Synthese, 35*(4), 397–421.
Valero, P. (2011). Re-interpreting students' interest in mathematics: Youth culture resisting Modern subjectification. In D. Di Paola & J. Díez-Palomar (Eds.), *63th Commission Internatinale pour l'Etude et l'Amélioration de l'Enseignement des Mathématiques (CIEAEM): Facilitating access and participation: mathematical practices inside and outside the classroom. Quaderni di Ricerca in Didattica (Mathematics)* (pp. 72–83). Spain: Barcelona.
Valero, P., & García, G. (2014). Matemáticas escolares y el gobierno del sujeto moderno (School mathematics and the governing of the Modern subject). *Bolema: Boletim de Educação Matemática, 28*(49), 491–515. doi: http://dx.doi.org/10.1590/1980-4415v28n49a02.
Valero, P., García, G., Camelo, F., Mancera, G., & Romero, J. (2012). Mathematics education and the dignity of being: Original research. *Pythagoras, 33*(2), 1–9.
Valero, P., & Knijnik, G. (2015). Governing the modern, neoliberal child through ICT research in mathematics education. *For the Learning of Mathematics, 35*(2), 34–39.
Verdine, B. N., Golinkoff, R. M., Hirsh-Pasek, K., & Newcombe, N. S. (2014). Finding the missing piece: Blocks, puzzles, and shapes fuel school readiness. *Trends in Neuroscience and Education, 3*(1), 7–13.
Wai, J., Lubinski, D., & Benbow, C. P. (2009). Spatial ability for STEM domains: Aligning over 50 years of cumulative psychological knowledge solidifies its importance. *Journal of Educational Psychology, 101*(4), 817–835.
Zañartu, M., Darrigrandi, F., & Ramos, M. (2012). *Matemática segundo año de Enseñanza Media. Guía didáctica del docente*. Santiago: Santillana.

Part IV
Disordering the Role of the Mathematics Education Researcher

Chapter 16
Disordered Order, Ordered Disorder: Threads, Folds, and Artistic Action

Peter Appelbaum

Abstract Alterglobal social movements and psychoanalysis are mined for actionable transformation of the mathematics educator, shifting the scholar of mathematics education, and offering phases for changing oneself in order to change the world: Identify (Deleuzian) nomadic terms; use these terms in nontraditional ways; craft the work with the new, nomadic topology; and study the points of time/space simultaneity within the new topology. Mathematics as public art is used to illustrate these phases. Alterglobal discourses act as tools of psychoanalytic understanding, functioning as hinges across the apolitical and the political, making mathematics education a part of changing social reality.

Opening: Not an Inchoate Project

I begin this chapter with a few questions about my role in this collective project, indeed *our* roles, phrased as a response to the introduction to the book you have in your hands. Do readers of this volume need to revitalize their individual and/or collective political imagination, to break with the alleged coherence or "order" of mathematics education as a field? Does this volume bring marginalized voices of mathematics education research to a center, or to destabilize presumed foundations of mathematics education practice, such as "mathematics for all," or "mathematics for social justice?" Do you expect this chapter to highlight problems, tensions and contradictions within existing research, or to present a well-developed, novel method that can be followed in order to merge the sociopolitical with localized school mathematics teaching and learning practices? Perhaps you hope to be entertained by what feels like a radical and re-politicized critique of already-existing research claiming sociopolitical relevance, or to witness a reshaping of what can be conceived as the practice of mathematics education in the first place? Following this book's introduction, you surely expect the chapter at hand to confront what has

P. Appelbaum (✉)
Department of Curriculum, Cultures and Child/Youth Studies, Arcadia University, Philadelphia, PA, USA
e-mail: appelbap@arcadia.edu

H. Straehler-Pohl et al. (eds.), *The Disorder of Mathematics Education*,
DOI 10.1007/978-3-319-34006-7_16

come to be commonly expressed as "main-stream mathematics education" with a sociopolitical context that is often neglected. An initial impression from that introduction might lead a reader to interpret the delightful collection of scholars brought together in this volume as claiming a shared, playful stance toward the apparent lack of consensus on any and all questions and concepts related to what might or might not be considered central or peripheral to mathematics education.

Situating themselves in such relation to a hypothesized "field" of mathematics education, these mathematics educators render themselves in some ways disordered as outliers to what would then be understood as the "mainstream." Such an interpretation would be premature, however, as the various contributors would never attempt to agree on such a reordering of the disorder of mathematics education: While some of my colleagues in this volume seek to "awaken" mainstream mathematics educators to the potential of the sociopolitical, to construct what might then be taken as a canon[1] of an alternative field of upstart, destabilizers of the old-fashioned, former canon[2]—and in the process to confuse and befuddle notions of insiders and outsiders of research fiefdoms, relevance and irrelevance to policy and practice, and bestowers of accolades for scholarly excellence—others would never work within that type of epistemological or ideological framework. This chapter falls in the latter category of scholarship: it does not present a new method or set of phases of work to be copied by the reader. Instead, it opens up possible stages of awareness that can be used to reexperience our roles in research, scholarship, teaching, learning, and other educational practices, in any ordering and/or overlap of the phases of awareness.

My work in alterglobalization and psychoanalysis applied to mathematics education would seem suitable to such a broad project. Alterglobal movements (Appelbaum & Gerofsky, 2013; Pleyers, 2010) help us theorize communities in flux and becoming that coalesce and take action without requiring fixed identities, clear goals, justification, or defined structures, while maintaining a strong commitment to ethical principles of inclusion, diversity, human recognition and dignity (Butler, 1997; 2010). The latter phrase, referring to recognition and dignity, may feel like familiar goals; however, in the context of globalization, it has become necessary to maintain a focus on these goals, as it becomes necessary to redefine the meaning of these terms and the means toward these goals, in the light of transnational business interests, global information flow, mass migration, refugee movements, and new uses of technology for political transformation.

The term "alterglobal" has come to represent the various forms of collective action that respond to the negative aspects of globalization—corporate personhood, the dominance of markets over ethics, the increasing need to understand diaspora

[1] The politics of mathematics education, mathematics education as enculturation and acculturation, postmodern critiques of knowledge and power, sociological challenges to assumptions of equity, power, diversity, and progress, public pedagogies, expansion of the institutions of mathematics education beyond the school, and so on.

[2] Cognitive and Social Psychology, teaching methods, analysis of mathematics content for ideal sequencing of curriculum topics, instructional organization, evaluation and measurement of outcomes.

identities, the changing nature of identity through social media, and so on—in ways that are not naively "anti-globalization," (which would treat globalization as something to work against). Alterglobalization should also be distinguished from counter-globalization (working in opposition to globalization), super-globalization (overcoming globalized interactions), or hyper-globalization (taking globalized interactions to an extreme).

Alterglobalization is a call for a renewal of political citizenship and activism. This means bypassing traditional ideas about how to make social change. For example, alterglobalization may reject traditional ideas of creating revolution, whether by peaceful or violent means, which usually assumes that people live within nation states and that change happens within the confines of national boundaries. Alterglobalization happens when people create opportunities for communities to have an impact on the course of things in the world. Social movements like the Green movement, various Feminist movements, Slow Food and Slow Clothing movements, and Animal Rights actions, all have in common the ways that they transcend country categories and work inside and outside of the marketplace.

Given the variety of contributors to this volume, we can add a particularly porous, transnational character to the work that is being highlighted here. What might in some contexts be considered globalization becomes a particular kind of "international trans-nationalism," as scholars from various regions of the world come together to speak across national boundaries, as well as through and around transnational projects that embrace and maintain an awareness of the ways that particular local cultures, national policies, and unique patterns of population shifts are related to a broader parallel set of experiences that share commonalities. Growing out of this international, trans-national project is the realization that the global dimension of research in mathematics education, from a sociopolitical and cultural perspective, is clarified in the potential for significant action at once locally and globally.

I study the ways that mathematics education communities, mostly outside of traditional schools, are bubbling forth, taking action, and supporting social justice movements, without being limited by or defined by boundaries of in-school/out-of-school, formal/in-formal, programmed/self-directed, etc. These communities are often found in surprising places for mathematics education: in contexts such as community arts groups (Spiral Q Puppet Theater "SparQ" groups, Philadelphia, USA), university course contexts (Undergraduate seminars in the US that are not situated in mathematics or in education), online hacker communities, and other social media spaces that provide networks for creative mathematics "educators." These people can only be so defined in retrospect, with labels that communicate the attributes of teaching, research, reflecting upon, or inventing mathematics and/or mathematics education strategies.

The second main area of my prior work, psychoanalytic theory, highlights how understanding forms of resistance to learning might be particularly useful in changing the language of instruction and curriculum design. The understanding of many forms of resistance would shift from problems to solve to signifiers of important learning; theories of transference and counter-transference, projection, and so on, could provide powerful tools for assessment, instructional decision-making, and teacher training.

For example, instead of interpreting certain behaviors as an avoidance of engagement, a teacher can interpret them as indicators of important learning in emergence. For psychoanalysis, resistance occurs when someone meets the otherness of their own unconscious knowledge. I don't want to say that teaching is the same thing as acting like a psychoanalyst, but there are many parallels. During the course of every psychoanalytic therapy, the patient will behave in ways that interfere with the progress of the treatment. This interference is called resistance. Because psychoanalytic therapy helps the patient to achieve freedom of thought and action by talking freely, the negative emotional forces that caused his symptoms manifest themselves as obstacles to the talking therapy. Similarly, we should not be surprised if students exhibit behaviors of resistance throughout every educational curriculum.

In mathematics education, too, we want students to speak about their thoughts, to share their ideas, beliefs and opinions; the difficulties of confronting new knowledge will manifest themselves in behaviors that delay learning. Like a patient undergoing analysis, students might become unable to talk any longer, feel they have nothing to say, need to keep secrets from their teacher, or display a variety of other behaviors (see Appelbaum, 2008, for a more extended list of possible behaviors). As many therapists recognize that their patient may *need* to resist, teachers could recognize this need in their students. Following such recognition, the teacher and students could together study their resistance to learning, and the meaning and purpose of this resistance. Resistance to learning is a part of learning itself. Our first interpretation of resistance is often clouded by ourselves feeling angry, annoyed or helpless, and the desire to do something about our own feelings. The trick is to remind ourselves that these behaviors are just as possibly signs that important learning is taking place. A critical response to resistance behavior requires that we internally monitor ourselves.

Phases of Arrival That Become New Points of Departure

The advent of the DOME project is an opportunity to rethink how one might combine alterglobal social movements with psychoanalysis in ways that are not about finding a new, supposedly better discourse or set of practices. These methods I employ are in no way disordered or radical or innovative, and the presumption that they might be is belied by the facts that those actors in alterglobal social movements include numerous teachers of mathematics, mathematicians, mathematics students, people who routinely employ mathematics as features of alterglobal educational programs, and so on. More personally, the inappropriately conceived inside/outside nature of any application of a method upon a field of study is demonstrated in this case by the reality that my uses of alterglobal strategies "outside of traditional schools" often function as components of teacher education schemes housed in traditionally sanctioned teacher training programs at a university (Appelbaum, 2015), by my very position as a tenured, full professor at a university in the United States, and by the presence of this work in publications in mathematics education venues, each of which indicate a relatively high level of complicity in what would otherwise be termed insider or mainstream within a broader field of practice.

What is needed, perhaps, is a way to theorize and capture in an open way the feeling of marginality that coexists with functionality. We might recognize the *disordered ordering* and the *ordered disorder* that defines the field, where we are all inside and outside at the same time as experiencing ourselves neither inside nor outside.

I am not advocating a shift in research methods nor in research topics, but instead in the ways of being a scholar of mathematics education. The shift is more in terms of the ways in which we construct ourselves for ourselves as researchers, a shift that bears more than casual resemblance to a common expression in political action, that one must change oneself in order to change the world (Appelbaum & Davis, 2013; Davis, Mausbac, Klimke, & McDougall, 2010). To take this on as a challenge more than a cliché is to seriously rethink our ways of being in the world as mathematics educators, and to change our practices of thinking, asking, working, exploring, fearing, hoping, pondering, and so on. What follows are several phases in a collection of activities I have used to take on this challenge:

- First is to actively pursue a collection of words that coexist in our old and new ways of being, and to struggle to invest our energy into using the words as part of a larger discourse that intentionally alters the meaning of these words to be that of the nonmainstream as much as possible.
- Second, is to work in ways that are parallel and coexisting with the mainstream but are simultaneously independent of that mainstream.
- Third, we use the new discourses, now functioning as alternative epistemologies, as a topology of subjectivity that brings together formerly unrelated points in time and space into recognizable events and relations, that is, to establish a subjectivity whose purpose is to construct and explore topologies.
- Fourth, we work to make these events or points of simultaneity stable long enough to analyze and study them, by using the new discourses as threads that sew together the fabric of mathematics education in time and space across the folds, which these events of simultaneity represent.

These phases appear in an ordered sequence in this chapter, due to the linear nature of writing. Yet they need not take place in such a sequence. One might use any combination of the approaches in any order, and some at the same time. The idea is to take on different subjectivities, to change oneself as a mathematics educator in intentional ways, and in the process to change the world of mathematics education.

- The chapter concludes with artistic practice as an illustrative example (or fifth, synthetic phase) that enables each of the previous four phases. Other enactments are of course possible, opening up a new world of mathematics education.

From Nomadic Epistemology to Nomadic Topology

Readers of (Deleuze & Guattari, 1987; Guattari & Deleuze, 1972) will hear in my last section echoes of the notion of a "Nomadic Epistemology," a distinct and independent set of concepts that can exist within hegemonic discourses, and that also live and function in a parallel universe of discourses not trapped by the dominant ideologies that are, in turn, repeatedly recreated by the hegemonic discourses. The metaphor evoked is that of the nomad, who is at once homeless yet carries their home with them, and at the same time also always at home yet without a permanent home.[3] The project would be to evoke terms that are analogously nomadic with respect to mainstream and marginal mathematics education practice. Possibly nomadic concepts include: fluid and self-defined identity and diasporic cultures; forms of collective action independent of nation-states that also take on local, regional, and national forms; and nonhierarchic conceptions of teaching and learning, within which it is unclear who is the one who knows and who is the one learning at any given time. These terms, at the heart of alterglobal social movements, would be simultaneously analyzed and applied within traditional school mathematics contexts, and also across national, cultural, class, religious, and traditional knowledge boundaries, to create a "nomadic" mathematics education discourse. There is much potential in the distinctions among work, labor, and action (Arendt, 1958; Biesta, 2006), the categories of youth leadership, voice, and participation (Appelbaum, 2008; Pitt, 2003), and the varieties of argumentative uses and to which models can be employed (Appelbaum, 2012; McElheny, 2007). I refer to these latter nomadic topologies in my discussion of other phases of work.

At the same time, Deleuze might have something more directly relevant to the project of defining ourselves as mathematics educators in his lesser-known work, *The Fold* (Deleuze, 1993). As we ask, Who are we? And who am I?—as a researcher, educator, researcher-educator, mathematics-educator, etc., we can make use of the Deleuzian notion of "the fold" as critiquing any sense of subjectivity that is premised upon exteriority and interiority, since the fold announces that the interior and exterior are merely folds of one another. The "disordered" scholar of mathematics education, placed in the obscurity of a special research group on the sociopolitical at an international conference, becomes a "presenter," or marginalized "outsider" in the interior of the group within the larger group of conference researchers, the fold of the fold of the conference. Indeed, this "folding" can take place in any of a variety of unlimited modalities, including ourselves and our bodies, the folding of time and chronology, memory, and semiotic analyses. In this way, our subjectivity as scholars of mathematics education, and indeed the subjectivity of any other groups, individuals, diasporic communities, social agencies, and so on, might be understood as an interactive topology of these different kinds of unfoldings within unfolding of chronology and geographic location. What began as a shift in epistemological categories turns out to be at once also a shift in the terrain upon which those epistemological perspectives are directed; this is one way in which we can change ourselves and in doing so change our world.

[3] Deleuze and Guattari evoke this conception of the nomadic as disrupting a static set of socially assumed ways of experiencing the world. How ironic, then, that one might perceive a nomad as the most static and immobile of all, as carrying one's home everywhere at all times, never changing. Similar to students resisting learning, I worry the nomad's "immobility" as a symptom of the hegemonic game of "change."

Working with Nomadic Epistemologies to Reexperience the Folds

The previous section of this chapter introduced some examples of nomadic concepts that work both in and out of "mainstream mathematics education," and also neither inside nor outside of "mainstream mathematics education"; this section briefly describes how these collective concepts might be better understood as the folds of subjectivity for mathematics education research, so that a nomadic epistemology can, if we choose, become associated with a nomadic topology for the field of mathematics education.

Nomadic Topology Example 1: Work, Labor, and Action

Hannah Arendt (1958) suggested a nuanced distinction of the types of activity that might take place in a group of people who are interacting in some way—perhaps with mathematics, or in such a way that we might say there are indications of learning about or with mathematics. "Work" is defined by Arendt as activity intimately related to the conditions of life. If we were to make a few chairs together so that we had places to sit while we meet, any activity that contributes to the creation of these chairs would be considered work. However, if we make a few chairs so that someone else can sell them to yet another person, Arendt suggests that our work is reduced to "labor," a kind of activity that removes the effort and time involved from the conditions of life and of being human beings. "Action," on the other hand, would be if we made the chairs so that we could meet as a group for some purpose, maybe even to specifically learn mathematics, or to use mathematics to learn something else. We would come together in action and form a community of chair makers, whereas work might have been done by isolated individuals.

Mathematics educators and mathematics education researchers might use the distinctions across types of activity from Arendt to analyze the forms that are manifested in their practices. They would be interested in the relationships and potential for mathematics to be a catalyst of action rather than merely work, and hope to encourage forms of activity that avoid the alienating effects of what would be described as labor (see Swanson in this volume). In classrooms, students would come together to create mathematics problems that they need to solve, rather than provide answers on tests that others score for no apparent purpose, disconnected from the classroom experience. In community mathematics circles, intergenerational members might form groups that study together, or groups that hope to change their neighborhoods in particular ways. These people would avoid tutoring sessions that help individuals complete school homework assignments unrelated to the goals of the group, because the production of answers to the homework problems by the tutees are unlikely to contribute in any direct way to something that the individuals would use in their lives (work), nor are they likely to contribute to the creation of the group (action); however, such tutoring (itself possibly characterized as labor by the tutors?) might become part of the broader efforts of a community group if it were designed as a sub-goal of a broader need for the group—for example, as apprenticeship in skills that members need for subsequent, collective action. Whether employed in traditional school contexts or in *avant garde* mathematics experiments, work, labor, and action make sense, and can be used to enrich the experience of those engaged in activity; in the nontraditional environments, they are especially rich in potential. Note that the focus is not explicitly on learning, but instead on the relationships that are established with others in the group and upon the nature of the activity (Biesta, 2006).

Nomadic Topology Example 2: Youth Leadership, Voice, and Participation

Sharon Todd (1997) writes about ways that differences can lead to norms, which clarify what is not the same and therefore highlight differences; norms subsequently construct disparities, the unequal in difference. A focus on disparities in material conditions that structure differences *differently* helps us to avoid collapsing diversity into an individualized, psychologized rendering (Pitt, 2003). It is in the struggles against the disparities of difference that desires are produced, mobilized and frustrated in the pedagogical encounter with difference. Desire can be a new term, referring to those things that ceaselessly circulate among the unsaid, and manifesting itself in expectations, hopes, visions, and fears. Desires are not merely handled or dealt with, writes Todd (1997), but are also produced and constituted. How might disparities and differences be used as nomadic epistemologies that enable a topology of practice? Alice Pitt (2003) proposed youth leadership, voice, and participation as discursive strategies for engaging with disparity and difference. In a school classroom context, teachers and students might together reflect on the ways that they are supporting, or can begin to support, those typically referred to as learners, in assuming leadership in response to disparity, to analyze how desires have been produced and are constituted, to consider how they might more successfully communicate with each other and to others beyond the classroom, and then to participate in their own education in ways that address disparity and difference (Appelbaum, 2007). Those previously thought of as "mere students" would now be considered community leaders in a neighborhood or regional context; and these leaders would make demands upon school personnel and others to provide the sorts of mathematics education and training that they need to address disparity and respond to desires more broadly construed. Whether in a school classroom context, or in a less traditional educational endeavor, the concepts of leadership, voice, and participation can contribute to assessment and evaluation of programs, might be used as strategies to effect successful educational change, or might be employed by facilitators and teachers to reflect on the nature of the experience in their groups. These concepts would replace the currently dominant, globalization focus on accountability, test scores, and national comparisons.

Nomadic Topology Example 3: Architects, Scientists, Artists Using Models

The sculptor Josiah McElheny (2007) has described different stereotypical ways that different professionals tend to use models in their work. Mathematics teachers and students often work with models of concepts, whether they are visual representations such as circle diagrams for fractions, physical objects, such as base-ten blocks for place value, or flow charts for complex algorithms. Given the association, we might learn from McElheny, who notes that an *architect* might use models in any of these ways, but often specifically uses a scale model to convince others of the strength of their proposed design, whereas a *scientist* mostly uses a model to analyze the relationships among the components of a model, and then to test out hypotheses about these relationships. An *artist*, McElheny suggests, might also use any of these approaches; but he notes that an artist has the option of using a model to challenge our assumptions or to raise questions rather than answer them. In a classroom, teachers and students might use models to convince others of a conclusion they have come to, to learn about new concepts, or to raise new questions about the mathematics (Appelbaum, 2012). Members of social circles, NGOs organizing projects, or political actors working to change public policy might similarly use mathematical and other models in a variety of ways. Whether in a traditional classroom, or not, this set of characteristics can be used to create variety of experience, to clarify the purposes of one's situation, or to address specific needs of the group (see Brown in this volume, for further discussion of students relating to different formations of self in order to model their mathematical/learning experience).

Folds of Subjectivity

In each of the above examples, new sets of categories function as nomadic epistemologies. They could be used in traditional school mathematics concepts, or to construct entirely new experiences in any social situation, whether face-to-face in a neighborhood, or through social media across diasporic communities, members of social movements, or moreover, to challenge ongoing research foci and public policy. In this way, they act as alterglobal discourses that are not in opposition to globalization yet might be used to create a different sort of globalization that is not characterized primarily by global markets, accountability rhetoric, and national identity formation. They also can act as tools of psychoanalytic understanding of experience. They begin as nomadic epistemologies, yet subsequently function as folds of interiority and exteriority, as hinges across potential apolitical and potential highly political activity, making a shared discourse that coexists in more than one parallel world of mathematics education activity. The ways that "we" as mathematics education scholars use the ideas—of work/labor/action, disparity/desire, youth leadership/voice /participation, and architect/scientist/artist to conceive of our work, to interact or not to interact with others as part of a broader community of researchers; or as a member of a broader community of social activists; as members of political parties; diasporic communities working for human rights, and so on—create forms of unfolding subjectivity that are never ordered nor disordered, instead always constructed at the core by folds that indicate difference. These differences are, in the words of Todd, related to disparities and desires, norms, hopes, and fears. And in this sense these epistemological differences establish in their folds and unfolding, that is, in our uses of them, an alterglobal, topological space of mathematics education.

Because of the ways that we are integrating epistemology with our topologies of mathematics education, we are able to recognize a kind of agency on our own part. That is, by changing ourselves, through our epistemologies, and then the uses of these epistemologies to construct folds and unfolding that establish topological spaces of mathematics education, we might significantly change the world of mathematics education. Any act of theorizing, reflection, research, etc., creates an agency associated with subjectivity, as described above. The fold is in one sense the name for this relation to oneself as a subject. In other words, another example of a fold is this creation of ourselves as the objects of study, as per the project of the book you have in your hands. The fold is the effect of the self on the self created by the act of looking at oneself, becoming outside of oneself, who is already inside. In this way we can see that a sociopolitical agenda can benefit from a "view from the fold," because any political struggle is going to necessitate a new form of subjectivity, that is, a new set of unfoldings.

Which set of categories do we take as the core folds for the moment? This creates a particular subjectivity. This also could be said to define a scale of observation. Are we at a micro level of a student in a classroom, or at a more macro level of the effects of mathematics curriculum policy on the patterns of refugee migration globally? Are we studying teachers in schools or members of social groups learning the skills of statistical analysis in order to effect change in labor laws? I suggest that the more powerful folds are related to those nomadic sets of terms that can be applied at the most variety of scales, and I further suggest that the three examples in this chapter are good ones for this reason. These potentially powerful folds can come to be "creases" in our fabric of mathematics-education/mathematics-education-scholarship; through repeated folding and unfolding of interiority and exteriority, the "crease" is what I understand as the lasting trace of such repeated folding. The more that researchers consistently carry out the same foldings, this ongoing construction of subjectivity, this application of nomadic epistemologies to change oneself in order to change the world, takes on a "character" analogous to those creases in a person's face that emerge over time along with their life history and personal commitments. What this means is that one's work as a mathematics educator is intimately connected to one's social, cultural, political and ethical commitments. If one hopes to improve the prospects of oppressed communities, or to rescue the planet from the devastation of irreversible climate change, or to make one's professional work consonant with one's views of privilege and disparity, then the character of one's creases are increasingly important.

Threads That Sew the Fabric of Mathematics Education Research and Practice

The metaphor of the fold is especially powerful in helping us imagine ways that seemingly distant points in time and space might be brought together to become one and the same through a simple fold. One can visualize this by randomly drawing two dots on a piece of cloth; no matter which two points on the cloth, and no matter what side of the cloth, a small number of twists and turns will easily make it possible to bring these two points together. If the fold is one of disparity across race, class, ethnicity, nationality, geographic region of the world, we can begin to understand how any two points of data or concern can join through this simple technique of folding. The choice to make these folds creates our subjectivity as a researcher. To focus on issues of fluid identity, diasporic communities that cross boundaries of nation, age, and skill level, or any other category of difference, and to connect with other social movements world-wide, is to work in an alterglobal subjectivity. If one is concerned about ways that women of color might work to gain control of their lives, or the ways that workers can avoid the dehumanization of trans-national corporate policies on labor relations, or global climate change, one can enact this subjectivity by using any or all of the nomadic epistemological discourses above to make a direct connection among the points of the alterglobal movement and, for example, the activity of mathematics teachers and students in schools, or the work of local political parties to affect the support for community agencies. Any or all of these points can directly or indirectly bring into simultaneity mathematical skills and concepts, assumptions and expectations regarding mathematical modelling, and so on. This is not to declare that such educational experiences are to be universally celebrated as "wonderful," but rather to understand, for example, that refugee children assisting adults in the use of mobile phone navigation systems, map reading, and procurement of energy for recharging those mobile devices are both sad examples of what globalization has wrought as well as potentially alterglobal ways that these children might be building communities of connection with others that may, if they are lucky enough to find a way to a new home, inform their lives in the future. The dehumanization is globalization; the parallel possibilities for humanization are alterglobal. And, any use of mathematical ideas to limit the free movement of such children and their families to more positive circumstances would be forms of dehumanizing globalization, while any use of mathematics as tools of enabling people to define themselves for themselves would be considered alterglobal.[4]

My concern is how easily the folds are unfolded, leaving no residual effects of having gone through the folds, that is, losing the trace of such effort, and hence leaving no

[4] De Certeau (1984) might have described trans-national corporations, government agencies, religious and other social institutions as using strategies of power, while individuals and ad hoc groups such as those involved in alterglobal movements would be described as employing tactics that circumvent the strategies. There is a way in which tactics are routinely coopted for strategic use, say, in the name of profit or accountability, while tactics are constantly reinvented in an ongoing alterglobal give and take.

creases. Many fabrics can be folded and unfolded over and over again and one would never know. Anyone who has played with a cloth napkin during a boring dinner has had this experience. On the other hand, an ironed crease on certain fabrics leaves a permanent change that will never go away. How can we fold and unfold so that the points that we bring together stay together long enough to be studied or long enough to work together toward a common goal? How can we pursue the creases of character? Our research projects sew together the fabric so as to keep the points stuck in one place. This generates a repetition of activity that can, if designed well, foster the kinds of creases that began as barely noticeable folds, and then are nurtured to support the ideals of alterglobalization.

To continue the metaphor, if a thread is stronger than the material that it is being used to join it to itself, and then, if seams are placed under strain, the material may tear before the thread breaks. Garments are usually sewn with a thread of a lesser strength than the fabric, so that, if stressed, the seams will break before the garment. Indeed, items that must withstand considerable stress, such as car seats, upholstery, and horse saddles, require very strong threads. A light weight of thread usually results in rapid failure; however, using a thread stronger than the material being sewn will end up causing rips and tears in the material before the thread gives way. The metaphor suggests therefore an important politics of research: if these nomadic epistemological concepts are stronger and more sustainable than the cloth of mathematics education, then the fabric of research and practice we have been pulling together with them will rip and tear before the threads used to sew them together. Instead of experiencing mathematics education differently, we may destroy the world of mathematics education in the process of making particular folds. For example, what might have originally been conceived of as hopeful, positive, nomadic categories of accountability and progress as measured through test scores of learning outcomes have torn the fabric of many school systems, leaving serious disparities in the social fabric (see, e.g., Jablonka and Bergsten in this volume). Can we repair this fabric with threads that are no stronger than the fabric itself? Can we explore the potential of nomadic epistemological discourses for their flexibility, strength, and ability to fray when necessary? Might Youth Leadership, Voice, and Participation or Models for Convincing, Analyzing, and Challenging Assumptions be just the right balance? Might there be other threads with just the right amount of braiding to hold points of interest together that would otherwise be very far apart? This chapter invites you to join in this research program, and to help figure this out, to bring a new world of mathematics education and mathematics education scholarship.

Enacting the Fabric in Artistic Action: An Illustration

The idea of the actions, inspired by the mission of the Spiral Q Puppet Theater of Philadelphia,[5] is to use mathematics to build strong communities characterized by joy, a can-do attitude, and the courage to act on their own convictions. In this work,

[5] http://www.spiralq.org/.

mathematics is understood as the "art" that is the center of the work of the Spiral Q, and as happens with the Q, mathematical/art techniques are learned, developed, explored, discovered, invented, and so on, as both the purpose and the peripheral features of the community activity. Nothing is first, central, or directive in this work, other than the ongoing community-building, focused on supporting action based on the convictions within the group. Arendt and McElheny come together here in advocating a particularly new perspective that artistic practices are unique and powerful, with McElheny's suggestion that the artist's use of models can challenge our assumptions or raise questions, rather than answer them. Viewing mathematics as political and conceptual "art" is in this manner a nomadic epistemology, independent of common mathematics education practices. That is, such a view of mathematics education might or might not be possible in traditional school mathematics classrooms—it can be a radically different view of a coexisting view with a school curriculum. The unfolding topology of mathematics education blossoms here with new discourses and possibilities that might conform in some ways with a school mathematics curriculum, yet also radically differ from and challenge such a curriculum. Once one understands mathematics as the art that co-constructs relationships in evolving communities, one employs concepts to map out what is going on in terms of mathematical art and audience, mathematical ideas and applications, models and metaphors for things that one wants to make sense of, and so on.

So if we ask, "How are people working with mathematics as a medium in this group effort?" or "How are the people here using mathematics as a tool to craft their performance?" then we are understanding our work through the lenses of Arendtian "action" as "artistic practice"—the artful mathematics action is primarily about community building, and secondarily about other realms of activity. At the same time, McElheny-ish use of models sculpts new worlds of mathematics/art—in contrast to the domination of mathematics as tool of argument (the metaphoric architect) or analysis and prediction (the metaphoric scientist). We are changing as mathematics educators because we are thinking (primarily) about different things: for example, when we are looking for mathematics as art that is building community, we are still concerned with artistic techniques and materials (mathematics skills, concepts, and facts, mathematical models, and so on), but only in the context of making an effort to promote mathematical action as opposed to work or labor. And when we are considering mathematics as art that uses models to challenge assumptions, we are exploring the relationship between our own individual attention to aspects of mathematical skills, concepts, and facts, and the ways that we share them through representations with others. When, where, and how do we witness mathematical action? Nelson Goodman (1978) is helpful here. Instead of defining art (which for us would be mathematical action), he used a Wittgensteinian, open set of characteristics not shared by all members of a category, but united by strands of common, "family resemblances," likely properties that I propose, do indeed make many types of mathematical action sound appropriately characterized by art: Art activities tend to contain symbols while also being symbols themselves. When is a thing a work of art? When is an ephemeral moment an aesthetic experience? These things tend to be *replete*. That is, more of its properties are important than when this thing is not functioning as a work of art.

Winner shares the example of a zig-zag line. If this is the printout of an electrocardiogram, the reader focuses on the ups and downs; if we are told it is the outline of a mountain in a landscape painting, we start to observe subtle properties of the line—thickness, brightness, color, and so on. If we pick up a stone from the ground, it is one kind of stone; placed in an art museum, a stone is experienced as a stone with shape, color, texture, location, size, etc., Another property from Goodman is sometimes referred to as *expression*. This refers to the metaphorical exemplification that is symptomatic of aesthetic things and events. Works of art express something other than what they are—emotions, interpretations, temperature, questions, and so on. A painting, for example, might suggest happiness, or social injustice, well beyond the actual, specific objects depicted; choice of color, hue, compositional juxtapositions, allusions to other well-known works of art, and so on, contribute culturally bound possibilities for expression. Art may not carry both of these symptoms, and we might identify some things that have these properties that are not art, but art tends to have these properties. Mathematics as experienced in and out of educational environments can also often be described with repleteness and expression. I claim that worthwhile mathematics experiences share these properties. In this way, I argue that mathematics can and should be understood as aesthetic experience (Appelbaum, 2012).

I conceive the educational space as studio, and the mathematical work in the community as a performative, participatory action, in an attempt to explore this potential of (nomadic?) arts discourses to be folds in a new subjectivity. In this discourse of the arts, we can accept the role of the model, in McElheny's sense, along with the place of the deception, as part of the aesthetic—we accept the absurdity in order to perform the parody, we use the model to carry out a sales meeting and convince someone of the value of our proposal—but for a serious alterglobal commitment to dignity and recognition, we need to question how the asymmetry of information within this aesthetic leads to things not consistent with this dignity and recognition. Teacher and student in a classroom? Hardly the only option. McElheny suggests two others! McElheny's options are taking place in an idealized, perfect democracy, where everyone has the right to be heard—I suppose we can be catalysts for ideal spaces in our own work, but the alterglobal project of making our world a better place is an ongoing struggle. The arts performance discourse highlights the role of "audience" in small pieces of this struggle, a potential market for what is produced, and a politicized distinction between what might be considered "good work" and "work that gets the audience it needs." It makes the role of the mathematician/artist one to reflect on as part of the mathematical/aesthetic work: are we explaining to an audience? Provoking our audience to action? Setting up distinctions between people who already know certain mathematical facts and procedures, and others who do not, that is, an asymmetry of skills, culture, or power? "In relationships with bigger stakes [that] exhibit the same asymmetric information distribution (confidence scams, are an obvious example), the inherent misrepresentation turns sinister." (Kim-Cohen, 2002, p.16). More to the point, such a lack of radical democracy, according to Rancière, undermines most prospects of both learning and taking action (Appelbaum, 2012; Ranciére, 1991, 2010). If we begin with the alterglobal commitments, we not only reframe mathematics education, but education, knowledge, power, mathematics, action, collaboration, and so on, in general.

Disordered Order in Ordered Disorder

What is important about the particular nomadic epistemological examples in this chapter is not so much their specific power to transform researcher subjectivity, but the ways that they model the potential application of any nomadic concepts. That is, the more general tactic is to work with concepts that are independent of insider-outsider discourses of mathematics education, to in turn create a nomadic topology of mathematics education that can bring together seemingly unrelated points of interest or moments that appear far apart through a folding of space and time. For example, in my more recent research/community development projects, I enter ongoing social projects, and use what we are collaborating on to support the work of that community, while also seeking to interrogate that very community for its potentially harmful forms of exclusion. The nomadic terms here are community development, exclusion, and social projects, which exist independently of traditional mathematics education practices. The particular example of artistic action is shared in this chapter as one illustration for how mathematics can become the art that supports the development of communities characterized by creativity, joy, and the courage to act on one's convictions. These community projects are sometimes explicitly educational, and some involve mathematically informed applications, but others are more appropriately described as aesthetic encounters or carnivalesque performances. Once we become comfortable with this simultaneity of embracing nomadic terms as coexisting, we can work with the nonmainstream nomadic epistemologies as topologies of a fascinating fabric of mathematics education that bring together formerly distant ideas, events, moments in time, space, and positions in theory or practice, into events of connection and simultaneity. The nomadic topologies hold our fabric in place long enough for us to study these events of simultaneity, so that we can change our own subjectivities as scholars, and in the process bring forth new worlds of research and understanding that are grounded in these folds of space and time.

It becomes possible, for example, to study the forms of interaction between a teacher's uses of models of concepts and students' forms of participation in communities of action outside of a school as related to the conceptions of leadership possible in a neighboring community's attempts to establish more action and less labor in their elementary curriculum. It can become routine to see policy-writers on mathematics education working across diasporic forms of dialogue to bring forth a program about, say, the mathematics of health education for elderly in Scandinavia, while the same policy-writers themselves may be focused on the forms of work and labor in the implications of their assumptions about school texts that emerge for employees of a paper factory in Ecuador. Rather than what mathematics should be taught and learned to serve social justice, we might instead ask, how might mathematics be involved in alterglobal social movements in ways that serve the movements' commitments to human dignity and recognition? These movements defer the definition of an individual to the ongoing absence of required categories. Contrary to this chapter's emphasis on our own sense of ourselves as scholars, alterglobal movements do not expect participants to become outside of themselves and define who or what they are, and they do not want to define people as subjects, agents and organizers. They instead accept people as who they are for themselves, and work together on joint projects.

There are, however, a couple of psychoanalytic issues that arise in this sort of work. First, one can identify forms of resistance, discussed earlier in this chapter, in oneself, as this atypical community research work takes place. We resist our new subjectivities and fall back onto traditional notions of teaching and learning, mathematics and non-mathematics, and so on. The powerful theoretical move from psychoanalytic theory is to recognize resistance as both the experience and the response to the experience, the problem and its solution. Resistance to new subjectivities fostered by nomadic epistemologies is simultaneously "irritation" and method. Suppose, for example, that we are working with a neighborhood community surrounding a school in the U.S. to prepare a parade and pageant inspired by celebrations that take place in the mountain villages of Bolivia. On the one hand, this is a seemingly unusual location for mathematics education. Yet, if we use photographs and video to document mathematics concepts and skills that are developing over time among the teachers and students, we are merely using what might have been new forms of pedagogy to re-inscribe traditional subjectivities, rather than shifting the important focus to the ways that mathematics is, for example, building community through skills and concepts. We are in such cases using nomadic terms to work within mainstream discourses, that is, to resist change, while actively and purposely attempting to work at the same time in ways that do not fall within mainstream ideological expectations, that is, to understand our resistance to change. This raises a second psychoanalytic concept of "regression to the norm," in which we fall back on prior, frequent forms of interpretation and action, in this case, proving that we are meeting others' objectives, rather than using our newer ways of being in the world. Yet another psychoanalytic concern is transference. Suppose a group of youth from an urban LGBTQ center ride a bus route together and design activities that would provoke passengers to use mathematics to increase awareness of gender and sexuality fluidity. The mathematics they use and learn about in the process might be thought of as the medium through which they build community and relationships, perhaps constituting "action" following Arendt. At the same time, the ways that mathematical concepts and processes are represented and the nature of the unfolding mathematical dialogue with the bus riders who find themselves on the same bus as these young mathematical artists might be understood as establishing models that do any or all of the following: convince others to support the activity or even to join in and participate (the architect), assist others in understanding gender and sexuality relationships in particular ways (the scientists), or open up challenges to riders' assumptions about gender, sexuality, mathematics, or anything else (the artist). Are the youth and their broader bus community applying mathematics learned in school, or creating mathematical understanding that might be brought back into school experiences, or something else? How we and the youth answer such a question has more to do with us and how we project our own mathematical understandings upon others than the responses of the unsuspecting audience on the bus.

Reprise: Two Points of Arrival That Become Points of Departure

Back to my opening paragraph: "our" project makes identity a weird thing, because identity only exists through and in action. The project messes up previous understandings of place because the meaning of "this place" changes in the ways that it is connected, physically, virtually, and semiotically, with other places, spaces, and constructs—as well as bringing any things, ideas, events, relationships, and so on, together "at the same place," folds bring any place together with any other place. The linear direction of research and practice, teaching and learning, and so on, even if located in a circular and reflexive form, is no longer "sensible" for grounding our work. Instead, we have complex networks of interconnected changes in links and nodes that form a crystallization of what is happening, but which never capture the lines of flight that are components of that crystallization. Two things made this possible: (1) an abandonment of expectations and assumptions regarding mathematics, mathematical action, teaching and learning, and their locations; and (2) a willingness to work with sociocultural and psychoanalytic theories outside of the mainstream of educational discourse. In the language of de Certeau (1984), one might say capitalist expansion made this interconnectivity possible via institutional *strategies* of trans-national corporations, imperialist popular Western cultures, nation-state politics, and so on, while our alterglobal *tactics* of mathematics education dream that a parallel, coexisting, other world of practice and politics is possible. But wait! Is it not bad form to suddenly introduce new concepts, like strategies versus tactics, in the final paragraph of a coherent chapter? And, would not a binary set of terms cry out for a nomadic alternative, terms that might coexist with the institutional/strategic-alterglobal/tactical dichotomy yet remain independent and "nomadic"? Perhaps, but this chapter is calling for points of arrival that at once become points of departure, and in this spirit, we are already on our way to a new nomadic journey. Like other alterglobal movements, we refuse to describe a clear set of goals or methods, other than the search for "another way."

References

Appelbaum, P. (2007). Afterword: Bootleg mathematics. In E. de Freitas & K. Noland (Eds.), *Opening the research text: Critical insights and in(ter)ventions into mathematics education*. New York: Springer.

Appelbaum, P. (2008). *Embracing mathematics: On becoming a teacher and changing with mathematics*. New York: Routledge.

Appelbaum, P. (2012). Mathematical practice as sculpture of utopia: Models, ignorance, and the emancipated spectator. *For the Learning of Mathematics, 32*(2), 14–19.

Appelbaum, P. (2015). Service-learning as teacher education. In U. Gellert, J. Gimenez Rodriguez, C. Hahn, & S. Kafoussi (Eds.), *Educational paths to mathematics: A C.I.E.A.E.M. sourcebook* (pp. 409–424). New York: Springer.

Appelbaum, P., & Davis, B. (2013). Curriculum for disobedience: Raising children to transform adults. *Journal of Curriculum Theorizing., 29*(1), 134–172. http://journal.jctonline.org/index.php/jct/article/view/420/pdf_1.

Appelbaum, P., & Gerofsky, S. (2013). *Performing alterglobalization in mathematics education: Plenary in the form of a jazz standard.* Opening Plenary Address. CIEAEM (International Commission for the Study and Improvement of Mathematics Education). Turin, Italy. Retrieved July 21–26, 2013.

Arendt, H. (1958). *The human condition*. Chicago: The University of Chicago Press.

Biesta, G. (2006). *Beyond learning: Democratic education for a human future*. Boulder, CO: Paradigm Publishers.

Butler, J. (1997). *The psychic life of power: Theories in subjection*. Palo Alto, CA: Stanford University Press.

Butler, J. (2010). *Frames of war: When is life grievable?* New York: Verso.

Davis, B., Mausbach, W., Klimke, M., & McDougall, C. (2010). *Changing the world, changing oneself: Political protest and collective identities in West Germany and the U.S. in the 1960s and 1970s*. New York: Berghahn Books.

de Certeau, M. (1984). *The practice of everyday life*. Berkeley, CA: University of California Press.

Deleuze, G. (1993). *The fold: Leibnitz and the baroque*. Minneapolis, MN: University of Minnesota Press.

Deleuze, G., & Guattari, F. (1987). *A thousand plateaus: Capitalism and schizophrenia*. Minneapolis, MN: University of Minnesota Press.

Goodman, N. (1978). *Ways of worldmaking*. Indianapolis, IN: Hackett.

Guattari, F., & Deleuze, G. (1972). *Anti-Oedipus*. New York: Continuum.

Kim-Cohen, S. (2002). *On cats in bags: Asymmetric information, the cohort, and the discourse of the magician*. London: The Royal Academy of Arts. http://www.kim-cohen.com/Assets/Texts/Kim-Cohen_Cats%20In%20Bags%20(2002).pdf.

McElheny, J. (2007). Artists and models. Video of March 2007 presentation at the Museum of Modern Art, New York. http://www.gvsmedia.com/video-2/w-lFT7CCuT8/JosiahMcElheny-presenting-at-MoMA. Last visited October 2008.

Pitt, A. (2003). *The play of the personal: Psychoanalytic narratives of feminist education*. New York: Peter Lang.

Pleyers, G. (2010). *Alter-globalization: Becoming actors in a global age*. Cambridge, UK: Polity.

Ranciére, J. (1991). *The ignorant schoolmaster: Five lessons in intellectual emancipation*. Palo Alto, CA: Stanford University Press.

Ranciére, J. (2010). *The emancipated spectator*. New York: Verso.

Todd, S. (1997). *Learning desire: Perspectives on pedagogy, culture, and the unsaid*. New York: Routledge.

Winner, E. (1982). *Invented worlds: The psychology of the arts*. Cambridge, MA: Harvard University Press.

Chapter 17
Reflections on Research Positioning: Where the Math Is and Where the People Are

David Wagner

Abstract Using positioning theory and functional grammar, I reflect on the way I position myself when my research in mathematics education is published. I consider the way my authorship addresses the field, focusing on the distinction between scholars who attend to the sociopolitical context and they who ask of this research "Where is the math?" I identify a range of discourses in which mathematics might be located. Throughout this reflection I draw on two of my publications for examples. Finally, I suggest some tools for reflecting on the positioning and discourses at play in a research situation.

Introduction

In this chapter, I reflect on the way I position myself when my research in mathematics education is published. My reflection has and will continue to guide my work but I share it here to offer possibilities for reflection that could be used by my mathematics education peers. My hope is that such reflection can help us develop perspectives beyond dominant ideological storylines, and thus add depth to our work and to our field.

To be reflexive, I draw on methods that I have used to analyse other people's discourse—primarily teachers and students of mathematics. Much of my work is rooted in positioning theory (e.g., Harré & van Langenhove, 1999) and functional grammar (e.g., Martin & Rose, 2005). For this reflection, I focus on positioning. Nevertheless, researchers using positioning theory often use critical discourse analysis tools, especially referencing functional grammar, to denaturalize structures that are otherwise opaque: "the orderliness of interactions depends upon taken-for-granted 'background knowledge' [which] subsumes 'naturalized' ideological representations, i.e., ideological representations which come to be seen as non-ideological 'common sense'" (Fairclough, 1995, p. 28). As researchers, we can identify our positioning among participants by attending to our language in our reporting and in our research interactions. This is a productive approach, though not completely

D. Wagner (✉)
Faculty of Education, University of New Brunswick, Fredericton, NB, Canada
e-mail: dwagner@unb.ca

H. Straehler-Pohl et al. (eds.), *The Disorder of Mathematics Education*,
DOI 10.1007/978-3-319-34006-7_17

straightforward. As Fairclough (1995, p. 71) noted, "it is not possible to 'read off' ideologies from texts. This is because meanings are produced through interpretations of texts and texts are open to diverse interpretation, and because ideological processes appertain to discourses as whole social events." Positioning theory helps with interpretation of texts.

I preface the reflection with an overview of key aspects of positioning theory. From this I reflect on the position of authorship in general and scholarly authorship in particular. This reflection focuses attention on what I consider to be a significant divide in the field of mathematics education. The divide becomes most evident when peers ask "where is the mathematics?" I suggest some possibilities for seeing this question in new light. This leads to consideration of the range of discourses (beyond mathematics), in which mathematics might be located, and questions about my responsibility as an author to locate myself in these discourses. Within each of these approaches, I draw on two of my publications for examples of applying the reflection. Finally, I suggest some tools for reflecting on the positioning and discourses at play in a research situation.

Positioning Theory

The common thread in this volume is a focus on the "disorder of mathematics education". I argue later that the idea of disorder is problematic. Nevertheless, there are different possible orders in any given discourse or context. Positioning theory, as elaborated by Harré & van Langenhove (1999, p. 1), is described as the "study of local moral orders". A central idea in the theory is that there is a range of *storylines* available for interpreting any interaction. Interactions tend to follow already established patterns of development, which are called storylines. These storylines can be shared culturally or invented as participants interact. Either way, they develop in local interaction. In this way, they are like Foucault's (1972, p. 52) sense of discourses: "practices that systematically form the objects of which they speak".

Each storyline suggests a moral order to its interactions and associated positionings, and these positionings carry "rights and obligations of speaking and acting" (Harré & van Langenhove, 1999, p. 1). Because storylines and positioning are contestable and may be negotiated, they can shift during an interaction. Also, multiple storylines can be at play at any time. Interlocutors may imagine different storylines and they may shift their orientation to storylines or positions, whether or not they explicitly talk about roles and responsibilities in the exchange.

In my view, the most powerful aspect of positioning theory is its radical focus on the immanent. It rejects the presence of external forces in an interaction. With reference to Saussure's distinction between discourse practice and discursive systems, Davies and Harré (1999, p. 32) claimed, "La langue is an intellectualizing myth—only la parole is psychologically and socially real." They meant that we ought not to focus on disciplinary forces in an interaction—no mathematics traditions, no mathematics education field. Instead, they promote a radical focus on the way actual human interlocutors interact. This willful ignoring of outside forces enables emancipation because nothing outside the interaction actually holds force within the interaction.

Along with Beth Herbel-Eisenmann (Wagner & Herbel-Eisenmann, 2009), I noted that many references to positioning in mathematics education identified how people were positioned in relation to mathematics. Such analysis does not fit the theory. Thus, we proposed a way of reconciling with the reality that discourse systems do manifest themselves in local interactions: systems of discourse are present but only as mediated through the people and artefacts involved in the interaction. We later provided further conceptual tools for identifying a wide range of discourses at play in any interaction (Herbel-Eisenmann, Wagner, Johnson, Suh, & Figueras, 2015). Too often researchers discussing positioning identify only one storyline to describe the interaction. Exceptional research has identified two competing discourses in an interaction (e.g., Esmonde & Langer-Osuna, 2013). But as we pointed out, there are many more discourses at play or potentially at play in any situation (Herbel-Eisenmann et al., 2015).

Positioning and Authorship

As a mathematics educator, I position myself both in face-to-face interactions and in my writing. In this chapter, I have chosen to reflect on my writing, though I recognise that the positioning of my writing connects to face-to-face interactions (mostly at conferences).

In her musings about authorship, acclaimed Canadian novelist Margaret Atwood (2002, p. 126) asked, "For whom does the writer write?". She concluded that any writer writes for the reader—"For the reader who is not Them, but You. For the Dear Reader. For the ideal reader, who exists on a continuum somewhere between Brown Owl and God" (p. 151). "Writing for Brown Owl" refers to her early experiences writing for a woman she loved—a known individual. "Writing for God" refers to an Isak Dinesen story in which a writer needed to escape the stifling demands of his readers. Either way, writing is driven by readers—to face them or to escape them. Atwood's musings are reminiscent of Barthes' (1968/1977, p. 148) essay "The Death of the Author". He directed attention toward the agency of the reader and away from the agency of the author: "Classic criticism has never paid any attention to the reader [...]. We are now beginning to let ourselves be fooled no longer by the arrogant antiphrastical recriminations of good society in favour of the very thing it sets aside, ignores, smothers, or destroys."

While Atwood and Barthes foreground the reader to counter the prevailing focus on authors, I argue that both writer and reader have agency. I interact with my readers in a complex way, which I want to explore below using positioning theory. Nevertheless, Atwood and Barthes caution writers to let our texts speak and to recognise that they are separate from our identities. "Writing is that neutral, composite, oblique space where our subject slips away, the negative where all identity is lost, starting with the very identity of the body writing" (Barthes, 1968/1977, p. 142). But what about the identity of the reader? Atwood used a poem by Emily Dickinson (1960) to raise this question: "I'm Nobody! Who are you? Are you—Nobody—too? Then there's a pair of us!" (poem #288).

Eco's (1994, p. 9) theorization of the relationship between author and reader may help. He described how texts create a "model reader—a sort of ideal type whom the text not only foresees as a collaborator but also tries to create." The text addresses the needs of a real reader enough to transform him or her into the reader imagined by the text. It is the text, and not the author, that imagines and addresses the reader because a text constructs a model reader regardless of the author's intent. Nevertheless, as a writer, I may design my texts with an intention to imagine and address a particular model reader. I write my model reader into being.

Eco (1979, p. 62) theorised different kinds of model readers. A *closed text* imagines and constructs a single reader. It recognises only one interpretation. By contrast, in an *open text* "the author offers [...] the addressee a work *to be completed.* [The author] does not know the exact fashion in which his work will be concluded, but he is aware that once completed the work in question will still be his own." The text invites the reader to choose from a variety of interpretations. Thus, as a writer, I could make choices in my writing to construct text that opens up multiple points of view or I could try to force or seduce my reader to a singular interpretation. I have examined such choices in the context of writing a mathematics textbook (Wagner, 2012).

Eco's closed and open texts relate to Bakhtin's (1975/1981) notions of unitary language and heteroglossia, respectively. Bakhtin described how unitary language acts like centripetal force by pulling meaning to a unified centre, while heteroglossia acts like centrifugal force by pushing out from shared meaning to say something new. While Eco and others (e.g., the research tradition of appraisal linguistics) have evaluated text as either open or closed, Bakhtin explained how both forces are always present: "Every concrete utterance of a speaking subject serves as a point where centrifugal as well as centripetal forces are brought to bear" (p. 272). In order for there to be meaning, a text has to align sufficiently with the reader's experience (the unitary force), and it is impossible to write something that does not do something new (pushing out).

The complementary nature of heteroglossia and unitary language is central to my positioning as I conduct research and write about it. I may wish to write text that will *push* my readers to see something new but that will not be possible without *pulling* them in. As elaborated by Barthes (1968/1977, p. 142), "the birth of the reader must be at the cost of the death of the Author." Atwood (2002, p. 140) addressed this idea by focusing on life instead of death:

> One of my university professors, who was also a poet, used to say that there was only one real question to be asked about any work, and that was—is it alive, or is it dead? [...] living things grow and change, and can have offspring, whereas dead things are inert. In what way can a text grow and change and have offspring?

Barthes proclaimed the death of the author. Atwood identified the life in good text. As noted in many religious and other traditions, death is a requisite for new life.

Who Is My Reader?

There is a technique I use in my writing that arose out of my reading of Eco's model reader theory over a decade ago. I am sure that the technique was elaborated through my conversations with others. I consider the venue of my writing and imagine the prospective empirical readers. I choose someone who represents for me the audience I want to address. I imagine that I see a miniature version of that person sitting on top of my computer monitor. And I write to that person. I try to write in a way that addresses the experiences of that person, pulling the person in, while also pushing this representative of a group to see something new. This technique often helps alleviate writer's block, but more importantly, it also shapes my language choices to address the kinds of readers I want to address. In this case, addressing them means to respond to them (pulling them in) while also hoping to open new possibilities for them (pushing them).

In most of my work I choose to address mathematics educators who do not work with critical theory. This is because an aim of my work is to expose power relations. My colleagues who cite critical theory tend to see these power relations already. I would rather focus my efforts on people who do not see these power relations yet. My intent is to bring disorder to the assumptions of people who valorize mathematics uncritically. And, I am well-acquainted with these assumptions because I was one of this relatively uncritical crowd before perspective-challenging experiences opened my eyes. My memory of this perspective is noted in this piece of autobiography I wrote many years ago, which I continue to use for self-identification on my website (http://davewagner.ca/):

> Prior to doing graduate studies, I taught grades 7–12 mathematics in Canada for six years and in Swaziland for two and a half years [...] It was the experience of teaching mathematics in Canada, then Swaziland, then Canada that alerted me to the highly cultural nature of mathematics teaching, which I had thought was culture-free and values-free. This experience prompted me to leave teaching to investigate the cultural nature of mathematics and the impact of mathematics teaching practices on individuals and society.

To provide an example of how I address relatively uncritical readers, I now consider two articles I wrote with Beth Herbel-Eisenmann (Herbel-Eisenmann & Wagner, 2010; Herbel-Eisenmann, Wagner, & Cortes, 2010). We used quantitative empirical research to illustrate how students' experiences are shaped by positioning in mathematics classrooms. I believe that the use of quantitative data may avert prospective complaints about subjectivity that are often directed toward sociocultural analyses. While I do not share this complaint, I recognise that many educators have the complaint, and in the past I had the same complaint. Thus, I take it seriously.

This turn to quantitative evidence buys into discourses that favour the so-called objective research. When we justified our methodology, we obscured our agency, and thus positioned our work as objective (Herbel-Eisenmann et al., 2010, p. 9):

> Our lexical bundle analysis in this article is distinct from the above studies because neither we, nor teachers, nor students identified the focus of analysis. Rather, lexical bundle analysis was designed to find patterns in a large set of transcripts, which are identified empirically using a special computer program that works on a corpus of texts. Lexical bundle analysis identifies what one may not otherwise notice about the mundane yet important language patterns.

I wonder whether our positioning choice undermines other sociocultural research by appealing to quantitative evidence or whether it supports that research with empirical evidence that inarguably demonstrates the pervasiveness of status-based authority structures and paucity of open dialogue in mathematics classrooms. It probably has both effects.

To provide another example of addressing readers, I consider this chapter, for which I imagine a different audience (whom I usually ignore in my writing). I write for you, "Dear Reader" (c.f. Atwood, 2002, p. 151)—someone who already pays attention to power relations. And I write for myself as one such person. I intend for my writing to pull us in and push us to new perspectives. To illustrate, the miniature person on top of my monitor as I write this chapter is someone who participated in the conference that formed the interactions that underpinned the volume. The miniature person sometimes changes with another conference participant every once in a while—it is not always an intentional choice for me. Sometimes, I imagine myself speaking to the group in that fancy board room in Berlin. By contrast, when I wrote with Herbel-Eisenmann about our quantitative research, the miniature person on my monitor was never someone who would initially care about the "disorder of mathematics education". And if they would care, they would likely identify different "disorders" than the ones identified in this volume.

Where Is the Mathematics?

The distinction between my imagined readers in the two works described above warrants attention to a split within the field, generally between scholars who attend to sociocultural aspects and they who do not. To illustrate, I recall an experience at a recent conference. A couple of days before the plenary panel discussion at the 2014 Psychology of Mathematics Education (PME) conference (Halai, 2014; Setati Phakeng, 2014; Valero, 2014; Wagner, 2014; Walshaw, 2014), the panellists sat under a tree to organise the discussion. Having already read each other's papers, we discussed how we would best allocate time for each panellist's presentation and responses to each other, while still recognising and responding to the questions and comments from our colleagues in the audience. Inevitably, there were tensions among our various approaches to the frame we had agreed to address, but we were more concerned about the "Where is the math?" crowd in the audience than with the differences among ourselves. We had been selected for the panel because of our work with equity and we all had experiences with people questioning the place of this work in the mathematics education research community. Indeed, even the conference organisers in communicating feedback from the PME International Committee had explicitly asked us to focus our panel discussion more on mathematics.

I think there were at least two things going on with our worry. First, it does not feel good to have one's work rejected, and we knew many in the crowd would quickly reject our work for not being sufficiently "mathematical". Second, having access to this crowd gave us an opportunity to catch their attention and convince them of the importance of sociocultural work in our field. The first of these concerns relates to our audience's acceptance of our work, whether or not we would be pulling them in. The second concern relates to the possibility of moving them, pushing them.

I have heard this "Where is the mathematics?" question often in informal discussions at conferences—both from scholars who ask the question and scholars who are frustrated by others asking the question. But the question is rarely addressed in formal settings. And so I use the question to ask, what is the state of order in mathematics education? Or, as positioning theory would ask, what is the moral order? How do we listen to each other?

As noted in the opening chapter of this volume, the field is not unitary and static: "Regardless of whether the attention to the sociopolitical dimension of mathematics education is to be rated as a "shift of paradigm", a "turn" or rather as the development of a new "branch", such dimension has been gradually recognised as an important part of mathematics education research. It is about to become institutionalised as a firm strand of mathematics education" (Straehler-Pohl et al. in this volume, p. 2). The claim is true in some respects but it is also problematic. Valero (2004) identified a sociopolitical turn in the field, but Gutierrez (2010, p. 4) identified a divergence: "[W]hile many mathematics educators are comfortable with including social and cultural aspects in their work, most are not so willing to acknowledge that teaching and learning mathematics are not politically neutral activities." Which is it? Has mathematics education as a group turned together, or have two subgroups turned away from each other? Either way, the opening chapter's claim that work within the sociopolitical paradigm "is about to become institutionalised" (see above) warrants some alarm bells. In Canada and perhaps elsewhere, when we say someone is institutionalised, we often mean that they are being sent somewhere against their will—e.g., a prison, nursing home, or psychiatric hospital. Indeed, the opening chapter warns about regime change.

The tension we recognised at PME is significant to the discussion in this volume about the disorder of mathematics education. As identified in the opening chapter, "Scholars with a 'disorder' can thus be humorously understood as those who appear not to function in the way they are supposed to; a way that is not aligned with some of the most unquestioned assumptions in mathematics education" (Straehler-Pohl et al. in this volume, p. 13). The editors of this volume have identified, for example, the assumptions that "the idea that mathematics is important for the daily life of people or the enticing goal of 'mathematics for all'" (ibid.). Significantly, they have invited contributors to identify other assumptions.

I want to take this consideration of disorder in another direction. There are various ways that "disorder" or "out of order" are used. As suggested in the opening chapter, *disorder* can refer to a condition that identifies someone as abnormal or diseased. Most literally, "out of order" means out of sequence, or in other words a mess—situations in which things are not in their correct place. There is also an expression in English in which we say something is "out of order" if it is not operable (i.e., it is not in working order). The literal meaning connects to position, but the other meanings connect to positioning theory's metaphorical use of position to illustrate moral orders—in one case judging against some idea of normal characteristics, and in the other case judging against some idea of healthy system operations.

Which of these numbers is out of order—2, 4, 8, 6, 10, 12? If you say 8, I would ask why not 6? If you say 6, I would ask why not 8? If you say 8 and 6, I would ask why not 2, 4, 10, and 12? Maybe the "correct" order is 12, 10, 8, 6, 4, 2, in which case 8 and 6 are the only numbers in order. What makes an order correct? Instead of picking out members of the sequence as being out of order, it may be more accurate to say that the whole sequence is out of order. It is in disorder. Or, better yet, I might say that I cannot identify an order in that sequence (acknowledging that someone else may identify an order), or that pattern and predictability are not a characteristic of the set. When I identify disorder, I am making a value-laden choice. I might try to identify the improper elements in a set to identify the disorder, I might point attention to the whole system as being in disorder, or I might claim that order is not appropriate to the system.

For my reflection on the state of mathematics education, I want to use positioning theory's rejection of norm language and its preferred approach to identifying competing storylines that could be at play in a situation. That is what I want, but it is hard because I am a player in the interaction.

Around the time that the *Journal of Research in Mathematics Education* (JRME) was soliciting and publishing papers for its special issue on equity, editor-in-chief M. Kathleen Heid (2010) published a controversial editorial entitled "Where's the math (in mathematics education research)?" She connected the question to a catchy television advertisement from the 1980s, in which fast-food restaurant patrons looked in their hamburgers and complained, "Where's the beef?" She then provided apparently positive examples of "findings related to mathematical understandings" (p. 102), which comprise research that focuses on "specific transition points in students' mathematical work" (ibid.), "global understandings that underpin students' mathematical thinking" (ibid.), "the effects of new instructional strategies" (ibid.), "theory to explain why some mathematical topics are notoriously difficult to teach" (ibid.), "mathematical issues that arise even in classrooms taught by mathematics experts" (p. 103), and "teachers' understanding of mathematics" (ibid.). Though Heid did not say so explicitly, her editorial frames a deficit-based assessment of mathematics education research, implying that something is missing in some (or possibly, many) submissions to the journal and other venues for dissemination within the field. She gave no examples of research with insufficient "mathematics" but she made it clear that her idea of good mathematics education research talks explicitly about mathematics.

There are other ways of thinking about the "Where is the mathematics?" question. The way one inflects the question has significant implications. If we emphasise the word *mathematics;* we are asking Heid's question, which seems to provide an excuse to ignore research that takes a sociocultural perspective. In the context of journal article review processes, the question may justify rejection of an article, in which case the question becomes a filter for membership in the group.

If we instead emphasise the word *where,* the question changes radically. A lot of research looks closely at the mathematics, the concepts within it, and how children develop these concepts. I, along with others, who use sociocultural/political theories, worry that this research does not recognise the sociocultural context of the mathematics under investigation. When we ask, "*Where* is the mathematics in your research?" we are complaining, "You are looking at the mathematics but telling us nothing about *where* it is." This complaint might be taken as our excuse to ignore research that does not consider the sociocultural contexts of the mathematics. We may reject research that does not properly situate its analysis. Thus, this form of the question is also a deficit-based evaluation of research.

The result of these deficit-based evaluations from the various strands of mathematics education is that scholars may be ignoring each other's work. This kind of relationship is a symptom of a disorder in the field or of a field in disorder. We might say it is poor communication—perhaps not for a lack of ability, but rather from a lack of interest.

There are other ways to answer the "Where's the math?" question. For example, I might answer as I did in my response to Paola Valero in the PME Plenary Panel discussion. This response relates to my way of seeing the presence of discourses/disciplines in positioning theory; the discourses exist only in the people that embody them. Valero (2014, p. 76), drew on her earlier research studies to show how "school mathematics practices govern children, effect classifications, and inscribe in them forms of reasoning about themselves." In response, I offered an example of someone who was inscribed by mathematics. I answered the question "Where is the mathematics?" by saying, "It is right here, in me." Mathematics is in all of us who were at the Disorder of Mathematics Education conference in Berlin, and in all of us who were at the PME conference in Vancouver. Mathematics is an army of billions trampling this earth. This army does some good things, but it also does some terrible things. When I reflect on the good and the bad, I am thinking of D'Ambrosio's (e.g., 1994) promotion of reflection on the good and bad of mathematics. The army is seductive. It is easy to think of the army as good. Most often we think of mathematics as abstract and free of moral questions (but I note that such unquestioned forces are the most dangerous). If we do think about its moral force, it is easy to think about the good things it does because that is what we have been told often in mathematics classrooms. This is like conventional armies, which do things we think of as good—for example, "protecting people" (otherwise described as protecting a certain social order). What are the options available to us within this army? We could aim to rise in the ranks without questioning its culture. We could try to desert, and abandon association and participation with mathematics. Or, we could stay within the army and try to change the culture. I think most of mathematics education researchers are doing this, including researchers who use sociocultural/political theories and researchers who do not. We are all trying to change the culture from within, but we are also trying to keep parts of the culture. It is like Bakhtin's (1975/1981) metaphor of centripetal and centrifugal forces—the unitary force has us trying to keep aspects of culture and the heteroglossic force has us trying to change the culture. Nevertheless, our approaches to this vary, and the extent that we aim to change the culture varies.

To provide an example of reflecting on the question about the location of mathematics, I return to the two articles I wrote with Beth Herbel-Eisenmann. We made conscious efforts to say how the work was unique to mathematics learning. This was easier in the first of our two articles because we were comparing the discourse in mathematics classrooms with discourse in other contexts (Herbel-Eisenmann et al., 2010). The second article (Herbel-Eisenmann & Wagner, 2010) required us to pay more explicit attention to mathematics. Because the quantitative data pointed attention away from mathematical thinking, we could not simply follow the data. We felt compelled to say over and over that the data defied our expectations about mathematical discourse. It felt oppressive to have the "Where is the math?" question hanging over us, but it made our writing stronger. The question helped us work to pull in the readers we sought to address and to push ourselves into new perspectives (at least this was the hope). The question helped us construct our model reader.

Identifying Storylines

Positioning theory reminds us that in addition to the discipline of mathematics, there are other discursive systems at play in mathematics education research, just as there are many in mathematics classrooms. It can be illuminating to see how the storylines intersect and thereby sustain each other. Now, I want to reflect on the range of discourses within which I am positioned in my research and writing.

Within research from a sociopolitical perspective, it is often expected for researchers to "position" themselves in their work. The editorial panel for the special edition on "Equity" in the *JRME* wrote an article that featured a conversation regarding "the role of a researcher's position in mathematics education" (D'Ambrosio et al., 2013, p. 11). Their idea of positioning meant that researchers should comment on their own identity and how it impacts their research design, political frameworks, participant selection, interactions with participants, and analysis. They asked whether researchers only need to position themselves when they study issues of identity and power. The answer was no; they agreed that all research is political. In wishing that all researchers would position themselves, they recognised that doing so could undermine the status of research in fields that favour objectivity. Related to this, positioning oneself as an author may be more difficult for early-career scholars, as noted by Foote and Bartell (2011), who grounded "positionality" in feminist and anti-racist literature and explored the experiences of novice mathematics education scholars.

Nevertheless, while the work on positionality is important, it does not address all the positioning that is taking place when we do research and report on it. There are many relationships at work in research. When I explored the positioning of Stephen Lerman's articulation of the social turn in mathematics education (Wagner, 2014), I did some analysis of the way researchers and others in the discipline tend to position scholarly work. A range of factors influenced the positioning of his research claims within the field, including the effects of race, gender, and generation demographics; as well as the researcher's service to the field's institutions and to individuals in small-scale interactions. As always, I remain unsatisfied with the extent of my analysis because there are yet other storylines in play.

For example, partisan politics in my country recently propelled my interest in discourses relating to neoliberalism, and thus, I begin to see the force of neoliberalism in mathematics education. Political scientist Jakeet Singh (2014) identified then Canada's Prime Minister's "vendetta against sociology" in the popular press: "In 2013, in response to an alleged plot against a VIA train, Harper remarked that we should not 'commit sociology,' but pursue an anti-crime approach." Singh pointed to other Harper actions that not only disparage sociology but also hamstring particular sociological research, and identified these moves as aligning with "a standard component of neoliberal ideology: that there are no social phenomena, only individual incidents." He argued that this rejection of sociology undermines a society's ability merely to recognise structural injustices, let alone to act on them.

The point I want to make as I strive to identify the complex positioning in mathematics education research is that neoliberalism is present in the mathematics classroom dynamic, in the discourse of mathematics education research (see Llewellyn in this volume), and in popular press discussion about mathematics education. Within our field, there are pressures to ignore sociocultural perspectives. Thus, neoliberal forces are important aspects of the positioning within our research. My research also exposes neoliberal forces in classrooms. Some of the same people who comment publicly from a neoliberal perspective are also commenting on mathematics education (without pointing to research that might back up their claims). Neoliberal storylines intersect with public education storylines, which intersect with mathematics education research storylines and so on. These discourses sustain each other.

Besides the neoliberal storyline, there are other storylines that permeate classroom mathematics and our research. Much research in mathematics education bears evidence of an interest in progress (e.g., "moving the field forward"), but I have rarely seen researchers situate themselves in the discourse of progress. Llewellyn (in this volume) provides a welcome exception to this in her investigation of the discourse of progress accompanied by her reflection on her own positioning in the discourse. Similarly, though I have heard mathematics educators situate themselves on a partisan political spectrum (e.g., embracing Marxism) in casual conversation, I have not seen this in their research publications. This raises the question about which storylines we feel the need to include in our reporting and which ones we choose to mask. For example, if I write about gender politics, does that compel me to situate myself in gender politics? If so, does situating myself in gender politics absolve me from the need to situate myself in other larger storylines?

A more difficult question arises if I do not write about gender politics but still choose to identify the genders of my research participants. Does naming gender obligate me to situate myself in gender politics? I have reviewed a lot of articles in which authors identified gender of their participants but do not say why they chose to highlight this characteristic. This practice of thoughtlessly identifying gender but not other characteristics reifies the stereotype that a person's gender is their most important characteristic. To resist this idea, in a recent article (Wagner et al., 2015) my collaborators and I chose gender-ambiguous pseudonyms for our participants, and we avoided personal pronouns that identify gender. We did this because there was no compelling evidence in our data that gender was significant in the interactions we analysed. We did not want the text to raise any questions about gender. It is difficult to write without gendered pronouns in English; social disruption is hard work.

I claim that it is the responsibility of scholars to identify our positionality within important discourses and storylines in our research. We have to make choices about which of these to report on in our writing. Whether or not we report them, we need to reflect on our positions that motivate our research actions. However, we tend to see only the discourses that we are accustomed to identifying. For example, English language and other practices (e.g., a tradition of aggregation in statistics) draw attention to gender. In contrast, our language and other structures do not so readily draw attention to partisan politics (e.g., neoliberalism vs. social responsibility), nor to other important discourses.

It is not easy to identify a storyline that I have not reflected on before. For example, if I obsess on gender, what would it take to prompt me to consider progress discourses in my work? This difficulty suggests to me that our field needs more venues for response, in which a scholar draws on a different storyline to reposition the work of another. For example, the journal *For the Learning of Mathematics* often features "Communications", in which readers respond to articles published earlier. Similarly, at conferences we often invite responses to plenary talks. I suggest that editors and conference organisers intentionally make space for such responses and select respondents who are likely to bring new storylines to the discussion.

Identifying Positioning in Human Relationships

Finally, a reflection on positioning requires the fundamental focus of positioning theory—the positioning itself. Within the storylines used to conceptualise interaction, what obligations and rights do the interlocutors have? When it comes to written publications, it would seem that the author has all the rights and obligations because the author even constructs the model reader. Nevertheless, it is worthwhile reflecting on how the text negotiates relationships. This is the kind of analysis that can benefit from functional grammar and other conceptual tools for language analysis.

I demonstrate a possibility for such reflection here with some further analysis of the articles I have referred to a few times above. I did some sentence-by-sentence analysis of my research with Herbel-Eisenmann, and share here my analysis of the second sentence of the abstract of our second article on lexical bundles (Herbel-Eisenmann & Wagner, 2010, p. 43): "We extend our analysis from a previous article (Herbel-Eisenmann et al., 2010), in which we introduced a concept from corpus linguistics—a 'lexical bundle,' which has been defined as a group of three or more words that frequently recur together, in a single group, in a particular register." This is my analysis of this sentence:

> In Sentence 2, we identify ourselves as "we" but make no explicit reference to our audience. We pointed to the field of linguistics as the source of our analytical method without identifying the people who comprise the field. We define a key term without saying who defined it this way. I would attribute this obscured agency to the academic genre of abstracts. It is complicated to reference a particular work within an abstract because it has to be cited with its full reference information. However, in this sentence, we had felt compelled to reference another article—the other one we wrote to set up this one. Thus, we did take the uncommon decision to reference another work in the abstract but did not go so far as to identify the linguists from whom we were drawing. This decision demonstrates the possibility of defying the general practice of avoiding citation, and thus implicates us as complicit in the scholarly tradition of obscuring agency in abstracts. Another significant choice here was to locate the work as coming out of linguistics. Does this add legitimacy? Perhaps this choice to foreground linguistics marks our work as special and cutting edge.

In this sentence and analysis, I drew attention both to the discourses that were identified and to some of the many that were not. The text of our article explicitly positioned the research as connecting to the field of linguistics. The article's location in ESM positioned it in mathematics education. These are both scientific fields. The text ignored, for example, the connections between the research and the political tensions between neoliberalism and social responsibility even though the data clearly pointed to the presence of strong social force in a learning context that purports to favour logic and individual work. The text oozes abstraction while claiming to be interested in human relationships.

This analysis draws to attention the difficulty with identifying omissions when we analyse text. How does one identify relevant actors who are not mentioned? Thus, I suggest another tool that arose out of my conversations with teachers about authority (Wagner & Herbel-Eisenmann, 2014). I draw a diagram of the people, their discourses, and their relationships to each other in my research. This helps me notice positioning that I might otherwise overlook.

Reflecting on my positioning research with Herbel-Eisenmann, I drew such a diagram. It was messy but the process raised some good questions. The diagram included all the people I found myself connecting with in the research: me, my research partner, participant students, participant teachers, other students, other teachers, other education stakeholders, article reviewers, journal editors, mathematics educators, linguists, and scholars in general. It included the relevant texts: oral interaction among participants, transcriptions of this interaction, publications, article reviews, and editor syntheses of reviews. The act of diagramming compelled me to physically position these agents and texts in relation to one another. It invited me to create groups, each of which implied a discourse system. It helped me identify who mediated relationships, and thus, to identify power and voice. Although I used connecting arrows to indicate relationships, those arrows do not necessarily depict the storylines that are at play in those relationships. Looking at my diagram made me realise that I needed to reflect further on each of the relationships, and consider the storylines at play. This also required consideration of who decides which storylines are in play. Each person in the messy (disorderly) diagram would have different reasons to be interested in particular storylines.

Synthesis

I close this exploration of ways to reflect on positioning in mathematics education research by citing a caution from Fairclough (1995, p. 42):

> It is quite possible for a social subject to occupy institutional subject positions which are ideologically incompatible, or to occupy a subject position incompatible with his or her overt political or social beliefs and affiliations, without being aware of any contradiction.

Indeed, I find that such inconsistency is likely. This is why I take seriously the quest for approaches that help me identify the positioning that can be associated with the text I generate and other actions in my research.

To summarise, here is a set of prompts that might be used to guide reflection on positioning in research and research writing:

1. List the relevant agents in my work, including research participants and people in my field.
2. Draw a diagram to show how these agents all relate to each other.
3. How do I position mathematics in my interaction with each of these people?
4. What other discourses (storylines) do I associate with each of these interactions?
5. Report my work in progress to others, including research participants, colleagues, friends, and opponents in the field.
6. What further discourses are identified in these relationships when I tell others about them? Which of these are most important/relevant and why?
7. In my writing, whom do I wish to address and why?
8. How can I use my text to construct a model reader? (How do I meet the real readers in their experiences, and how do I help them see new perspectives?)

References

Atwood, M. (2002). *Negotiating with the dead: A writer on writing*. Cambridge, UK: Cambridge University Press.

Bakhtin, M. (1975/1981). *The dialogic imagination: Four essays*. (M. Holquist, Ed.; C. Emerson & M. Holquist, Trans.). Austin, TX: University of Texas Press.

Barthes, R. (1977). *Image, music, text*. (Trans. S. Heath). New York: Hill and Wang.

D'Ambrosio, U. (1994). Cultural framing of mathematics teaching and learning. In R. Biehler, R. Scholz, R. Sträßer, & B. Winkelman (Eds.), *Didactics of mathematics as a scientific discipline* (pp. 443–455). Dordrecht: Kluwer.

D'Ambrosio, B., Frankenstein, M., Gutierrez, R., Kastberg, S., Martin, D. Moschkovich, J., … Barnes, D. (2013). Positioning oneself in mathematics education research. *Journal of Research in Mathematics Education*, *44*(1), 11–22.

Davies, B., & Harré, R. (1999). Positioning and personhood. In R. Harré & L. van Langenhove (Eds.), *Positioning theory: Moral contexts of intentional action* (pp. 32–51). Blackwell: Oxford.

Dickinson, E. (1960). *The complete poems of Emily Dickinson* (Ed. T. Johnson). Boston, MA: Little, Brown.

Eco, U. (1979). *The role of the reader*. Bloomington, IN: University of Indiana Press.

Eco, U. (1994). *Six walks in the fictional woods*. Cambridge: Harvard University Press.

Esmonde, I., & Langer-Osuna, J. (2013). Power in numbers: Student participation in mathematical discussions in heterogeneous spaces. *Journal for Research in Mathematics Education, 44*(1), 288–315.

Fairclough, N. (1995). *Critical discourse analysis: The critical study of language*. London: Longman.

Foote, M., & Bartell, T. (2011). Pathways to equity in mathematics education: How life experiences impact researcher positionality. *Educational Studies in Mathematics, 78*(1), 45–68.

Foucault, M. (1972). *The archaeology of knowledge* (A. M. Sheridan Smith, Trans.). London: Routledge.

Gutierrez, R. (2013). The sociopolitical turn in mathematics education. *Journal for Research in Mathematics Education, 44*(1), 37–68.

Halai, A. (2014). Social justice through mathematics education: Skilling youth for a societal participation. In P. Liljedahl, C. Nicol, S. Oesterle, & D. Allan (Eds.), *Proceedings of the Joint Meeting of PME 38 and PME-NA 36* (Vol. 1, pp. 67–71). Vancouver, Canada: PME.

Harré, R., & van Langenhove, L. (1999). *Positioning theory: Moral contexts of intentional action*. Oxford: Blackwell.

Heid, M. (2010). Where's the math (in mathematics education research)? *Journal for Research in Mathematics Education, 41*(2), 102–103.

Herbel-Eisenmann, B., & Wagner, D. (2010). Appraising lexical bundles in mathematics classroom discourse: Obligation and choice. *Educational Studies in Mathematics, 75*(1), 43–63.

Herbel-Eisenmann, B., Wagner, D., Johnson, K., Suh, H., & Figueras, H. (2015). Positioning in mathematics education: Revelations on an imported theory. *Educational Studies in Mathematics, 89*, 185–204.

Herbel-Eisenmann, B., Wagner, D., & Cortes, V. (2010). Lexical bundle analysis in mathematics classroom discourse: The significance of stance. *Educational Studies in Mathematics, 75*(1), 23–42.

Martin, J., & Rose, D. (2005). *Working with discourse: Meaning beyond the clause*. London: Continuum.

Setati Phakeng, M. (2014). The calculus of social change—mathematics at the cutting edge. In P. Liljedahl, C. Nicol, S. Oesterle, & D. Allan (Eds.), *Proceedings of the Joint Meeting of PME 38 and PME-NA 36* (Vol. 1, pp. 55–60). Vancouver, Canada: PME.

Singh, J. (2014, October 18). The ideological roots of Stephen Harper's vendetta against sociology. *The Toronto Star*. Retrieved November 14, 2015, from http://www.thestar.com/opinion/commentary/2014/08/26/the_ideological_roots_of_stephen_harpers_vendetta_against_sociology.html.

Valero, P. (2004). Socio-political perspectives on mathematics education. In P. Valero & R. Zevenbergen (Eds.), *Researching the socio-political dimensions of mathematics education: Issues of power in theory and methodology* (pp. 5–24). Dordrecht: Kluwer Academic Publishers.

Valero, P. (2014). Cutting the calculations of social change with school mathematics. In P. Liljedahl, C. Nicol, S. Oesterle, & D. Allan (Eds.), *Proceedings of the Joint Meeting of PME 38 and PME-NA 36* (Vol. 1, pp. 73–77). Vancouver, Canada: PME.

Wagner, D. (2012). Opening mathematics text: Resisting the seduction. Educational Studies in Mathematics, 80(1-2), 153-169.

Wagner, D. (2014). Privileging local cultures and demographics in the mathematics classroom. In P. Liljedahl, C. Nicol, S. Oesterle, & D. Allan (Eds.), *Proceedings of the Joint Meeting of PME 38 and PME-NA 36* (Vol. 1, pp. 61–66). Vancouver, Canada: PME.

Wagner, D. (2015). A speech act in mathematics education—the social turn. In P. Gates & R. Jorgensen (Eds.), *Shifts in the field of mathematics education: Stephen Lerman and the turn to the social* (pp. 75–87). New York: Springer.

Wagner, D., Dicks, J., & Kristmanson, P. (2015). Students' language repertoires for prediction. *The Mathematics Enthusiast, 12*(1–3), 246–261.

Wagner, D., & Herbel-Eisenmann, B. (2009). Re-mythologizing mathematics through attention to classroom positioning. *Educational Studies in Mathematics, 72*(1), 1–15.

Wagner, D., & Herbel-Eisenmann, B. (2014). Mathematics teachers' representations of authority. *Journal of Mathematics Teacher Education, 17*(3), 201–225.

Walshaw, M. (2014). Raising political, psychoanalytic, and cultural questions of a proposed educational intervention. In P. Liljedahl, C. Nicol, S. Oesterle, & D. Allan (Eds.), *Proceedings of the Joint Meeting of PME 38 and PME-NA 36* (Vol. 1, pp. 79–83). Vancouver, Canada: PME.

Chapter 18
Urban^Boundaries^Space. Disturbing Choices and the Place of the Critical Research/Researcher in the Capitalist Wile

Mônica Mesquita

Abstract This chapter is a deep outburst! This outburst first materialized as a presentation during a small international meeting of mathematics education researchers—DOME (Disorder of Mathematics Education), in early 2015 in Berlin. Mathematic education researchers committed to Critical Social Theory made up the core participants at the DOME. The central aim of this chapter is to propose some thoughts regarding our situationality as critical researchers and considering the position that our research occupies in the capitalist wile. The systematization of the presented arguments came as a result of a reflexion on my recent path as mathematics education researcher and educator, self-labelled as critical. The discourse begins with a brief introduction of some points of view over our choices, questioning our freedom, going through the political flows of our survival, and enquiring the role of our production, reproduction, and contradictions as critical researchers—passing to another stage as critical researchers.

Introduction

Grounded on the bold assumption of individual freedom of choice, the global hegemony of the neoliberal system provides us with the possibility to change our posture to make choices. Simultaneously, the possibility of choice is framed within the precondition that choices do not disturb deeply the balance established by the dominant ideology. Hereby, the acts of different postures are "in the same bag"—*choice as bondage*. The concept of choice is linked to the concept of freedom and it can be assumed as a "rationalized" blind position. According to Žižek (2001), we are living in the era of the "escape from freedom," as we choose to refrain from choosing under the banner of an alleged rationality. I will take this reflections as a point of departure in order to rethink the opposition between "formal" and "actual" freedom

M. Mesquita (✉)
Policy and Governance, MARE - Marine and Environmental Sciences Centre, Caparica, Portugal
e-mail: mbmesquita@ie.ul.pt

H. Straehler-Pohl et al. (eds.), *The Disorder of Mathematics Education*,
DOI 10.1007/978-3-319-34006-7_18

(in Lenin's sense), leading to questions regarding our needs and our desires; they could develop into questioning "the wound of reality" (Žižek, 2006) or even the possibility of "free" decisions.

This chapter is a work in progress; it assumes a radical posture, which focuses on the disturbing collective choices while it invites (fellow) researchers to critically rethink the relations among their (our) needs, desires, and obligations, establishing a common ground by contextualizing them in our social-economic nets. The reflection about the threat (to rethink the opposition between "formal" and "actual" freedom in Lenin's sense) comes from inside, as Žižek (2002, p.154) points out: "from our own laxity and moral weakness, loss of clear values and firm commitments, of the spirit of dedication and sacrifice." Considering this chapter as self-collective exercise, it proposes walking from the victimization posture to a free collective subject.[1] The recognition of the human as an autonomous collective subject is central to humankind. To be free is then to be reasoning about the human being's closure and, accordingly, to create collective strategies to live with dignity. *Disturbance is the strategy I suggest to regain such freedom.*

It is interesting to realize that in the neoliberal system, in which we operate, only one form of knowledge has been fully recognized and validated—formal knowledge. Nowadays, the increasing presence of research in the academy exploring informal or nonformal knowledge—like in the case of indigenous studies, urban studies, ethno methodologies, or more specifically modelling or ethnomathematics in mathematics education—could provide a strong contribution to change this "unidirectional way" approach followed until now. However, this current process within the academy is still rooted in understanding behaviors arising from plural contexts with the aim to reorganize our current political system but not to validate the informal and nonformal knowledge; that is, to enable these knowledge and the people who develop them, to become an active part of the local political decision-making process.

Some movements of recognition and validation of informal and nonformal knowledge have been practiced, but they need to be always linked to some institution of formal nature in order to be socially recognized and validated—and even so, still maintaining the form of ghettos (i.e., nonformal knowledge recognized and validated by the academy will still be called nonformal knowledge and it will always be separated from formal knowledge). This is just one illustration for the tendency towards the normalization of society—a society that today is maintained by the exotic of the other, by the charm of the social exclusion (Demo, 1998). Hence, diversity becomes part of the "normal"—as a non-part (Rancière, 1995), feeding the present political system and securing the current economic apparatus.

[1] The autonomous collective subject is like the anarchist concept of "self": socially constructed, embedded in and constitutively social and without free will (an ideological idea, a myth without scientific basis) but with the natural right to freedom (a political concept), both socially and materially real. We cannot freely choose to speak a new language all of a sudden nor can we freely choose to fly without technological apparatuses (this would be the extreme implication of free will). We know, however, the difference between being in a prison (as institution) and not being in a prison, even though we might want to speak about society metaphorically or even to depict it in theory as a prison.

The subject that does not have her/his validated knowledge, or have it validated by and to minorities' groups, is kept on the margins of society, with no conditions to live a self-determined life in dignity within the current political and economic apparatus, but only surviving! A "local" community that does not celebrate this kind of knowledge collectively will be—I maintain—confined to mere survival within capitalism, similar to the individual subject. As an exercise of the present economic apparatus, society assumes these "local" communities to be centered in and defined at the local level. But quite the contrary, these communities are situated within society as a whole.

My own research identified dialogues and cultural clashes that were created by our dominant system of governance (Mesquita, 2014; Mesquita, Restivo, & D'Ambrosio, 2011). Making these clashes visible also unveiled the oppressions that were executed by externally imposing voices into the local communities. Making visible these clashes also enabled the communities to express their voice and thus, collectively, develop critical reasoning among all members involved in the research through an exercise of praxis. The extermination of a community begins with the act of silencing their voice and, consequently, their collective thought. The struggle for voice and for collective thought thus must produce disturbances. Disturbances provoke mechanisms of defense.

A Personal History of Disturbing Choices

In the following, I illustrate how disturbing choices provoke mechanisms of defense from the dominant system of governance by sharing and reflecting three research moments during my own research biography. Each of these moments is a testimony of how trying to be together—that is living with and being inserted in a community—with the members of the local communities has been another "big problem"[2] for the local elites.

The research moments are not only just another report of instances of oppression, but signify how creating disturbances, and being perceived as such, contributes to a freedom that emerges from a reasoning of closure resulting in the development of collective strategies.

In a first research, with two groups of children in street situation in the city of São Paulo from 1990 to 2001, I could realize how our daily and constant encounters were not welcome by the local authorities (Mesquita et al., 2011). However, at that moment, the most important analysis of the research practices was to realize how the academic environment does not recognize this encounter. With the extreme support

[2] The big problem consists in a small act—to be with the other, which is in an excluded position in any context (political, economical, social, cultural, religious, etc.). This act has been considered as a personal affront to the apparent linearity in the societies where I was living/working and the treatment has been done based on the exclusion as I discuss in this chapter. In fact, this big problem, or this small act, can be considered as a break in the vicious cycle of the capital (the food of its own decadency), which maintains the victim's condition so that it is the wide of who maintains it. The big problem here is that, with a small act, is possible to cut the big food.

of two respected researchers, internationally renowned in the mathematics education environment, the knowledge of these children could be systematically discussed through a doctoral work. The encounter of our knowledges promoted learning for all of us and it brought new perspectives about the concept of (urban) space. The movement of survival of these children shows, through ethnographic images and actions, that some mathematical knowledge (what we—academic view—identify as) is developed in the context of their labor, social, and affective relations. To assume that they use the Pythagorean theorem, or the Euclidian Parallel Postulate—the fifth one (as you prefer), in their labor process was considered by the mathematics department of the oldest and most famous university of Portugal as an affront and it was formally stated that the study should be disregarded by the scientific community.

In a second research, with mathematics teachers of different ethnic groups from Brazil, from 2000 until now, I could see the infinite care that we must have towards each other, in the learning process (Jesus & Mesquita, 2000). This finding had already been made during the first research and during my 15 years, until then, as a public high school mathematics teacher. However, at that moment, I could feel it even more, and I think it was because of facing ethnic groups that had (and still have) a very different culture from mine—so urban, until then. The sensibility to each other and the significance of their knowledges in their daily life, especially regarding the concept of space, is a posture. It is a worldview that reminds us—urban humans—how important the process of learning as political tool is, and how this tool can be useful to construct a way to solidary and equitable society if the central focus is autonomy, respect, and civility. Such importance can be attributed to the different conceptions of the own local mathematical structures developed in the context of labor, social, and affective relations. Finite, infinite, set, compactness, connectedness, neighborhood, separability,[3] empty, and unity are some concepts that have different definitions to these different ethnic groups. These differences revealed themselves to be fundamental for the relations that these ethnic groups have with the learning processes—a relation founded in cultural values. In this case, the cultural values are directly linked with the local political decision-makings. With this situation, and leaving the naivety aside—which is not permitted in the urban life—this research process allowed me to realize how education, as a political tool, is in our urban society, oppressed, shaped, and related to the economical values; which are directly linked with our cultural world view. In the urbanity, the learning process is shaped by the hegemony, acting as a capitalist bait, and so, currently, the learning process focuses on the maintenance of the economic crisis.

The third and last research occurred in Portugal and was shared with members of two local communities from the city where I live—Costa de Caparica—in small portions (Mesquita, 2014). Since 2009, when we found each other in a collective sense, some members of these three communities have done a collaborative educational work, which focused on the socioeconomic space existent in this city, more specifically, in the fishing and *bairro* (a multicultural slum) communities. The daily work, which developed in the communities, allowed us to experience, discuss, and

[3] In the sense of Hausdorff space.

rethink our ethnographic images and actions, and created the desire to construct a more systematic way of being together. The search for such systematization was important since, during the 3 years that preceded the project, we found a set of legal obstacles making it too difficult to establish an open and emancipatory collaboration. The local authorities—town council, church, and police, did not welcome our encounter. The destruction, by the local government, of a cultural center built by the Urban Boundaries Movement—how we identify ourselves as an independent local group—as well as the constant raids that members of this group suffered by the local police are two of many other examples. Based on the fact that we could not be together—after all we lived (and live) under a local camouflaged feudalism,[4] where "things" happen without a *locus* of responsibility for these facts—we decided to find a legal way to face these obstacles. As some of us were linked with the academic community as researchers and university teachers, a research proposal made possibilities (at the same time with a cynical seasoning) to legalize our lives and place. All of us spent some time discussing whether or not we desired to submit our struggle through this option—being directly linked with one formal elitist group—the academy. However, we could find in our images and actions a sense of transformation, which is supposed to be the central core to start an inside-out struggle in this closed and upstart system, or even blind by convenience. Our particular focus, while academics, was to reappropriate the signifier of research, researching our spaces as community, and, consequently, the researcher's role in this process.

These experiences of learning processes, supported (in their base) by affectivity, also reinforced some ideas in me, making me think about the academic coterie and sharpen my desire to continue studying the role of research in its space: an elitist urban space, and also to think more and more about our role as researcher. These three experiences in my life as an ethnographic researcher, i.e., my posture as researcher until now (questions, methods and methodology, objectives, and chosen "study objects"), have caused me discomfort and many obstacles in my professional life. Some obstacles are: (1) maintaining myself on the outside of a college "co-status"—being difficult to find positions as a professor to survive; (2) denigrating my image as researcher—not being taken seriously and being categorized as an activist; (3) limiting my research to the academic boundaries. Disturbing choices have built my research path. Like me, some colleagues, from different fields of knowledge, and some people outside of the academic environment have demon-

[4] A social and political local organization based on servo-contractual relations that appears on the socioeconomic relations in the Costa de Caparica, namely within the fishing and *bairro* communities. Both fishermen and the residents of the *bairro* community—the servants, have their home because the local government gave them (implicitly or explicitly). The right to a piece of land (house/apartment) comes from their labor relations, not always paid, with (1) the owners of the boats, (2) the owners of the agricultural lands, where the *bairro* is located, or even (3) the religious or public local institutions. These local owners can be understood as the feudal lords. The servants, when they need to transgress the limit of their houses, they need to pay (fines) both because the fishermen are fishing in places where the local government determines that they cannot as because the members of the *bairro* are walking in the city without a residence permit.

strated the same reflection around the situationality[5] of the critical researcher. Following these thoughts, I propose a discussion around the academic apparatus as a factory of the dominant ideology to (de)form consciences. The central point here is: for whom do we (critical researchers) produce knowledge?

The Production of Knowledge: The Boundaries Among the Research, the Critical Researcher, and the Political Flow in Their Survivals

Current thinking on the production of knowledge presents different ways. From a critical point of view, as David Harvey shared with us in the first pages of his most recent book *Seventeen Contradictions and the End of Capitalism*, published in 2014, this thinking points out to the fact that "something different in the way of investigative methods and mental conceptions is plainly needed in these barren intellectual times if we are to escape the current hiatus in economic thinking, policies and politics" (p. xiii). Barrenness, in this case, is not a singular causality; it is a fruitful tool to maintain things as they are. We see that, as the movie Children of Men (2006) suggests, barrenness can be analyzed as a process of infertility, in which the new cannot be produced on a natural historic way—the socially unplanned creation may cause disturbance within a system of hegemonic thought, shaking the existing comfort in a dominating elite. In fact, and as Žižek (2006a) argues in his critical view of the movie, the "infertility is the very lack of meaningful historical experience. It's a society of pure meaningless historical experience." Some academic historical acts—i.e., some production of systematic knowledge—disappear. They are there but they are deprived of the capitalist wile, they become invisible. The rhetoric, the servile, and the mainstream discourses are the food of bareness lived today.

This work would not be unusual if it analyzed and discussed (1) the relationship between domination and knowledge, or (2) the relationship between the intellectual and the university as an institution located in the dominant discourse. Following the focus of this work—the disturbing collective choices (highlighted in page 2), the central point here resides on the analysis and discussion of the relationship between (1) domination and critical knowledge, (2) the critical intellectual and the university, while being in the space of the dominant discourse, (3) the relationship of knowledge, university, and their roles in the capitalist wile, and, more specifically, in the (4) relationship between "doing" and "thinking." Mauricio Tragtenber (1990, 2002) categorizes the old separation between "doing" and "thinking" as being "one

[5] Men, as beings "in a situation" find themselves rooted in temporal-spatial conditions, which mark them and which they also mark. They will tend to reflect on their own "situationality" to the extent that they are challenged by it to act upon it. Men are because they are in a situation. And they will be more the more they not only critically reflect upon their existence but critically act upon it (Freire, 1970, p. 100).

of the diseases that characterize academic delinquency—the analysis and discussion of relevant issues in the country is a political act, a form of action, inherent to the social responsibility of the intellectual" (p. 4, own translation).

The current motto of the capital wile has been "realize yourself"! How can critical researchers "realize themselves"? What does it mean to people that are normally very compromised with the complexity of life, analyzing critically human relations, people that feel that to "realize yourself" is to realize a set of things that are transversal to the economic collective condition, including their own? Returning to the historical absence, previously discussed, what does it mean to have critical theories? Do they only work when signaling a certain world in which creation (production of knowledge) is allowed only to those who, without being forced by the academic political flow, accept the bureaucratic control of its production? When is this control something more than comparing, classifying, watching, or punishing? Or when is this control also practiced through pleasures, gatherings, innovations, even "disorders" like our own? Researchers that bring disturbing studies to the academic order are controlled by institutions (academic and research) under invisible, or masked, evaluations—under the false alibi of the evaluation criteria.

The conditioned knowledge production is rooted in the academic space, which operates it and ensures its own space as cloister—trying to maintain its pureness without boundaries, being the house of true knowledge, of the intellectual subject and being, and of the capital moral following the ideological despair of capitalism: how to generate knowledge to regulate life. Other critical researchers such as Ubiratan D'Ambrosio and Sal Restivo have developed their theories around that—around the closed and regulatory conditions of the production of "true" knowledge. Based on these theories, I have argued (Mesquita, 2014, 2015) that, in a topological ontology, the boundary is the essential space for the production of knowledge, developed in the encounters, breaking the hegemony of the systems and creating, dialogically, new conceptions. However, being on the boundaries is a risky exercise, i.e., to be on the academic boundaries is to be in the boundaries of the life, economically and socially speaking. How can a critical researcher survive if critical researchers need to eat (and, therefore, they need to reproduce the normality required by the cultural environment where they work)? The critical researchers need to be "inside" in order to have economic power, which allows them to maintain their intellectual life and, in this dichotomist movement, allows them to "think" but not "do." The capitalist wile develops this sense of freedom, creating a free subject and making impossible the development of the collective subject—the socially unplanned creations. In that sense, maybe we can discuss that critical researchers do not produce knowledge for the humanity. In fact, the capitalist wile opens spaces to produce critical research as one of its own tool of reproduction. Yet the intellectual barrenness produced in this process of production of knowledge is perceived by many critical researchers not as a weakness, but as a movement of empowerment, i.e., as intellectual fertility.

The Crisis, the Reproduction, and the "Land of the Lost"[6]

A collection of encounters with Etienne Balibar[7] allowed me to reconstruct the idea of the "land of the lost" when thinking about our current political time. Balibar, talking about the current European crisis, provides the following Gramsci's thoughts as an argument: "The crisis consists precisely in the fact that the old is dying and the new cannot be born," made me reinforce this previous idea. We cannot be born! We are, while critical researchers, within the dominant ideology. In fact, even though we perceive ourselves as critics, we are slaves of our own reality and, at the same time, food to the barren intellectual times—so important in maintaining the academic world order; "the land of lost" of the neoliberal system. According to Žižek (2014), "when we think that we escaped into our dreams, at that point we are within ideology. Ideology is not simply imposed on ourselves. Ideology is our spontaneous relationship to our social world" (00'37"; online video).

The reproduction of the urban system where we operate has as main gear welfarism—the great virtue of the democratic person. It is undeniable that democracy has been portrayed, in an academic way, as the way of social equality. However, in a certain sense, we can question ourselves about our welfare as critical researchers. To bring new conceptions developed from a critical view over this reproduction, through academic thought, even if this position of criticism comes from the boundary, is an exercise embedded on the democratic politics that lives in the critical researcher. The great challenge here, to the critical researcher, is the urban boundary (with the other and not to the other) that promotes disturbing choices, i.e., to propose a global historical analysis, a local genealogy, and the collective rethinking of our current fears; to lay bare the gear of this hegemonic system; to unravel its complexity and understand the consensus by the way it is engineered.

To live the "land of the lost" inside of our margin of freedom to decide, to make choices, outlined by the neoliberal politics of the urban system, is to assume that we are living the "urban land of the lost." This act can be considered as an elevation of our critical research into our current social and political urban life. This position assumes the slave role of the urban system and shakes us inside our symbolic deadlock. How to assume our position of opposition without serving the situation? How to survive if our position disrupts the urban social order? How to assume our role of mediation if we see our position as revolutionary? When has our position made a difference? What do disturbing choices mean to a critical research?

In basic Marxist terms, any social reproduction, including the reproduction of knowledge, can be identified by its production, distribution, and consumption.

---

[6] I am using this term thinking about the television series "Land of the Lost," created by David Gerrold and produced by Sid and Marty Krofft, who codeveloped the series with Allan Foshko in the 1970s.

[7] In his book *Politics and the Other Scene* (2002/2011), in his interview *Europe is a dead political project* (2010) or even in his discourses in Birkbeck Institute, UK; during a summer course 2010 or in Bogazici University, Turkey; and during a public lecture called *In Globalization and the Crisis of Cosmopolitan Idea*.

Embedded in the social reproduction of knowledge, to be a critical researcher means to be critical in this cycle (which is the source of the determinism of our social being), i.e., to institute a critical cycle. It is not enough to "think" without "doing." The praxis, in this case, is essential to the critical reasoning. However, this cycle of the social life in our urban hegemonic system is fed by the economic cycle, which imposes a drain on production, a barrier on distribution, and a normalizing frame on the consumption. The junction between "thinking" and "doing," in the academic community, presents a body rooted into the "academic dream": mass production (paper fabricaction, project fabricaction, and so on...), distribution in high-rated journals (which offer blind reviews, regulatory evaluation, and so on...), and technical and quantitative consumption (what is important is the quantity of knowledge and not the quality of the process of knowledge—to understand, to discuss, to analyze, to infer, and so on). In the current academic cycle there is no space to be in the boundary, to produce from the practical experiences, to produce the new on the natural historic way—an academically unplanned creation that may be disturbing within the academic world order, shaking the existing comfort of the "academic dream." The intention here is not simply to conflate two dimensions: knowledge and economy, but to reinforce its intrinsic relation and, in this neoliberal system, its symbiosis as one of the main feed.

Do critical researchers want to change the academic world order that: (1) corroborates the hegemonic system and the neoliberal government; (2) self-supports the academic comfort—neutrality in the form of democracy; and (3) holds the intellectual infertility? What have we (critical mathematics education researchers) done beyond our desires, our wanting, and our possibilities from our comfortable zone?

The proposal, here developed, of a collective discussion may have its end in a cul-de-sac, walking by a discourse of impossibility guided by rhetoric: "what can we do against the academic world order, or even, against the (urban) 'land of the lost'?" However, the relevant point to this collective discussion is to question our own consensus about the way the academic world order is engineered: after all the "academics world" is us! This emergent exercise should not allow our critical position to take the course that Žižek (2001, p. 2) identifies as the "perfect example of impassivity," acting outside of an urban non-space, being located as "an act WITHIN the hegemonic ideological coordinates."

The "land of the lost" is found, in fact, more commonly than we think; it is indeed the mask of the urban ideology, in what we live, what we eat and what we are food for; in what we feel comfort and freedom; in how we exercise our bondage behind our choices and, at the same time, in how we can realize, as critical researchers, the contractions between our thinking and our doing.

Contradictions, Boundaries, and Urban Ideology

> I repeat that there is a politics of space, because space is political (Lefebvre, 2009 [1970], p. 174).

Henry Lefebvre outlines some aspects of space: as historical product, as stake of political struggle, and as ideology. As the critical researchers we are, it is particularly relevant to realize the anatomic description that exists in the academic space, especially considering these three aspects: historical, political, and ideological. In a topological view, the critical discourse that we are developing is rooted in these three aspects: (1) a reading of questionings, letting us inside the boundaries of the academic space and, at the same time, inside the boundaries of ourselves; (2) by reasoning about our comfortable and necessary position as the "politically correct"; and, (3) in contrast, a defense against our own innermost identification. When we study critical social theory, we (often) come upon revolutionary conceptual notions under different schools of thoughts, which, to maintain our economic survival, we subtly translate to attend our desires—the desires of the urban ideology, the dominant one.

In an urban architecture, we can compare the critical research with social edifications—social housing; in the field to attend and to maintain the urban space ideology, even if it has been developed in rural, *caiçara,*[8] or indigenous areas. The consistency of our scientific production is hybrid and makes a conurbation with the normalized scientific society, permitting us to be part of that. In our case, being critical researchers in (mathematics) education, we know about the "political project on the part of reformist bourgeoisie to create a 'respectable' working class that would refrain from riot and revolution and succumb to the blandishments that capital could offer" (Harvey, 2014, p. 183). We, as educators (in a Freirian' sense) and as critical educational researchers, have a commitment to capital's demand for ideology conformity, even though we are conscientious (falsely or not)—i.e., even though our research subject is in itself the process of bondage that the empowerment of the mass education provokes. We live in contradiction, questioning the schooling process and, at the same time, giving empowerment to investments in education and teacher formation—according to Harvey (2014, p. 184), "a sine qua non for capital's competitiveness." Education and research development can be considered as a profitable business sector, in our current time.

Though it could be considered obvious, I reinforce that it is incredible how researchers developed, for example, the so-called human capital theory. It seems to be that we (researchers) often theorize about the other, and about the workforce of certain working classes, feeling as if on a stage ourselves. However, we seem to forget that, in our urban system, the working classes have suffered the conurbation[9] movement, as have the big cities and their small neighbors. Researchers, as well as teachers/educators, are totally inserted in the low level of the current labor pyramid. Nowadays, as the fisherman ensures the fish and the slum ensures nonurban spaces oriented to welfare, the researchers, teachers, and educators ensure standardized

[8] *Caiçara* is a Brazilian word, from the *Tupi* language family, to refer to the inhabitants of the coastal areas.

[9] It is the moment that an expanding body encounters other(s). In urban studies and architecture the "conurbation" can be defined as an expansion movement extensive urban area resulting from the expansion of several cities or towns so that they coalesce.

thinking. In fact, we (critical researchers) are nothing more than the complementary part, with different performances, of other researchers, and vice versa; the neoliberal system creates illusory dichotomous positions. The freedom of the exercise of conurbation is nothing more than a current political tool to imbricate the mass behind labels of freedom: "yes, we can"; of equality: "from the mass, to the mass"; of unity: "we are all equals"; of fraternity: "we must develop laws and systems of education to recognize and tolerate the other."

Since the 1980s, Žižek has discussed the current cynical position mentioned above, which he recognizes as a form of ideology that dominates our political moment.

> Peter Sloterdijk puts forward the thesis that ideology's dominant mode of functioning is cynical, which renders impossible—or, more precisely, vain—the classic critical-ideological procedure. […] The formula, as proposed by Sloterdijk, would then be: "they know very well what they are doing, but still, they are doing it." Cynical reason is no longer naïve, but is a paradox of an enlightened false consciousness: one knows the falsehood very well, one is well aware of a particular interest hidden behind an ideological universality, but still one does not renounce it. (Žižek, 1989, p. 28–29)

I have questioned myself whether this cynical position in which researchers (including myself) have, in my opinion, assumed is a question of courage or a question of blindness, or if it is both—after all, as critical researchers, we know perfectly well what we are doing but we are still doing it. To think about the process of our situationality (about how we occupy the urban spaces, contributing to the regulation of society), is to think about (1) the level of bravery it would take to understand the complexity of human relationships and their contradictions and (2) the level of blindness it takes to survive in the current economic system, developing political tools to do so.

On the one hand, in a short movement, courage is not relevant to develop our convictions, following Nietzsche's thoughts in his studies about genealogy of moral, but it is relevant "to attack one's convictions" (1878/1996, §630), shaking our contradictions. On the other hand, blindness, following Santos's (2001, p. 2) thoughts in his studies about the epistemology of blindness, is a forced exercise we are performing in our current urban system "while unveiling the blindness of others." The small gap between these two closely linked postures (bravery and blindness) is in fact filled with contradiction and can be one of the spaces where we find the possibility of radical social change. Going a little bit further, this space can be categorized as being another example of urban non-space—we know it, we transit it, we construct it but we do not act (on) it—a boundary urban space. Contradictorily, this game inside this urban non-space can only be reasoned under critical thoughts; only by radical social thinkers—the critical ones, who can do the radical social changes.

A critical view regarding the current "reign of cynical reason," as Žižek (1989, p. 29) defines our social moment, shows us that we, critical educational researchers, are not a flaw and that our research, i.e., our critical thoughts, should not be treated as an accident of a soft and perfect functioning system. We must treat ourselves as a local symptom of a much larger phenomenon (Žižek, 2006, p. 242)—the phenomenon of dehumanization through the fear politic. Today, we have fear as our chain balls. The careless process of globalization (is it *démodé* to argue about globalization today?) has brought, among other things, this new, normalized, passive, egocentric culture, which has contributed to the destruction of our humanity—and which, just to remind ourselves, we are part of. Some current philosophical praxis have invited us to shake our comfortable position, reversing "our concept of what is possible and what isn't; maybe we should accept the impossibility of omnipotent immortality and consider the possibility of radical social change" (Žižek, 2010, online).

Let us be intolerant with ourselves and "pass on to another stage". (Balibar E. 2002, 2011; p 21)

Acknowledgments Lia Laporta, Liz de Freitas, David Swanson, and Alexandre Pais ... thank you very much for the discussion times! Special thanks to Hauke Straehler- Pohl for the warmth and professionalism with which he edited this work. This chapter was developed in the context of Post-Doctorate in Science Education, funded by the *Fundação para a Ciência e a Tecnologia*, under internal identification: SFRH/BPD/87248/2012.

References

Balibar, E. (2002/2011). *Politics and the other scene*. London: Verso.

Balibar, E. (2012). *Classes notes* (Summer course: Social critical theory). London, UK: Birkbeck Institute/London University.

Balibar, E. (2014). In *Globalization and the crisis of cosmopolitan idea*. Public Lecture. Turkey: Bogazici University. Retrieved May 6.

Balibar. E. (2010). Europe is a dead political project. In *The Guardian*. Retrieved May 25, from http://www.theguardian.com/commentisfree/2010/may/25/eu-crisis-catastrophic-consequences.

Children of Men. (2006). Dir. and co-author Alfonso Cuarón. Perf. Clive Owen and Julianne Moore. Universal Pictures. DVD.

Demo, P. (1998). *O Charme da Exclusão Social*. Campinas: Autores Associados.

Freire, P. (1970). *Pedagogy of the oppressed*. New York: Herder and Herder.

Harvey, D. (2014). *Seventeen contradiction and the end of the capitalism*. London: Profile Books.

Jesus, C., & Mesquita, M. (2000). *Relatório do Curso de Matemática da IV Etapa do Curso de Formação de Professores Indígenas*. Brasília: FUNAI.

Lefebvre, H. (1970/2009). Reflections on the politics of space. In N. Brenner & S. Elden (Eds.), *State, space, world: Selected essays, Henri Lefebvre*. (pp. 167–184). Minneapolis: University of Minnesota Press.

Mesquita, M. (2014). Fronteiras Urbanas. Sobre a humanização do espaço. In M. Mesquita (Ed.), *Ensaios sobre a humanização do espaço* (pp. 19–31). Anonymage: Viseu.

Mesquita, M. (2015). Urban boundaries and love: the rebirth of political thought within the disturbing education. *International Journal of Žižek Studies, 9*(2).

Mesquita, M., Restivo, S., & D'Ambrosio, U. (2011). *Asphalt children and city streets. A life, a city, and a case study of history, culture, and ethnomathematics in São Paulo*. Rotterdam: Sense.

Nietzsche, F. (1878/1996). *Human, All too Human*, trans. In R. J. Hollingdale. New York: Cambridge.

Rancière, J. (1995). In J. Rose (Ed.), *Disagreement: Politics and philosophy*. Minneapolis: University of Minnesota Press.

Santos, B. (2001). Towards and epistemology of blindness—why the new forms of "Cerimonial Adequacy" neither regulate or emancipate. *European Journal of Social Theory, 4*(3), 251–259.

Tragtenber, M. (1990/2002). Sobre Educação, Política e Sindicalismo. In Revista Espaço Académico, Ano II, n° 14. Maringá: Universidade Estadual de Maringá Ed. http://www.espacoacademico.com.br/014/14mtrag1990.htm.
Žižek, S. (1989). Cynicism as a form of ideology. In *The sublime object of ideology* (pp. 28–30). London: Verso.
Žižek, S. (2002). *Welcome to the desert of the real*. London: Verso.
Žižek, S. (2006). *The parallax view*. Cambridge, MA: MIT Press.
Žižek, S. (2001). Lenin's choice. In *Repeating Lenin*. Lacan.com. http://www.lacan.com/replenin.htm.
Žižek, S. (2006a). Children of men: Comments by Slavoj Žižek. In Cuaron, A. (Ed.), *Children of men*. DVD. Universal City, CA, USA: Universal Studios.
Žižek, S. (2010). *The end of nature*. Retrieved December 2, from http://www.nytimes.com/2010/12/02/opinion/global/02iht-GA12zizek.html?_r=0.
Žižek, S. (2014). *O Guia da Ideologia do Ódio*. Retrieved November 4th, from http://vimeo.com/110715167.

Index

H. Straehler-Pohl et al. (eds.), *The Disorder of Mathematics Education*,
DOI 10.1007/978-3-319-34006-7

Q

R

S

T

MIX
Papier aus verantwortungsvollen Quellen
Paper from responsible sources
FSC® C105338

If you have any concerns about our products,
you can contact us on
ProductSafety@springernature.com

In case Publisher is established outside the EU,
the EU authorized representative is:
Springer Nature Customer Service Center GmbH
Europaplatz 3, 69115 Heidelberg, Germany

Printed by Libri Plureos GmbH
in Hamburg, Germany